PMMA Circularity Roadmap

I0131698

Also of interest

Industrial Green Chemistry.
Kaliaguine, Dubois (Eds.), 2025
ISBN 978-3-11-138340-8, e-ISBN 978-3-11-138344-6

Circular Plastics Technologies.
Chemical Recycling
Knauer, 2024
ISBN 978-1-5015-2328-1, e-ISBN 978-1-5015-1561-3

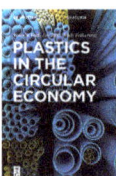

Plastics in the Circular Economy
Voet, Jager, Folkersma (Eds.), 2024
ISBN 978-3-11-120029-3, e-ISBN 978-3-11-120144-3

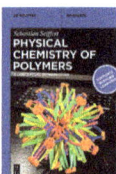

Physical Chemistry of Polymers.
A Conceptual Introduction
Seiffert, 2023
ISBN 978-3-11-071327-5, e-ISBN 978-3-11-071326-8

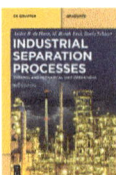

Industrial Separation Processes.
Thermal Unit Operations and Mechanical Unit Operations
De Haan, Eral, Schuur, 2025
ISBN 978-3-11-106363-8, e-ISBN 978-3-11-106381-2

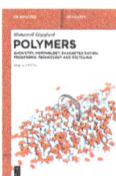

Polymers.
Chemistry, Morphology, Characterization, Processing,Technology and Recycling
Elzagheid, 2026
ISBN 978-3-11-158565-9, e-ISBN 978-3-11-158573-4

PMMA Circularity Roadmap

—

Industrial Practice and Academic Insight

Edited by
Simon van der Heijden, Pascal Lakeman and Jean-Luc Dubois

DE GRUYTER

Editors

Simon van der Heijden
Heathland B.V.
KVK 17246130
Arkansasdreef 8
3565 AR Utrecht
The Netherlands

Pascal Lakeman
Trinseo Netherlands BV
Sustainability and Engineered
Materials R&D
Innovatieweg 14
4542 NH Hoek
The Netherlands

Jean-Luc Dubois
Altuglas International / Trinseo
Tour CB21
16 Place de l'Iris
92400 Courbevoie
France

ISBN 978-3-11-107683-6
e-ISBN (PDF) 978-3-11-107699-7
e-ISBN (EPUB) 978-3-11-107742-0

Library of Congress Control Number: 2025933602

Bibliographic information published by the Deutsche Nationalbibliothek
The Deutsche Nationalbibliothek lists this publication in the Deutsche Nationalbibliografie;
detailed bibliographic data are available on the internet at http://dnb.dnb.de.

© 2025 Walter de Gruyter GmbH, Berlin/Boston, Genthiner Straße 13, 10785 Berlin
Cover image: Jean-Luc DUBOIS
Typesetting: Integra Software Services Pvt. Ltd.

www.degruyter.com
Questions about General Product Safety Regulation:
productsafety@degruyterbrill.com

Contents

List of contributing authors

Jean-Luc Dubois
Trinseo France SAS
Altuglas International SAS Tour CB21
16, place de l'Iris
92400 Courbevoie
France

Simon van der Heijden
Trinseo
Heathland B.V.
Arkansasdreef 8
3565 AR, Utrecht
The Netherlands

Nikolaj Garnitsch
Trinseo
Heathland B.V.
Arkansasdreef 8
3565 AR Utrecht
The Netherlands

Mariya Edeleva
Laboratory for Chemical Technology (LCT)
Ghent University
Technologiepark 125
Zwijnaarde 9052
Belgium

Pablo Reyes
Laboratory for Chemical Technology (LCT)
Ghent University
Technologiepark 125
Zwijnaarde 9052
Belgium

Rudinei Fiorio
Centre for Polymer and Material Technologies
(CPMT)
Ghent University
Technologiepark 130
Zwijnaarde 9052
Belgium

Yoshi W. Marien
Intelligence in Processes
Advanced Catalysts and Solvents (iPRACS)
Faculty of Applied Engineering
University of Antwerp
Groenenborgerlaan 171
2020 Antwerp
Belgium

Eli K. C. Moens
Laboratory for Chemical Technology (LCT)
Ghent University
Technologiepark 125
Zwijnaarde 9052
Belgium

Freddy L. Figueira
Laboratory for Chemical Technology (LCT)
Ghent University
Technologiepark 125
Zwijnaarde 9052
Belgium

Kevin M. Van Geem
Laboratory for Chemical Technology
Ghent University
Technologiepark 125
Zwijnaarde 9052
Belgium

Marek Blahušiak
Inprocess Technology and Consulting Group
S.L Carrer de Pedro i Pons 9
08034
Barcelona

Juraj Hrstka
RWE Technology International
Altenessener Straße 32
45141 Essen
Germany

https://doi.org/10.1515/9783111076997-203

Fady Boutros
Speichim
Allée des Pins
Saint-Vulbas 01150
France

Michiel Van Melkebeke
Laboratory of Environmental Toxicology and
Aquatic Ecology
Ghent University
Gent B-9000
Belgium
And
Department of Green Chemistry and Technology
Kortrijk B-8500
Belgium

Clément Lemenu
Certech
Rue Jules Bordet 45
Seneffe B-7180
Belgium

Philippe De Groote
Certech
Rue Jules Bordet 45
Seneffe B-7180
Belgium

Maria Savina Pianesi
Delta Srl
Via Tambroni Armaroli 2
Montelupone MC 62010
Italy

Tommaso Compagnucci
School of Science and Technology
ChIP Unicam Research Center
University of Camerino
Via Madonna delle Carceri
Camerino
Italy

Dagmar R. D'hooge
Department of Materials
Textiles and Chemical Engineering

Technologiepark 125
Zwijnaarde 9052
Belgium

Mark Bierens
Delta Glass
Deltaweg 4
RX Tholen 4691
The Netherlands

Jean-François Devaux
Arkema France
Rue Henri Moissan
Pierre Bénite F-69491
France

Juanjo Bermejo
ProCoat
Avda de la Industria 4
Castellgalí 08297
Spain

Pascal Lakeman
Trinseo
Innovatieweg 14
Hoek
The Netherlands

Jonathan Ouziel
Quantis
Rue de la Gare de Triage 5
Renens 1020
Switzerland

Andrea Corona
Quantis
Rue de la Gare de Triage 5
Renens 1020
Switzerland

Elnaz Zeinali
Laboratory for Chemical Technology
Ghent University
Technologiepark 125
Zwijnaarde 9052
Belgium

Jacopo de Tommaso
Ecole Polytechnique de Montréal
Polytechnique Montréal
2500 ch. de polytechnique
Montréal, Québec
Canada

Gregory Patience
Ecole Polytechnique de Montréal
Polytechnique Montréal
2500 ch. de polytechnique
Montréal, Québec
Canada

Bruno de Caumia
Pyrovac Inc.
176-2 Damase-Breton
Saint-Lambert-de-Lauzon
Québec G0S 2W0
Canada

Benjamin Dubois
Pyrovac Inc.
176-2 Damase-Breton
Saint-Lambert-de-Lauzon
Québec G0S 2W0
Canada

Gloire Justesse Adolphe Mbou
UNIDO/Centre d'Excellence d'Oyo pour les
Énergies Renouvelables et l'Efficacité
Énergétique
République du Congo

David Laplante St-Martin
Pyrovac Inc.
176-2 Damase-Breton
Saint-Lambert-de-Lauzon
Québec G0S 2W0
Canada

Gaétan Mercier
Pyrovac Inc.
176-2 Damase-Breton
Saint-Lambert-de-Lauzon
Québec G0S 2W0
Canada

Hooshang Pakdel
Pyrovac Inc.
176-2 Damase-Breton
Saint-Lambert-de-Lauzon
Québec G0S 2W0
Canada

Christian Roy
Pyrovac Inc.
176-2 Damase-Breton
Saint-Lambert-de-Lauzon
Québec G0S 2W0
Canada
croy@pyrovac.com

Ahmet Turhan Ura
M-D2 Mühendislik Danışmanlık Dönüşüm / M-D2
Engineering Consultancy Recovery Systems
ODTÜ-Ostim Teknokent
Uzayçağı Blv. 1308. Sok. No:06/ZK
13–14
Ostim/ANKARA
06370
Turkey

Ezgi Güzel
M-D2 Mühendislik Danışmanlık Dönüşüm / M-D2
Engineering Consultancy Recovery Systems
ODTÜ-Ostim Teknokent
Uzayçağı Blv. 1308. Sok. No:06/ZK
13–14
Ostim/ANKARA
06370
Turkey

Yusuf Uludağ
Middle East Technical University D-112
METU-Department of Chemical Engineering
06800
Ankara
Turkey

Sara Babo
LAQV-REQUIMTE
Department of Conservation and Restoration
NOVA School of Science and Technology
Universidade NOVA de Lisboa

2829-516 Caparica
Portugal

Anna Micheluz
Conservation Science Department
Deutsches Museum
Museumsinsel 1
80538 Munich
Germany

Eva Mariasole Angelin
Chair of Conservation-Restoration
Art Technology and Conservation Science
Technical University of Munich
Oettingenstr. 15
80538 Munich
Germany

Susanne Brunner
Professorship of Recent Building Heritage
Conservation

TUM School of Engineering and Design
Technical University of Munich
Arcisstraße 21
80333 Munich
Germany

Joana Lia Ferreira
CIUHCT—Interuniversity Center for the History
of Sciences and Technology
Department of Conservation and Restoration
NOVA School of Science and Technology
Universidade NOVA de Lisboa
2829-516 Caparica
Portugal

Marisa Pampolona
Conservation Science Department
Deutsches Museum
Museumsinsel 1
80538 Munich
Germany

Jean-Luc Dubois and Simon van der Heijden

1 Preface: the genesis of the MMAtwo project

Poly(methyl methacrylate) (PMMA) has been depolymerized back to its monomer soon after its first commercialization. From the initial "dry distillation" process, in which the polymer is heated to high temperature, the depolymerization technology evolved to enhance heat and mass transfer. One of the most efficient processes is the "Molten-Metal Bath" process, in which quite often molten lead is used to transfer the heat needed for the depolymerization. Although performing well, the process is not the most appropriate to recycle low-quality scraps that could generate a lot of solid residues.

Back in 2016 and 2017, Altuglas (at that time part of Arkema) and Heathland convened to investigate the possibility to initiate a collaborative project on regeneration of methyl methacrylate (MMA). Since then, Arkema's PMMA business – Altuglas International – was divested to TRINSEO, and Heathland was acquired also by TRINSEO. This was just before the Chinese ban on imported plastic wastes, and at that time, most of the PMMA scraps were still exported from Europe to Asia and elsewhere to be reprocessed. But we felt that the market would change and request more and more recycled products and that there would be a need for new depolymerization capacities in Europe.

At that time, we had identified very few depolymerization plants that are still in operation in Europe. Arkema (Altuglas) had closed its molten lead depolymerization plant a decade earlier, like ICI (which became Ineos Acrylics, then Lucite, and later part of Mitsubishi Rayon and now Mitsubishi Chemical). Evonik had a plant in Austria, under the name of ParaChemie, also using the molten lead process, but closed it definitively at the end of 2017. The rest of the PMMA business of Evonik was also divested and became Röhm. So, the three main PMMA players on the EU market at that time had once a depolymerization plant, and all decided to shut down those units for various reasons. The remaining players were smaller companies and we estimated that altogether only about 7,000 tons of PMMA were reprocessed in Europe, while a significant share of the regenerated MMA was for captive use.

In these conditions of large wastes supply and small recycling capacity, only the best scraps are processed. Even outside of Europe, where all kinds of scraps would be in demand to fill the recycling facilities, recyclers need to be selective and cannot accept all kinds of PMMA wastes. For example, the molten lead bath process is not appropriate

Jean-Luc Dubois, Trinseo France SAS – Altuglas International SAS Tour CB21,16, place de l'Iris, 92400 Courbevoie, France
Simon van der Heijden, Heathland B.V., Arkansasdreef 8, 3565 AR Utrecht, The Netherlands

https://doi.org/10.1515/9783111076997-001

to process scraps that would generate a lot of solid residues that would otherwise be contaminated with lead.

We concluded that there is a large volume of PMMA waste which is out there, but which is not collected yet and not processed. In fact, out of the 300,000 tons of PMMA produced in Europe every year, only about 30,000 tons were collected, and most of it was exported outside of Europe. We wanted to address all the other PMMA wastes, including Post-consumer and Post-industrial wastes, which would be sufficient to justify several new recycling plants and to change the public perception of PMMA.

To process those wastes, we needed to have a technology that would be flexible enough to adapt to the various scraps. PMMA exists on the market under several different grades, shapes, and colors. To make it simple, there are cast-grade sheets and blocks with high average molecular weight, extrusion-grade sheets, and injection grade with lower molecular weight but also comonomers. End-of-life PMMA wastes are challenging, because in many applications, like car tail-lights or construction materials like parapet and soundproof walls, the product's life expectancy can be of several decades. This means that the product arriving in the recycling loop corresponds to formulations that were allowed at the time of commercialization; but might include additives that are no longer allowed and must be removed by the recycling process. In addition, over such a long period, one lost track of the initial producer in most cases. Postproduction wastes are easier to trace, but when collected along the value chain, all grades, producers, and colors would be mixed together, and the mainstream is a blend of materials.

The recycling process should not only be flexible to the feedstock, but should also be a technology that could be adopted by recyclers. It means that it should give them enough confidence that they could operate such a process, and that in case the market flips, they could still use the assets to reprocess another polymer and the investment would not be lost. Ease of operation was another important criterion. Dry distillation processes typically involve a labor-intensive cleaning process of potentially toxic char. For wide adoption of the process, it should be very safe and relatively easy to operate for the employees involved. The recycling process being implemented in Europe, the regenerated MMA should reach a product quality which is high enough to fit with European standards. Some of the impurities present in regenerated MMA are process related. It means that the amount of impurity depends on the process used for the depolymerization. So, one more selection criterion for the process was to be able to lower the amount of some selected impurities by the selection of the depolymerization process. Finally, we wanted to select a technology which was mature enough to be able to build a commercial plant after 4 years. And that is how we came up with the selection of the Japan Steel Works technology. Through Japan Steel Works Japan we got in touch with their European Branch and quickly organized a meeting. JSW had already patents in the field and experience with PMMA.

Once PMMA is depolymerized, it has to be purified to reach the quality needed to close the loop. This means that we wanted to be able to make the best optical transparency and reach virgin-like materials. Several companies and research organizations were contacted to assemble the appropriate technologies and that is how Speichim, Process Design Center, Certech, and later Suster (now Demcon Suster) joined the consortium. It was also important for us to demonstrate that the combined processes would be able to deliver convincing data to demonstrate the superiority of the recycling over other end-of-life alternatives. We approached Quantis, which had been a partner in previous projects also, to take care of the life cycle analysis and life cycle costing part of the project, not only to assess what our recycling process could deliver but also to benchmark the virgin and other recycling processes.

End users were also needed, in various applications, to demonstrate that the regenerated MMA could be used in a large panel of applications. Besides Altuglas and Arkema, which were virgin MMA producer and end users, we also approached Delta Glass to make a demonstrator that can speak to everybody. Delta Glass produces PMMA sheets for caravan windows. This is a demonstrator which is both an interior and an exterior part, which has to satisfy several technical properties, such as transparency, odor, mechanical, and surface finish. Later, during the project, other end users like Plados Telma and Procoat joined the project for composite kitchen sinks and technical coatings, respectively. In addition, after the divestment of Altuglas by Arkema, Trinseo also joined the project for the last year. Besides being a producer of sheets, Trinseo, through Altuglas, is also a producer of PMMA raw material for injection molding. It therefore helped to broaden the pallet of applications for the regenerated MMA. Plados Telma replaced AKG, which unfortunately went bankrupt few months after the start of the project. AKG had originally been in the consortium to investigate uses for by-products from the depolymerization process, such as glass fibers. Luckily, we had invited Plados Telma to join the project Advisory Board since the project inception, and they were fully aware of the project objectives and could join rapidly.

To complete the consortium, we wanted to address the PMMA waste which is in selected applications. Comet Traitements, which deconstructs a large volume of cars, could supply PMMA waste from taillights and other parts but also investigate the best way to collect those streams. Ecologic is a French Extended Producer Responsibility scheme accredited by the French government for waste electrical and electronic equipments (WEEE) collection and treatment on the national territory. It was the perfect fit not only to investigate the supply from flat screens deconstruction but also to communicate with similar organization throughout Europe and pass the message that PMMA should be collected, sorted, and recycled.

Although the consortium has a strong industrial focus, there were still some research questions to be addressed. Specifically, we wanted to understand to which extent we can influence the depolymerization chemistry and speed up the reaction, and also how to counterbalance the impact of various grades of PMMA on the regenerated MMA (rMMA) rMMA purity. Ghent University was then invited to join the project and also to assist in organizing dissemination and training materials in order to attract talented people to the recycling opportunities.

Last but not least, we needed professional assistance to manage the project. Alma Consulting Group, which became Ayming, and currently Benkei, was included to organize the day-to-day management of the project, and all the reporting to the EU. In project construction, this should not be underestimated as management activities and interactions with the European commissions are time-consuming and need to be very efficient. To manage the project, Simon van der Heijden agreed to take the lead as Coordinator, and Jean-Luc Dubois took the Chair of the Executive Committee. This organization was selected to demonstrate a strong commitment to implement the project results by a small company (Heathland) in a future commercial plant, and also the project was backed by a larger organization more familiar with the management of large collaborative projects.

Finally, the Advisory Board, which is not mandatory but strongly recommended, was built looking for people who would have additional expertise than the project partners, and also who could replace one of them in case of failure. The Advisory Board has been a living group, which was invited in several periodic events. Every six months, the project organized special events on invitation basis, to debate of different topics and to disseminate selected results. The Advisory Board included Philippe Salemis and Marcelo Vollmann (CEFIC, Methacrylate sector group), Peter Kelly (Polycasa), Andrew Bragg (Lucite), all four representing the MMA/PMMA major producers; Savina Pianesi and Antonio Bugiolacchio (Plados Telma, who later joined the project) representing regenerated MMA end users; H.R. (Bob) Fowler for his expertise in recycling; Prof. Dimitris S. Achilas (Aristotle University of Thessaloniki), Prof. Martin Olazar (Universidad del Pais Vasco), and Dr. Helene Wiesinger, researcher at ETH Zurich, Chair of Ecological Systems, for their expertise in PMMA depolymerization and for the impact of polymer additives in recycling value chain. We hereby take the opportunity to thank them for helping us to make from MMAtwo a very successful project.

The book has been a piece of work by itself. It became a masterpiece, with a lot of information and technical details on recycling, in general, and MMA regeneration, in particular. It was well received by both stakeholders new to PMMA recycling and by companies active in PMMA depolymerization for many years. At the end of the project, we also took the commitment to prepare a second edition two years later, and here we are. This new edition includes several new chapters, but the landscape has also significantly changed: Trinseo has started its demonstration plant in Rho, Italy. The plant integrates the best technological improvements on PMMA depolymerization developed during the MMAtwo project, and was inaugurated on June 25, 2024.

The project also received recognition with the "Innovation Team Best Practices" award in 2023, and it was also selected as "Project of the Month" in April 2024 by the Cordis platform of EU projects. The book (first and second editions) would not have been possible without the support of the European Union's Horizon 2020 research and innovation funding scheme (grant agreement no. 820687) as illustrated by the contribution of most of its partners in writing and preparation of the books.

We hope that the readers will enjoy it as much as we enjoyed working in the project, and that it can serve as a guide to recycle both PMMA and other polymers alike.

Simon van der Heijden, Jean-Luc Dubois, and Nikolaj Garnitsch

2 Vinyl polymer waste market: opportunities for PMMA chemical recycling

2.1 Applications defining the PMMA market

The PMMA market can be segmented by the end-use industry. The key industry segments accounting for the bulk of PMMA demand are advertising, appliances, automotive, construction, and electronics including optical media. Together these five end-use market segments account for about 90% of the global demand. The market size per field of application is illustrated in Table 2.1.

Table 2.1: Major applications for PMMA market[a].

Application	Characteristics	Estimated annual growth (%)	Market size (M US$)
Signs and displays	Durable and shatterproof, facilitating applications such as advertising, visual communication, and instructions	5	1,700
Construction	Usefulness in skylights, partition sheets, shockproof- and protective glass	5	1,000
Automotive	Lightweight automotive parts are relevant for PMMA material developments	5	1,300
Electronics	Screens of various sizes, light guide panels, cover panels, cassette doors, and appliances	4	500
Lighting fixtures	With more desired design, PMMA is a good alternative to fix lightening structures	6	200
Others	Healthcare (surgeries and dental practices) and sanitary ware applications	5	1,000

[a]Numbers have been rounded for a general overview.

Of course, looking broader at (P)MMA markets, its applications in coatings and resins should also be included. For this chapter, we shall mainly focus on end-uses of PMMA, that can be collected as a waste, either somewhere in the production and distribution value chain, or as a finished product at the end if its lifespan. Most applications in-

Simon van der Heijden, Nikolaj Garnitsch, Heathland B.V., Arkansasdreef 8, 3565 AR Utrecht, The Netherlands
Jean-Luc Dubois, Trinseo France SAS – Altuglas International SAS Tour CB21,16, place de l'Iris, 92400 Courbevoie, France

https://doi.org/10.1515/9783111076997-002

volve the use of PMMA in the production of sheets and/or injection molded parts. A casting process is used to produce sheets and large blocks, while an extrusion process is used to produce resins and sheets. The resins can in turn be used for injection molding. While signs and displays are typically made from sheets, automotive applications like taillights are made by the injection molding process.

New markets in which PMMA is used are composites. "Solid surface" (kitchen sinks and cladding) has been a growing business for years; and using methyl methacrylate (MMA) resins as an alternative for polyester in windmill blades and boat hulls is gaining traction. Recyclability of these composites, however, comes with its own set of challenges.

Having a good overview of PMMA markets is important. First, it indicates where regenerated PMMA (rPMMA) and regenerated MMA (rMMA) could be sold. Potentially, given a high enough quality of rMMA, all markets for virgin MMA could also be considered markets for rMMA. Second, knowing what PMMA quantity and quality markets use will make the targeted collection of PMMA (production) waste easier. For example, PMMA is used to produce automotive taillights. Therefore, companies producing automotive parts are a potential source of high-quality PMMA production scrap. Third, waste collectors need to know what types of products are made from PMMA, in order to be able to identify as much as possible end-of-life (EoL) PMMA waste. Collection schemes can be instigated to efficiently draw PMMA waste. An example would be the knowledge that PMMA is often used to produce sky domes. Therefore, the deconstruction of a shopping mall could be an opportunity to collect EoL sky domes.

2.2 Variation of PMMA waste and feedstock

Any case for PMMA waste recycling starts with PMMA waste feedstock. After all, if there is no feedstock, there is no reason to justify an investment for a recycling capacity and to set up a value chain. Therefore, the presence of sufficient waste feedstock is key to any endeavor in the field of PMMA recycling. It should be noted that, in general, waste markets for which there is available viable recycling technology are typically sellers markets. This can be explained by the fact that the creation of waste is a result of production and abandonment of products, and not a goal in itself. Waste, as such, is not deliberately "made" and is therefore available in limited supply. For good quality waste, demand often outstrips supply. Such equilibrium is also seen in PMMA waste markets where good quality waste typically has a positive value. This, in turn, supports the collection of waste: efforts to obtain contaminated or difficult to collect EoL waste may prove to be rewarding.

Each PMMA manufacturer has its own process technology for making PMMA products, leading to many variations of PMMA formulations. The inclusion of a comonomer minimizes or eliminates detrimental properties, such as poor temperature resistance and shock resistance. Comonomers can be required to be able to process PMMA pellets properly in extrusion or injection processes, avoiding undesirable depolymerization. The properties of PMMA, or processability of PMMA, improve with a small percentage of a one comonomer, such as ethyl acrylate, usually between 1% and 5%, although the proportion could be exceptionally as high as 30% depending on specific properties required. These formulations are necessary to improve thermal stability during the molding and extrusion steps, but they will affect the depolymerization and recovery of high-purity monomer. Sulfur compounds are also added in some PMMA grades as chain transfer agents during the polymerization and/ or to provide higher heat stability (saturation of the terminal double bonds of the polymer chains) and are suspected to contribute to the odor issues in depolymerized MMA and derived PMMA products.

The waste can be generated at various stages in the value chain of the PMMA industry before the actual postconsumer stage. Some wastes will originate from the producer of PMMA sheets and resins. This waste could include PMMA/poly(vinyl chloride) waste from the casting process; PMMA off-cuts and dust from cutting and sawing process; PMMA purging from cleaning the equipment; PMMA color changing material; and off-grade material which can have defects for a variety of reasons. Then, once sheets or resins are supplied to the customers of the original producers, there will be other wastes from cutting, rejects, injection molding process, and contaminated waste with additives and composites.

The final product is positioned at its "final destination," like signage at a petrol station. At the end of its life, the signage is renewed, and this waste also finds its way to the waste collector. This EoL waste (or postconsumer waste) is much more difficult to collect than production waste, because it has been scattered to final consumers. Ironically, the amount of EoL waste that could be potentially sourced is around 10 times more than postproduction waste. Newly developed methods of waste collection, however, provide access to this great source of waste for depolymerization.

The waste from PMMA will differ from preconsumer to postconsumer, in grades (cast/injection/extrusion), in colors, in the level of contamination, in additives, in age, and in size. This makes recycling challenging. Components locked into the PMMA polymer matrix could pose a challenge. An example is the heavy metal cadmium, which used to be an ingredient of pigments for PMMA sheets. As all locked substances will be unsealed in depolymerization, harmful components can pose a serious risk to operators and result in solid residue that must be treated with great care. However, dedicated depolymerization offers the opportunity to remove these contaminants that are no longer acceptable in products. For illustration purposes, Figure 2.1 provides some examples of the various types of PMMA waste described earlier.

2.3 When to select the chemical recycling route for PMMA waste

A challenge for every waste stream, after proper collection, sorting, and pretreatment (see details in Chapter 3), is to identify whether a mechanical recycling (principles in Chapter 4 for a general feedstock) or chemical recycling route (reaction and reactor principles in Chapters 5–7 selecting PMMA as feedstock) is best suited.

If both routes are an option, and the final product can be identical, then mechanical recycling should be the preferred method, because the energy consumption and CO_2 emissions are lower and the yield is higher.

Figure 2.1: PMMA waste types.

After all, mechanical recycling requires only transformation of PMMA waste into new products via heating (energy consumption) once or twice to a temperature range of 180–250 °C. An example of heating only once (single energy consumption) is PMMA waste that is grinded into 10 mm particle size and directly re-extruded or used for injection molding. An example of heating twice is pelletizing the waste regrind in an extruder, then cooling down and using the pellets in re-extrusion or injection molding.

Chemical recycling, in contrast, requires at least two heating steps, if not more, to retrieve a high-quality purified product. PMMA is first heated in the range of 350–400 °C, to chemically unzip polymer into monomer. Then, the monomer needs to be heated again in order to repolymerize in the presence of chemical initiator. The polymer may require once more heat to use it for extrusion or injection molding. Upon using the monomer for sheet casting, at least two heat/energy consumption steps must be included (see Chapter 10).

Despite the advantages of mechanical recycling of PMMA, the chemical route has four main advantages:

1) Most PMMA waste is cast grade or a mixture of cast and extrusion grades, and generally cannot be sorted because it became mixed very early in the value chain. Cast-grade PMMA is unsuited for mechanical recycling because of too long polymer chains (too high average molecular weight). It means that any waste, which has (a mixture of) cast grade inside, needs to be recycled by means of the chemical recycling route.

2) Whenever PMMA waste is mixed with other polymers or special additives, in a mechanical process, it will immediately result in quality deteriorations, for example surface imperfections, spots, color changes, and lesser mechanical properties, while the chemical process is much more flexible, due to various purification steps after depolymerization. Aiming for high-end applications like automotive or signage businesses, mechanical recycling can only be applied to very well-sorted and high-quality PMMA waste; and even then, usually only if blended with virgin PMMA. Chemical recycling of PMMA, combined with the right purification steps, will result in an rMMA raw material of the same quality than virgin MMA, making it suitable for high-end (optical) applications. The chemical recycling process can be repeated without significant loss of material quality.

3) The chemical recycling method will, after purification, result in a transparent MMA that can be repolymerized and turned into any color needed, while removing all additives that were present in the waste. The mechanical method will result in limited color possibilities and restrict the material to final uses that require the exact properties as present in the waste. Clear transparent material suitable for mechanical recycling can in some cases be reprocessed into clear transparent product or color of choice after adding a masterbatch. However, most waste is not transparent; and, once there is a mixture of colors, the options are very limited with usually black and dark gray solid colors.

4) Heavily contaminated wastes, typically EoL PMMA waste, or PMMA waste filled/contaminated with glass fiber, fillers, and composites, are generally unsuitable for mechanical recycling and must be chemically recycled. Glass fibers and other inorganics can be separated at the depolymerization phase and be potentially reused.

Hence, for PMMA waste, there is a clear market potential and relevance for chemical recycling technology. A question then arises which chemical reactor technology is most suited and which operating and downstream conditions should be utilized, as addressed in the main part of this book. The main technologies for PMMA recycling, still in use in Europe today, are unsuitable for many types of PMMA wastes, in particular for contaminated and EoL PMMA wastes. Therefore, when working on a solution for PMMA waste available in Europe today, the requirement for developing better technology is evident. The resulting technological solution, in turn, will propel the collection and recycling of challenging PMMA waste.

2.4 Products generated during PMMA depolymerization

A criterion which makes the difference between the various depolymerization technologies are the impurities which are generated during the depolymerization process. Some impurities are related to the PMMA formulation and others are related to the depolymerization process itself. Among the first ones, we find comonomers such as methyl and ethyl acrylate which are commonly used in injection and extrusion-grade formulations; but also styrene, which is a common comonomer but which can also come from additives or polymer contamination. One would also find molecules such as isobutylene which is a decomposition product of PMMA. Furthermore, one can find decomposition products from the PMMA formulation, such as lubricants, plasticizers, additives, or polymerization initiators. The other family of impurities is related to the depolymerization process conditions: technology, temperature, pressure, separation. To some extent these impurities could be avoided or reduced.

Pyrolysis of PMMA (also called thermolysis) and other acrylate polymers has been studied in laboratory scale micro-pyrolysis coupled to gas chromatography-mass spectrum analysis [2]. This technique gives already a good insight of the impurities generated during the pyrolysis of various PMMA samples (Table 2.2). The PMMA sample used is a very low molecular weight PMMA, so naturally it will tend to generate more impurities. For a 15,000 g/mol molecular weight, the average chain length is 150 MMA molecules. It also means that two of those are at the end of the chain. If we just assume that the activation will generate side products and that the polymeric molecule generates two side products by chain, that's already more than 1% of impurity. The longer the chain, the lower the impact of those impurities.

Isobutene is produced by thermal degradation of MMA. MMA decomposes into methyl formate and a radical, which recombines with the CH_3 of the methyl formate, releasing isobutene and CO_2. The reaction mechanism corresponds to the transfer of the methyl group from the ester to the tertiary carbon, and release of CO_2. Similar mechanisms generate also other impurities. Among the reaction products detected there are several saturated molecules: methyl propionate, methyl isobutyrate, dimethyl glutarate, and so on. Although the reaction is done on a very small amount of material, in conditions where we can assume that there were no mass and heat transfer limitations, these saturated molecules are produced. This is a sign that the hydrogen transfer reactions which are taking place during the depolymerization are very fast reactions occurring between the depolymerizing chain and other intermediates. Dimers and trimers, as well as their hydrogenated forms, are also detected. The temperature at which the depolymerization takes place is high enough to be above the boiling points of those products. As soon as they are produced, they can evaporate, especially since there is a continuous flow of helium which ensures a low partial pressure. Methyl acrylate was probably produced because the PMMA sample was containing it as comonomer. Methanol is also produced either by hydrolysis of PMMA/MMA or through the degradation of the ester groups, releasing simultaneously CO. The isoprene fragment is more surprising, but it could correspond to a hydrocarbon fragment generated by the degradation of the polymer backbone, or it could be due to a comonomer used in the PMMA sample.

This analysis also shows that there are multiple degradation products with similar physicochemical properties than MMA, which are going to make the purification more difficult.

This simple analysis which can be reproduced for many different PMMA samples already illustrates the complexity of the depolymerization process, and of the purifications that have to be considered. Therefore, it is very important to properly select the types of waste, the technologies, and the process conditions to minimize the impurities generated.

Table 2.2: Pyrolysis products detected in micro-pyrolysis of PMMA with a molecular weight of 15,000 g [2].

Compound/IUPAC name	Common name	CAS no.	MW	Area (%)	Reference
Carbon dioxide	CO_2	124-38-9	44	1.29	[3]
2-Methylprop-1-ene	Isobutene	115-11-7	56	Trace	[4]
2-Methylbuta-1,3-diene	Isoprene	78-79-5	68	Trace	[5]
Methyl formate	Methyl formate	107-31-3	60	Trace	[6]
Methanol	Methanol	67-56-1	32	Trace	[7]
Methyl propanoate	Methyl propionate	554-12-1	88	1.18	[8]
Methyl 2-methylpropanoate	Methyl isobutyrate	547-63-7	102	1.31	[9]
Methyl prop-2-enoate	Methyl acrylate	96-33-3	86	Trace	[10]
Methyl 2-methylprop-2-enoate	Methyl methacrylate	80-62-6	100	58.58	[11]
Methyl 2-methylidenebutanoate	2-Methylene butanoic acid methyl ester	2177-67-5	114	Trace	[12]
Methyl 2-methylbut-3-enoate	3-Butenoic acid 2-methyl methyl ester	51747-33-2	114	0.39	[13]
Methyl 2,4-dimethylpent-4-enoate			142	0.59	[14]
Methyl cyclohex-1-ene-1-carboxylate		18448-47-0	140	0.43	[15]
1,5-Dimethyl pentanedioate	Dimethyl glutarate	1119-40-0	160	0.29	[16]
Dimethyl 2,4-dimethylpentanedioate		2121-68-8	188	0.41	[17]

Table 2.2 (continued)

Compound/IUPAC name	Common name	CAS no.	MW	Area (%)	Reference
1,5-Dimethyl (2E)-pent-2-enedioate	Dimethyl-pent-2-enedioate	5164-76-1	158	1.35	[18]
2-Methylprop-2-enoic acid	Methacrylic acid	79-41-4	86	Trace	[19]
1,2-Dimethyl cyclopropane-1,2-dicarboxylate		702-28-3	186	1.86	[20]
Dimethyl 2,2-dimethyl-4-methylidenepentanedioate	(Dimer)	71674-93-6	200	1.80	[21]
Dimethyl 2,2,4-trimethylpentanedioate	(Hydrogenated dimer)		202	1.50	[22]
2.4.4-Trimethyl-2-pentenedioic acid dimethyl ester	(Dimer)		200	Trace	
Methyl 2-[2-(methoxycarbonyl)-2-methylcyclopropyl]propanoate			200	0.53	
Methyl 2-[2-(methoxycarbonyl)-2,2-dimethylcyclopropyl]propanoate			214	Trace	
1-Methyl-3-cyclopentene-1,3-dicarboxylic acid dimethyl ester			198	0.39	
Dimethyl 4-(methoxycarbonyl)-2,4,6.6-tetramethyl-2-heptene-1,7-dioate	(Trimer)		300	0.85	
Dimethyl 4-(methoxycarbonyl)-4.6.6-trimethyl-2-methyleneheptane-1,7-dioate	(Trimer)		300	4.49	

The pyrolysis was done on 0.4 mg of material, at 600 °C in a flow of helium and at 20 °C/ms. Some of the peak assignments given in the publication are tentative as no references were available. IUPAC names have been used when available and CAS numbers have been added. A reference for more data on the compound is also provided. MW: molecular weight (g); area (%): Peak area from the GC analysis – peaks due to contamination with another polymer have not been included in the table for clarity.

2.5 Market potential of recycled projects

The market for rMMA in Europe is not yet mature, as is evident from a current production capacity of 7 kt/year of rMMA versus 700 kt/year production capacity of virgin MMA in Europe [1]. With such a small supply of rMMA, it is not surprising that the European market is fully organized to use virgin MMA. After all, which producer would rely on rMMA in case supply is short and unstable, knowing that virgin MMA can be produced with large capacity? Add on top of that potential variations in monomer purity. rMMA can only be produced based on availability of the waste and a good recycling technology. Since waste cannot be produced on purpose, there is a delicate line to balance supply and demand. A very different situation had existed in China, with an estimated absorption of 150 + kt annually of rMMA until the waste import ban in 2018.

Importantly, the demand in Europe seems now to be present. Governmental incentives and programs, which stimulate the use of recycled raw material, like the End-of-Life Vehicle regulation [23], support this. More regulation is in the making, potentially enforcing the use of sustainably derived raw materials and limiting further the possibility to export waste to destinations outside of Europe [24]. Furthermore, a consumer demand for "green" and sustainable products has raised interest in rMMA as an environmentally sound feedstock.

PMMA is typically processed into high-end products, which have a lifespan of several years to decades. The taillights of cars, for example, are made from PMMA. PMMA is not typically used for packaging or single-use applications with a short lifespan. Plastics used in packaging are facing increased resistance in the European Union. PMMA, however, with its advantages over glass (weight and strength) makes it a material that can be, and will be, used in a wide spectrum of applications. It is expected that the consumption of PMMA (and therefore MMA) will only increase in the coming years.

Key markets for rMMA are:

1. Sheets (rMMA can be used directly to cast sheet, but in order to use it for sheet extrusion or injection molding, it should be polymerized first (indirect use))
2. Injection molding (indirect use)
3. Solid surface material
4. Composites, for example, as a replacement of polyester with glass fiber
5. Paints and coatings
6. Adhesives
7. Resins

Though much regulation is aimed at increasing the use of regenerated raw material, other regulation such as REACH makes the adoption of rMMA more challenging. In the field of recycling, REACH is still in development, and uncertainties resulting from this may result in short-term resistance to adoption of regenerated raw material in the market.

Chemical recycling of PMMA paves the way for closed-loop recycling systems, in which suppliers of waste can also be the end-users of the regenerated raw material, without significant loss of quality, and allowing for replicability. This key advantage of PMMA unveils the potential of chemical recycling for this specific polymer.

2.6 Trinseo's PMMA depolymerization demonstration plant

Since the first edition of this book, there have been several announcements of new plants and pilot operations:
– Sumitomo Chemical in Japan announced that its pilot plant was completed at the end of 2022 [25].
– Biorennova in Italy was acquired by Maire Tecnimont/Nextchem, which announced a new plant to be built by 2025 [26, 27].
– Mitsubishi Chemical and Microwave Chemicals collaborate on a demonstration plant [28].
– But most importantly, the technology developed during the MMAtwo project has been implemented by Trinseo, in its plant in Rho, Italy [29].

Trinseo's depolymerization facility has been constructed within the Rho plant which is already producing virgin MMA. It then benefits not only of the expertise of the personnel but also from the local environment to produce the monomer. The plant construction was announced end of 2023 and was inaugurated on June 25, 2024. Figure 2.2.

Figure 2.2: Inauguration of Trinseo's PMMA depolymerization demonstration plant.

References

[1]   IHS Markit. MMA report 2019.
[2]   Moldoveanu SC. Analytical Pyrolysis of Synthetic Organic Polymers, vol. 25, 1st Edition, Elsevier, January 20 2005, ISBN: 9780444512925.

[3]   https://echa.europa.eu/fr/substance-information/-/substanceinfo/100.004.271.
[4]   https://echa.europa.eu/fr/brief-profile/-/briefprofile/100.003.697.
[5]   https://echa.europa.eu/fr/brief-profile/-/briefprofile/100.001.040.
[6]   https://echa.europa.eu/fr/brief-profile/-/briefprofile/100.003.166.
[7]   https://echa.europa.eu/fr/brief-profile/-/briefprofile/100.000.599.
[8]   https://echa.europa.eu/fr/brief-profile/-/briefprofile/100.008.238.
[9]   https://echa.europa.eu/fr/brief-profile/-/briefprofile/100.008.118.
[10]   https://echa.europa.eu/fr/brief-profile/-/briefprofile/100.002.274.
[11]   https://echa.europa.eu/fr/brief-profile/-/briefprofile/100.001.180.
[12]   https://echa.europa.eu/fr/substance-information/-/substanceinfo/100.283.590.
[13]   https://pubchem.ncbi.nlm.nih.gov/compound/Methyl-2-methyl-3-butenoate.
[14]   https://pubchem.ncbi.nlm.nih.gov/compound/557840.
[15]   https://echa.europa.eu/fr/substance-information/-/substanceinfo/100.038.468.
[16]   https://echa.europa.eu/fr/brief-profile/-/briefprofile/100.012.980.
[17]   https://pubchem.ncbi.nlm.nih.gov/compound/Dimethyl-2_4-dimethylpentanedioate.
[18]   https://echa.europa.eu/fr/brief-profile/-/briefprofile/100.210.011.
[19]   https://echa.europa.eu/fr/brief-profile/-/briefprofile/100.001.096.
[20]   https://echa.europa.eu/fr/substance-information/-/substanceinfo/100.313.511.
[21]   https://pubchem.ncbi.nlm.nih.gov/compound/10035672#section=Canonical-SMILES.
[22]   https://pubchem.ncbi.nlm.nih.gov/compound/11424214.
[23]   End-of-Life Vehicles – European Commission (europa.eu), https://environment.ec.europa.eu/topics/
       waste-and-recycling/end-life-vehicles_en#publications, last accessed July 10th 2024.
[24]   https://environment.ec.europa.eu/topics/waste-and-recycling/waste-shipments_en, last
       accessed November 18th 2024.
[25]   Sumitomo Chemical Completes Construction of Pilot Facility for Acrylic Resin Chemical Recycling
       Accelerating the development of a circular system for plastics through cross-industry collaboration
       | Business & Products | SUMITOMO CHEMICAL (sumitomo-chem.co.jp), https://www.sumitomo-
       chem.co.jp/english/news/detail/20221223e.html, last accessed July 10th 2024.
[26]   NEXTCHEM (MAIRE) signs EUR 4 million grant agreement with the EU Co . . . | NextChem,
       https://www.nextchem.it/en/newsroom/press-releases/detail/nextchem-maire-signs-eur-4-million-
       grant-agreement-with-the-eu-commission-under-the-innovation-fund-for-the-scale-up-of-its-
       proprietary-chemical-recycling-technology-nxre-pmma/, last accessed July 10th 2024.
[27]   MAIRE'S NEXTCHEM ACQUIRES CONTROL OF CATC, AN INNOVATIVE PLASTIC CH . . . | Maire
       (groupmaire.com), https://www.groupmaire.com/en/newsroom/press-releases/detail/maires-
       nextchem-acquires-control-of-catc-an-innovative-plastic-chemical-recycling-technology/, last
       accessed July 10th 2024.
[28]   Microwave Chemical and Mitsubishi Chemical have agreed to build a demonstration plant to
       commercialize acrylic resin chemical recycling | News | Microwave Chemical Co., Ltd. (mwcc.jp),
       https://mwcc.jp/post_en/press_release/2638/, last accessed July 10th 2024.
[29]   Trinseo – Trinseo Opens Next-Gen Depolymerization Facility in Italy, https://investor.trinseo.com/
       home/news/news-details/2024/Trinseo-Opens-Next-Gen-Depolymerization-Facility-in-Italy/default.
       aspx, last accessed July 10th 2024.

Mariya Edeleva, Pablo Reyes, and Simon van der Heijden

3 Collection, sorting, and pretreatment

3.1 Plastic material waste

Approximately a century ago, humankind endured a revolution in the field of materials, as synthetic polymers, so-called plastics, started to be produced for a handful of applications. As plastic science evolved, polymers have become close to an ideal material due to the ease of tuning up its properties [1]. A downside is the large amount of generated plastic waste. Plastic packaging, especially single-use packaging, is the primary source of plastic waste [2, 3]. Even if its production carbon footprint can be less than that of other materials such as glass and metal, as supported by the diagrams in Figure 3.1, and its use importantly reduces the waste of food, there is a significant concern on optimizing plastic recycling rates [1, 4]. Furthermore, public opinion to some extent has turned on plastics, with movements in certain countries calling for a "plastic diet." While plastics, especially ones used for non-packaging purposes, have a key role in our life, it is important to address these concerns and showcase the excellent recyclability of certain polymers, such as PMMA.

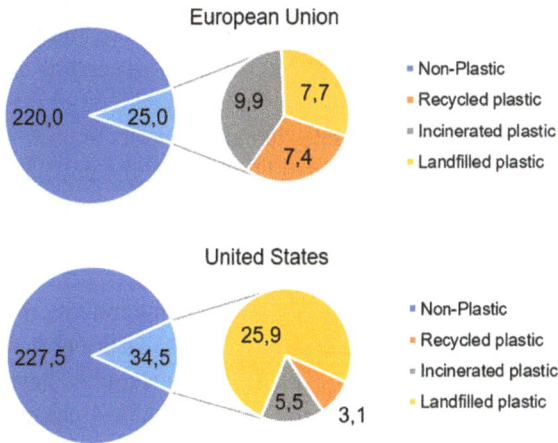

Figure 3.1: Breakdown of the plastic waste disposal expressed in million tons collected in 2015 in the United States and Europe, in comparison with the total amount of solid waste (US 262 million tons and EU 245 million tons) [3, 5, 6].

Mariya Edeleva, Pablo Reyes, Laboratory for Chemical Technology (LCT), Ghent University, Technologiepark 125, Zwijnaarde 9052, Belgium
Simon van der Heijden, Heathland B.V., Arkansasdreef 8, 3565 AR Utrecht, The Netherlands

https://doi.org/10.1515/9783111076997-003

Per every ton of mixed plastic that is effectively recycled, an estimated energy saving equivalent to seven barrels of oil is made [5], which makes evident the importance of plastic recycling. Virgin PMMA, for example, has a significant production carbon footprint, making an appealing case for recycling this polymer (see Chapter 12). As recycling policies advance worldwide specifically aiming for higher recycling rates for household plastic waste, the challenge falls on collection strategies, sorting methodologies and technologies, policies, and plastic market economy dynamics [7, 8]. In what follows, emphasis is first on the main collection principles, then the main sorting principles, and finally the main pretreatment routes.

3.2 Collection principles

Waste collection has become part of urban life that responds to local regulations. In Europe, several countries follow a collection model which might be comprised of a combination of the following activities: (1) curbside collection by the local authorities, (2) operation of recycling collection centers for household waste, and (3) the introduction of convenience collection points at high-circulation urban locations [7, 9]. Each of these activities has a different contribution regarding the quality of the streams for recycling and also the total amount of plastics that are collected [7]. From the curbside collection yields, by far the highest quantity of plastics was collected [7, 10], and this becomes enhanced with a fortnightly frequency of collection [10]. Also, depending on the model of collection, the waste could range from fully commingled with other materials to (in the limit) fully sorted at the source. This selection affects both the quality of the waste to be recycled and the total quantity collected [7]. From all the possibilities, either a postseparation system or a curbside collection system combined with a policy of unit-based pricing are the best options to increase the quantity of plastics collected and to optimize the cost-efficiency [9].

The quality of the collected plastic materials is affected most dominantly by the presence of contaminants and the presence of other materials, either plastics or not, which cannot be processed alongside the targeted plastics [10]. Some studies have shown that only about half of rigid plastics and less than 5% of film-shaped plastics are collected from household sources [4].

One should realize that recyclable plastics could be disposed with the mixed residual waste, which could end up being incinerated, land-filled, littered, misallocated with other wastes, all of which could affect the recycling rates. Additionally, the quality of the collected material could cause rejection at the processing facilities, also negatively affecting the recycling rates. Increasing the recycling rates could be challenged by problems with the sorting done at households [4]. The differences in the models applied from place to place originate from somewhat complex sorting rules which might be confusing and therefore impact collection. Also, from the circular model, some proposals might enhance the collection rates in case implemented throughout the supply chain, for example, standardization of plastic containers, marking the materials to facilitate sorting [4].

It is interesting to point out that collection has a very high cost throughout the recycling process. Besides, both the fuel consumption of collection vehicles and the consistent use of plastic bags for the handling of solid household waste pose an additional environmental impact of considerable importance [11]. This is relevant as a positive environmental balance must be obtained after a circular plastic cycle. A broad share of knowledge throughout the plastic recycling value-chain is therefore important. Once stakeholders are informed about the (im)possibilities of plastic recycling, and the value created, a collaborative effort can be made toward efficient plastic waste collection.

3.3 Sorting principles

The efficiency of recycling is dependent on the performance of both sorting and pretreatment [5, 12] and the amount of knowledge of downstream markets and material requirements, as to avoid treating waste in a way that reduces its value. As shown in Figure 3.2, plastic waste might undergo mechanical recycling, leading to reuse by reshaping the material into production of the same products (circular production) or new different products without further recycling (material recovery) [3]. Reshaping the plastic via thermomechanical processes can lead to degradation of the material properties [5]. Therefore, plastic from mechanical recycling must sometimes be mixed with new plastic to keep up with the (societal) specifications or expectations of the products. Energy recovery is the last option because of the implicit CO_2 emissions [3]. Chemical recycling is an appealing option that aims to generate valuable chemical components. It is not yet developed to be a primary option for most polymer waste streams, but shows potential for complex streams. In the case of PMMA recycling, chemical recycling is often the only option available, and the preferred option for most PMMA waste streams. Overall, though, mechanical recycling remains momentarily the best option in hand [1, 3].

The sorting (and pretreatment) processes need to provide an input feed to recycling that is as pure as possible. Mixed materials, foils, contamination, and multilayer packaging are just among some of the factors that can cause the rejection of a plastic waste batch [1, 12]. In the remaining part of this chapter, focus is on the most important design considerations for sorting, the concept of manual sorting, the principle of sorting based on material properties, (semi)automated sorting, and other separation methods.

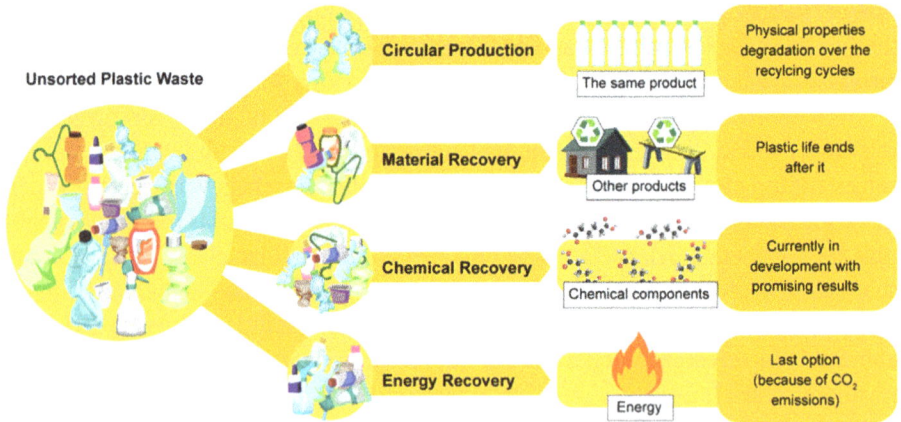

Figure 3.2: General overview of plastic recycling according to the process: (1) circular production, (2) material recovery, (3) chemical recovery thus chemical recycling, and (4) energy recovery [1, 5]. Final disposition of plastic that is in no way recycled or reused, for example, landfilling or incineration, is not considered in this figure.

3.3.1 Considerations for designing a sorting process

Different final applications or targets have different requirements. Therefore, sorting is a process that must be designed specifically for the implemented recycling process in mind. Perhaps, the simplest case is mechanical recycling of post-industrial plastic waste, for example, from a failed production batch, as it can simply be rerun in the company [3]. This case requires no sorting and no pretreatment, aside from the typical crushing of production waste into manageable particles, as the stream contains one single (defined) material, and it has not highly unexpected contaminants. In contrast, chemical recycling is broad in the number of options available. Some chemical processes require the absence of foreign materials such as metal, wood, or paper. The size of the plastic pieces in the feedstock to be loaded in the reactor can be a limiting factor either for the reactor or for the mechanical feed to it [12]. Mechanical recycling beyond the own company also needs sorting to process the material according to its composition. The streams processed by mechanical recycling should have high purity to keep the quality of the material as close as possible to the material produced from raw stock. In many cases, these streams have to be rectified, for example, by mixing it with new plastic or additives.

In general, material is sorted according to the shape, density, size, color, or chemical composition, among other properties [3]. The most important parameter is the chemical composition of the plastic. The five most common polymers found in the market are low-density polyethylene (LDPE), high-density polyethylene (HDPE), polypropylene (PP), polyvinyl chloride (PVC), and polyethylene terephthalate (PET). LDPE,

HDPE, PP, and PET are mostly used in the packaging sector, and PVC is most common in the building and construction sector [13] (Figure 3.3). PMMA is a bit of an outlier, with its excellent recyclability not widely recognized, but with a large recycling potential.

3.3.2 Manual sorting

Typically, the sorting of postconsumer plastic waste in developed countries is conducted via (semi)automated equipment. However, some products such as electronic equipment and large household appliances such as televisions, microwave ovens, refrigerators, and washing machines are usually dismantled before recycling, and a coarse manual sorting of different plastics is still a common practice [14].

Since the manual sorting of plastic waste is very laborsome, lower value waste is exported to low-wage countries for sorting. This practice has been reduced due to more stringent regulation that is expected to evolve further https://environment.ec.europa.eu/topics/waste-and-recycling/waste-shipments/plastic-waste-shipments_en.

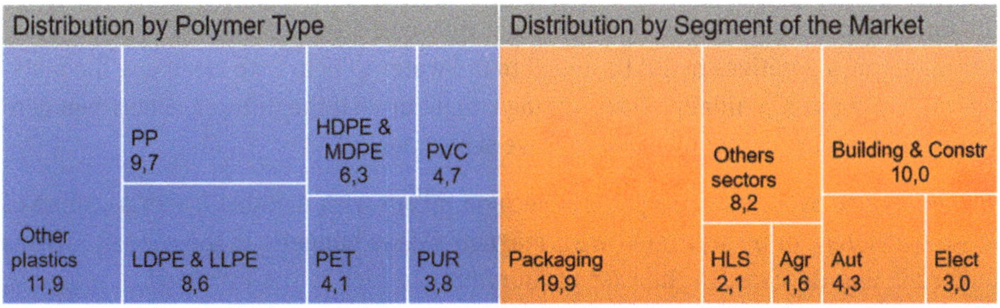

Distribution by Polymer Type				Distribution by Segment of the Market				
	PP 9,7	HDPE & MDPE 6,3	PVC 4,7			Others sectors 8,2	Building & Constr 10,0	
Other plastics 11,9	LDPE & LLPE 8,6	PET 4,1	PUR 3,8	Packaging 19,9	HLS 2,1	Agr 1,6	Aut 4,3	Elect 3,0

Figure 3.3: Demand of plastic materials in the European Union in 2020 (49.1 million tons), expressed in million tons and classified by polymer types and by segment of the market [13]. LLPE, linear low-density polyethylene; MDPE, medium-density polyethylene; PUR, polyurethane; Const, construction; Elect, electrical and electronics; Aut, automotive; HLS, household, leisure, sports; Agr, agriculture.

3.3.3 Sorting by material properties

By size: Materials can be primarily sorted by means of sieves. An example is the mechanical separation made in progressive rotary drums in which the materials are fed and then forced to move through it by a helix mounted fix to the inside [3]. The first purpose of these kinds of drums is to wash the material, and these drums might also be equipped with several types of sieving devices such as perforated plates, wire screens, or fabrics. As the drum rotates, the solids tumble and move along, with a fraction falling through the corresponding openings according to their size. Opening sizes are selected to certain types of plastic waste, like polyolefin bottle caps, which are normally small. Rotary drums can handle large volumes being fed continuously and have a simple mechanical design. Their energy consumption is low, and their maintenance is rather straightforward.

By density: Density variations are also useful at sorting facilities by allowing to remove easily low-density elements, such as films, paper, and labels. An example of this operation is the air density separator, or wind shifter, or wind sifter [3]. The core section of a wind sifter equipment is a chute that is fed from above with the waste stream to be separated. The feed falls through the chute while a controlled stream of air is blown in the opposite direction and then diverted to a lateral exit. The principle is that a heavy fraction passes while the remaining light fractions are blown away. Wind sifters operate at low cost and have a high capacity. In comparison to rotary drums, wind sifters witness a dry operation.

Density allows separating plastics not only from contaminants but also from each other. Examples are float-sink separation tanks that, besides the separation, can perform washing since they operate with water [3]. Sink-float separation tanks are simple as they often use water as a means to separate different plastics. The working principle is that any material with a density higher than the one of water will sink, while the lower density materials will float. It is a very cost-effective method of separation, and sometimes additives might be added to the water to help with cleansing the material from some contaminants. Disadvantage might arise if the difference between densities is not enough to perform an effective separation [3].

By magnetism: Whenever the plastics become mixed with metallic fractions, some of these fractions can be separated with magnets [3], as happens in the case of ferrous materials like tin cans. In the case of aluminum, which is not attracted to magnets, devices like the eddy current separator can induce a magnetic field on it and effectively separate it from other materials [3].

Conventional magnetic separators come in different configurations, but the basic principle remains. One part of the equipment is magnetic, attracting ferrous materials and leaving the remaining fraction to be diverted. Magnets can be installed above a conveyor belt to attract ferrous material and divert it, while the nonferrous fractions remain to be transported to another process (phase). Another possible configuration is letting the mixed material fall through a vertical shaft equipped with one or more magnetic drums that capture and then divert the ferrous fraction and leave the rest to fall. Conventional conveyor belts can be equipped with magnetic roller heads, so that in case the mixed material reaches the roller head (at the end of the conveyor belt), nonferrous material can be projected forward, while ferrous material keeps rotating with the belt until it passes the magnetic field regions of the roller head (from below) and then falls.

Eddy current separators operate under a somewhat different principle. They are equipped with magnetic poles that can rotate at high speed. As nonferrous metals pass over the rotating magnets, the rapidly changing polarity induces the so-called eddy currents. Material such as aluminum develops for an instant its own magnetic field, matching the polarity of the rotating magnets. This creates a repulsion force allowing to expel the nonferrous metal away, separating it from the remaining fraction on the separator.

By rigidity: Rigid and flexible materials can be separated from each other based on their difference in material response. A good example is ballistic separators, which by using alternating moving paddles can separate flexible material from the rigid fraction [3]. These ballistic separators consist of a deck in which the waste stream is fed from above. The bottom of the deck comprises several long moving flat paddles, lying with a certain angle of inclination, next to each other. These paddles oscillate in an alternating manner. The working principle is based on the difference in material properties. Light and flexible material will be prone to bounce up the deck, while the heavier and more rigid fraction will have the tendency to tumble down. That behavior makes the separation possible. Additionally, paddles can be provided with a screening surface to let fine material to pass through and fall, providing a third outlet stream.

Another equipment that performs this kind of separation is star screens. Star screen separators are somewhat different in the operating principle. In a star screen, rotating shafts lay across the equipment, and their rotation conveys the material along the equipment. Every shaft is equipped with star-shaped forms made of rubber, and in some cases metal. As the stream is conveyed along, the larger and more rigid materials keep being transported on top of the stars, while flexible and small material falls through. These star screen separators may suffer entanglement of soft flexible materials such as films and bags, which may be an operational issue in some cases. However, some models address this weakness by adjusting the geometry of the shafts or the stars. Another operational aspect is that rubber stars wear off more rapidly than metal pieces, increasing the need for maintenance.

By melting point: If solid contaminants remain in the plastic material before being processed, melt filtration can be used to separate the targeted plastic from other materials. The material is melted and extruded through a sieve that retains particles of glass, metal, polymers with higher melting point, among others [3]. The contaminants to be removed via melt filtration should ideally not melt or melt only at a higher temperature, so they can remain solid before passing the filtering element. Melt filters normally operate as a final separation before the streams are thermomechanically reformed. They can play a very important role, assuring the quality of the final product which otherwise would be affected if the contaminants remain in the stream.

Melt filters are operations that imply maintenance as the contaminants will be unavoidably trapped in the filtering element. There are options that range from stopping the process and changing the filtering element to the use of units that automatically monitor pressure drop, change filtering elements, clean them, and keep the process running. The actual choice must be always evaluated and selected according to the needs.

3.3.4 (Semi)automated sorting

Sorting of plastics can be done in semiautomated or fully automated ways. In these approaches, the identification of the different plastic types is commonly achieved by spectroscopic techniques. Fourier transform infrared (FTIR) spectroscopy is usually applied. Other techniques used are Raman spectroscopy, laser-induced breakdown spectroscopy, and hyperspectral imaging methods [15]. The use of one of these techniques, or a combination of them, is very beneficial for recycling companies, increasing the sorting yield and the quality of the sorted plastics [14].

FTIR spectroscopy: Depending on the chemical structure of the plastics, certain infrared (IR) wavelengths are absorbed or reflected. Since plastics show characteristic chemical compositions, the identification of the types of plastics by the absorbance/ reflectance of specific IR wavelengths allows their identification. Among the IR techniques, near-infrared (NIR; wavelength 780–2,500 nm; wavenumber 12,800–4,000 cm^{-1}) is commonly used by the industry. Mid-infrared (MIR; wavelength 2,500–16,667 nm; wavenumber 4,000–600 cm^{-1}) is also applied in some cases [14].

The spectrometric identification of plastic residues by NIR can be conducted by applying IR light on a conveyor belt. The plastic parts located on the conveyor belt are identified according to their absorbance and reflectance of IR. Then the particles are expelled from the belt into specific collecting bins according to their chemical structures. One disadvantage of NIR is that it cannot identify dark-colored plastics. However, these plastics can be identified by MIR [15].

Table 3.1: Advantages and disadvantages of the NIR and MIR technologies [14].

Technology	Advantages	Disadvantages
NIR	– Commonly used in the industry – Fast, noncontact technique – Suitable for use with conventional optical fibers – Portable dives are also available (for quality control)	– Limited information in the near-IR range – Not suitable for the identification of dark-colored plastics – The presence of carbon black affects the absorption and scattering of NIR light
MIR	– Ability to detect characteristic absorption bands (fingerprint region in the MIR range) – Higher accuracy and information than NIR – Suitable for the identification of dark-colored plastics – Established technique	– Very sensitive to the surface characteristics of the sample – Difficult remote analysis – Requires special optical fibers, more fragile, and expensive – Normally the analysis time is higher than that for NIR

A summary of the main advantages and disadvantages of NIR and MIR is shown in Table 3.1. Specifically, an important limitation to the FTIR sorting is that it relies on detecting a surface, which can be a label or another material (as in multilayer packaging) or even dirt, all these leading to false identification [3].

Optical color recognition: This technique uses light detection to sort material according to their color [3]. Optical sensors capture the signal, and this signal is compared to a standard that rules the sorting. It is useful in sorting products colored differently but from a single type of material.

3.3.5 Other separation methods

Specialized techniques exist to separate mixtures of polymers. For example, triboelectric (electrostatic) separators can be utilized. These work more efficiently on binary mixtures, and in some lesser extent with mixtures of more than two components. They can induce a charge in either one or two of the plastic materials, creating the conditions to separate them by repulsion forces [3]. This process requires that the mixture is comprised of nonconductive and noncomposite species, shows different triboelectric behavior, and is dry and clean with fineness in size of the materials.

Froth flotation is an alternative technique not yet widely implemented but being in development. It aims to separate polymers with densities lower than that of water by using a stream of bubbles in a tank with water and modifying the surface properties of the plastics. That modification should be done before feeding the stream in the tank [3]. Another density-based separation technique is magnetic density separation (MDS), which is similar to traditional density separation in water, but the liquid used contains iron oxide, and its density can be varied by applying a special magnetic field. MDS allows separating multiple fractions, but it still fails to be able to separate fractions with close density values [3].

3.4 Pretreatment principles

The feedstock to either the recycling process or the sorting process might need to be pretreated to suit the needs of the case. Examples are size reductions and the removal of contaminants, for which different methods have been developed.

Size reduction: Some reactors for recycling require particles of even size. This might be because of the need of a bed to be effectively fluidized, or because the thermodynamics rely on a uniform size distribution [12]. This operation is mostly performed with shredders. These shredders commonly consist of a hopper-shaped chamber over one or more rotating shafts possessing sharp blades. These blades crush and cut the materials into a preselected sized, designed specifically for the processes downstream the shredding operation.

Cleansing: Dirt or some other materials can be removed from the waste stream by physical separation or by washing. Some options are rotating drum washers, friction washers, and so on [3].

Rotating drum washers can separate some solid contaminants such as glass, metals, and rocks, while streams of water along the drum can provide washing to remove other kinds of dirt. The waste is loaded in this equipment which consists of a long horizontal cylinder, possessing an inner helix which will make the materials move alongside due to the rotational movement. The particles and object will tumble as they advance, being eventually exposed to the liquid. This cleansing liquid, most commonly water, can be applied by immersion, which allows for better reaching of cavities. It can also be applied by pressurized spray which reaches better the outer surfaces.

Friction washers are mostly designed for shredded material. The equipment consists of an inclined casing which is loaded from the lower end. Inside, there is a rotating longitudinal shaft, normally provided with paddles. Around the shaft one encounters a secondary inner case, normally made of perforated metallic plates. As the material is fed, the shaft rotates at high speed, throwing the material against the inner casing. At the same time, water is applied to wash dirt off. As the shaft rotates, the paddles keep throwing the material against the casing, creating friction between them, and making it easier for the water to remove the contaminants. The inclination helps increasing the residence time, enhancing the cleansing.

Drying: It might be required in some cases to remove water which is most effectively carried out with a mechanical dryer [3]. Mechanical dryers work with different principles. High-spin dryers possess a rotating mechanism forcing the material against a perforated barrier or a mesh. This kind of equipment requires high rotational speeds to effectively remove water. Hot air dryers are also common and operate by letting the waste material fall through a hopper while hot air is circulating in counter-current absorbing the water. Air dryers can, instead of having a hopper, contain a rotating drum which could enhance the contact between the solids and the hot air.

References

[1]   Vollmer I, et al. Beyond mechanical recycling: Giving new life to plastic waste. Angewandte Chemie International Edition Sep 2020, 59(36), 15402–15423. 10.1002/anie.201915651.
[2]   Dahlbo H, Poliakova V, Mylläri V, Sahimaa O, Anderson R. Recycling potential of post-consumer plastic packaging waste in Finland. Waste Management Jan 2018, 71, 52–61. 10.1016/j.wasman.2017.10.033.
[3]   Ragaert K, Delva L, Van Geem K. Mechanical and chemical recycling of solid plastic waste. Waste Management 2017, 69, 24–58. 10.1016/j.wasman.2017.07.044.

[4]   Burgess M, Holmes H, Sharmina M, Shaver MP. The future of UK plastics recycling: One bin to rule them all. Resources, Conservation and Recycling Jan 2021, 164(September 2020), 105191. 10.1016/j. resconrec.2020.105191.

[5]   Thiounn T, Smith RC. Advances and approaches for chemical recycling of plastic waste. Journal of Polymer Science May 2020, 58(10), 1347–1364. 10.1002/pol.20190261.

[6]   European Union. Municipal waste by waste management operations. *Eurostats*, 2022. https://ec.eu ropa.eu/eurostat/databrowser/view/env_wasmun/default/table?lang=en (accessed Feb. 01, 2022).

[7]   Hahladakis JN, Purnell P, Iacovidou E, Velis CA, Atseyinku M. Post-consumer plastic packaging waste in England: Assessing the yield of multiple collection-recycling schemes. Waste Management 2018, 75, 149–159. 10.1016/j.wasman.2018.02.009.

[8]   Ventola V, Brenman E, Chan G, Ahmed T, Castaldi MJ. Quantitative analysis of residential plastic recycling in New York City. Waste Management & Research May 2021, 39(5), 703–712. 10.1177/ 0734242X211009968.

[9]   Dijkgraaf E, Gradus R. Post-collection separation of plastic waste: Better for the environment and lower collection costs? Environmental and Resource Economics 2020, 77(1), 127–142. 10.1007/s10640-020-00457-6.

[10]   van Velzen EUT, Brouwer MT, Feil A. Collection behaviour of lightweight packaging waste by individual households and implications for the analysis of collection schemes. Waste Management Apr 2019, 89, 284–293. 10.1016/j.wasman.2019.04.021.

[11]   Fernández-Braña Á, Feijoo-Costa G, Dias-Ferreira C. Looking beyond the banning of lightweight bags: Analysing the role of plastic (and fuel) impacts in waste collection at a Portuguese city. Environmental Science and Pollution Research Dec 2019, 26(35), 35629–35647. 10.1007/s11356-019-05938-w.

[12]   Qureshi MS, et al. Pyrolysis of plastic waste: Opportunities and challenges. Journal of Analytical and Applied Pyrolysis Nov 2020, 152(March), 104804. 10.1016/j.jaap.2020.104804.

[13]   Plastics Europe. Plastics – The Facts 2021, an Analysis of European Plastics Production, Demand and Waste Data, Brussels, Belgium, 2021.

[14]   Andrady AL. Plastics and the Environment, vol. 148, John Wiley & Sons, Inc., Hoboken, NJ, USA, 2003.

[15]   Parameswaranpillai J, Rangappa SM, Rajkumar AG, Siengchin S. Recent Developments in Plastic Recycling, Springer Singapore, Singapore, 2021.

Rudinei Fiorio, Mariya Edeleva, and Simon van der Heijden

4 Mechanical recycling principles and challenges

4.1 Plastic waste management with key role for mechanical recycling

Although today's world is difficult to imagine without plastics, these materials have only been around for a hundred or so years. At the time of their first synthesis, plastics were seen as extraordinary materials, due to their remarkable properties such as durability, lightweight, and easy processability [1].

Initially, the variety of polymers commercially available as well as their production was limited. After the Second World War, the development of new plastic or polymeric materials increased substantially, leading to continuous growth in their production [1, 4]. Figure 4.1 (black symbols) shows the worldwide global annual production of plastics from 1950 to 2018 [2]. For example, in 2019, the global production of plastic was almost 370 million tons [5]. As one could expect, the generation of plastic waste followed the same trend as its production. Figure 4.1 (red symbols) shows the estimated global plastic waste generation also from 1950 to 2015 [3]. Considering the interval 2000–2015, the approximated plastic waste generation is close to 90% of the yearly plastic production. The cumulative quantity of plastics manufactured from 1950 through 2015 is estimated as 8,000 Mt, and all plastic waste ever generated from primary plastics is around 6,000 Mt. Hence, it is estimated that ca. 2,000 Mt of plastic is currently in use as products.

Unfortunately, some of the major advantages of plastic materials such as durability and resistance to (nonforced) degradation make them difficult to be assimilated by nature, accumulating and causing pollution in different ecosystems [3]. Despite the fact that plastics are being used for more than a century, reports of plastic pollution in the ocean in the scientific literature date from the early 1970s [6], in agreement with the beginning of the exponential increase in the plastic waste generation shown in Figure 4.1.

Rudinei Fiorio, Centre for Polymer and Material Technologies (CPMT), Ghent University, Technologiepark 130, Zwijnaarde 9052, Belgium
Mariya Edeleva, Laboratory for Chemical Technology (LCT), Ghent University, Technologiepark 125, Zwijnaarde 9052, Belgium
Simon van der Heijden, Heathland B.V., Arkansasdreef 8, 3565 AR Utrecht, The Netherlands

https://doi.org/10.1515/9783111076997-004

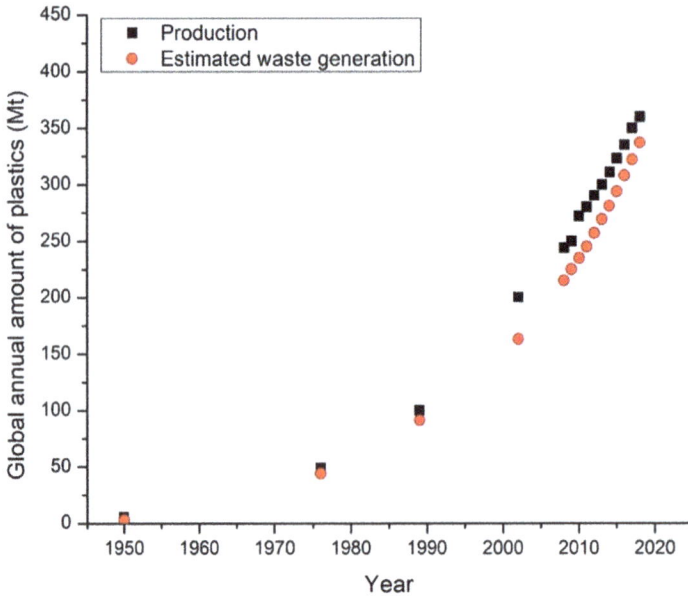

Figure 4.1: Global annual amount of plastics/polymers with production (■) [2] and estimated waste generation (●) [3].

With the increase in plastic waste generation, adequate management of this waste is necessary. Plastic waste can have several destinies. They may be reused in the manufacturing of other items, for example, introducing plastic packaging residues into construction materials [7]. Furthermore, these residues can also be recycled, generating either new plastic products (mechanical recycling route) or high-quality chemicals allowing for repolymerization or other modifications (chemical recycling route). A specific route is thermal degradation (e.g., pyrolysis) with subsequent generation of fuels but one can also apply incineration, which aims at recovering energy. In principle, one can also consider discarding in a controlled manner in sanitary landfills [3]. Recycling is the preferred solution for plastic waste management, since it requires in many cases less energy than manufacturing virgin materials [8, 9]. However, only approximately 12% of the plastic waste generated worldwide is currently recycled [10].

Recycling of plastics began to become relevant starting from the 1980s [3]. Aiming for the easy identification of plastics as well as stimulating plastics recycling, the Society of the Plastics Industry assigned different codes to be used during the manufacturing of products [11]. These codes, according to the resin identification code informed in the ASTM D7611 standard [12], are shown in Figure 4.2. Numbers 1–6 identify the most consumed thermoplastics: 1, poly(ethylene terephthalate) (PET or PETE); 2, high-density polyethylene (HDPE); 3, poly(vinyl chloride) (V or PVC); 4, low-density polyethylene (LDPE); 5, polypropylene (PP); and 6, polystyrene (PS); the number 7 is applied to all other plastics [12].

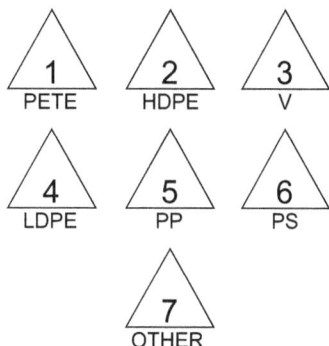

Figure 4.2: Resin identification code according to ASTM D7611 standard – option A [12].

If one desires to directly obtain a recycled polymeric product, one needs to pay attention to sorting, collection, and pretreatment (cf. Chapter 3), and one needs to exploit mechanical recycling, the actual transition of waste polymer into recycled polymer (e.g., via extrusion). In this chapter, the general principles of mechanical recycling are addressed, further addressing a concise comparison with chemical recycling and a case study selecting a typical group of vinyl polymers suitable for the mechanical recycling route. The group of vinyl polymers covered in case studies in this chapter does not include polyacrylics, for which chemical recycling is much more preferred, as explained in full detail in Chapters 5–11.

4.2 Mechanical recycling principles

Regarding plastic recycling, four classifications are often listed: primary recycling and secondary recycling, both involving physical reprocessing of plastics; tertiary recycling, concerning chemical modification, which is also known as chemical recycling; and quaternary recycling, concerning the recovery of energy from plastic waste [11].

Figure 4.3 schematically represents the aforementioned different types of plastics recycling. This chapter focuses on physical or mechanical recycling, namely, primary and secondary recycling. Depending on the source of the waste, mechanical recycling is subdivided into primary recycling, which is applied to post-industrial (or pre-consumer) residues (Figure 4.3a), and secondary recycling, which refers to the reprocessing of post-consumer plastics (Figure 4.3b) [13].

4.2.1 Primary recycling

Primary recycling consists of the reutilization of industrial residues; thus, pre-consumer wastes such as scraps and defective parts are generated during the manufacturing cycle. Therefore, primary recycling is a closed-loop recycling approach, in which the recycled material is used to produce the same items made using virgin plastics, as shown in Figure 4.3a. Despite the fact that (limited) degradation occurs during the recycling step, this operation involves the physical transformation of the material, which can also be known as remolding (or re-extruding). Usually, the manufacturer of plastic parts applies primary recycling internally, reusing its own wastes. These pre-consumer wastes are normally uncontaminated and easily identifiable materials, simplifying the recycling process [11].

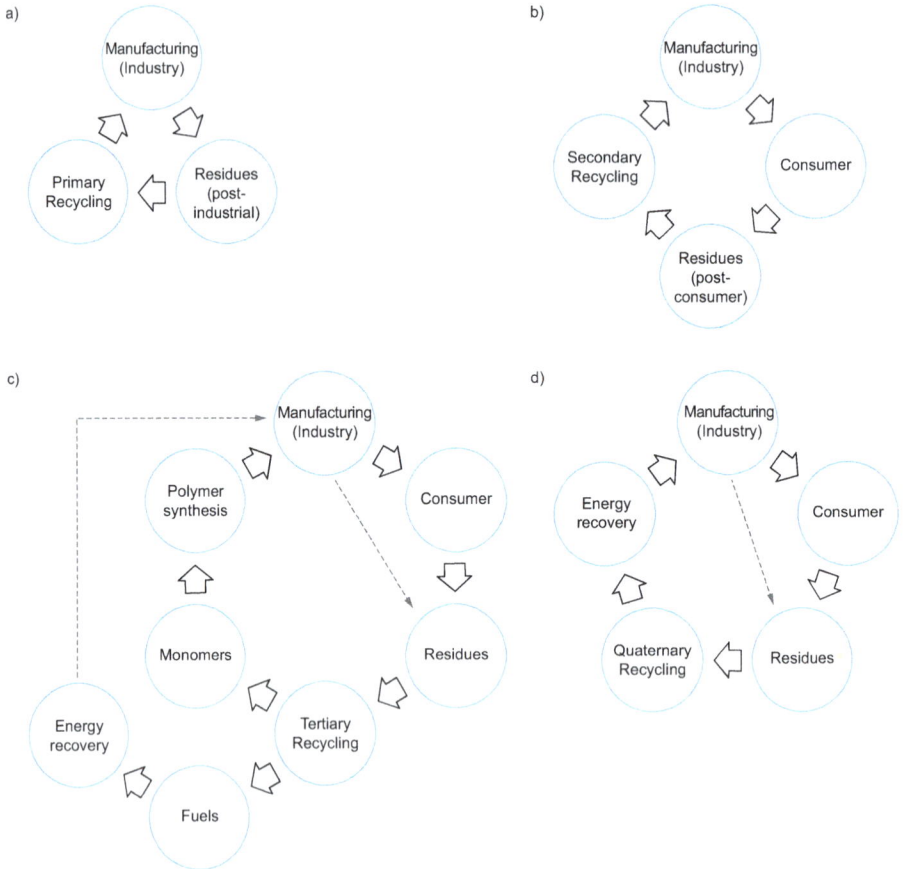

Figure 4.3: Types of plastics recycling: (a) primary recycling (mechanical); (b) secondary recycling (mechanical); (c) tertiary recycling (chemical); and (d) quaternary recycling (energy recovery).

4.2.2 Secondary recycling

Secondary recycling also consists of the physical remolding of plastic residues. However, these residues have a diverse source than those of primary recycling. They are postconsumer plastics, obtained from discarded products by the end user or, less commonly, from industrial postprocessing operations, for example, scraps from cutting and machining plastic parts. As shown in Figure 4.3b, they consist of a mix of several plastics [8]. Since most of the plastics cannot be blended due to their incompatible chemical structures, sorting is a fundamental step in recycling.

The final application of mechanically recycled plastics strongly depends on the quality of the material. According to its quality, the recycled material can be used in three different ways. It can be used to substitute the virgin material in the manufacturing of the same type of product, which is so-called closed-loop recycling. It can also be employed for higher end applications compared to the original one, which is known as upcycling. Under such circumstances, the properties of the recycled material are equivalent or better than those of the virgin polymer, which is the ideal situation. Here, properties can be improved during recycling by additives and blending. In contrast, if properties of the material decrease due to recycling, the recycled material is used for the manufacturing of low value-added products, called downcycling [14]. Currently, the latter approach is the most dominant one.

4.2.3 Sorting and collection

As explained in the previous section, mechanical recycling heavily depends on sorting and collection. Postconsumer plastics require proper pretreatments before their reprocessing, usually presenting contaminants. They also exhibit a certain level of degradation that occurred during their lifetime, leading to lower quality products upon comparison to goods made from virgin plastics.

Common pretreatments and sorting operations applied in postconsumer plastics before their reprocessing are shown in Figure 4.4. According to this figure, after a coarse sorting, the waste stream is shredded and non-plastic fractions, such as metals, are removed. Next, light products such as foams, fine particles, and textiles are taken away. Then, low-density plastics are separated from high-density plastics, by placing the mixed plastic waste in liquids with known density so that the plastic fraction showing lower density than the liquid will float, and the denser fraction will sink [15]. The plastics fraction can be further sorted by, for example, spectrometric methods. These spectrometric methods can also be applied in earlier stages of the sorting process depending on the waste source, for example, electrical-electronic equipment and automotive wastes.

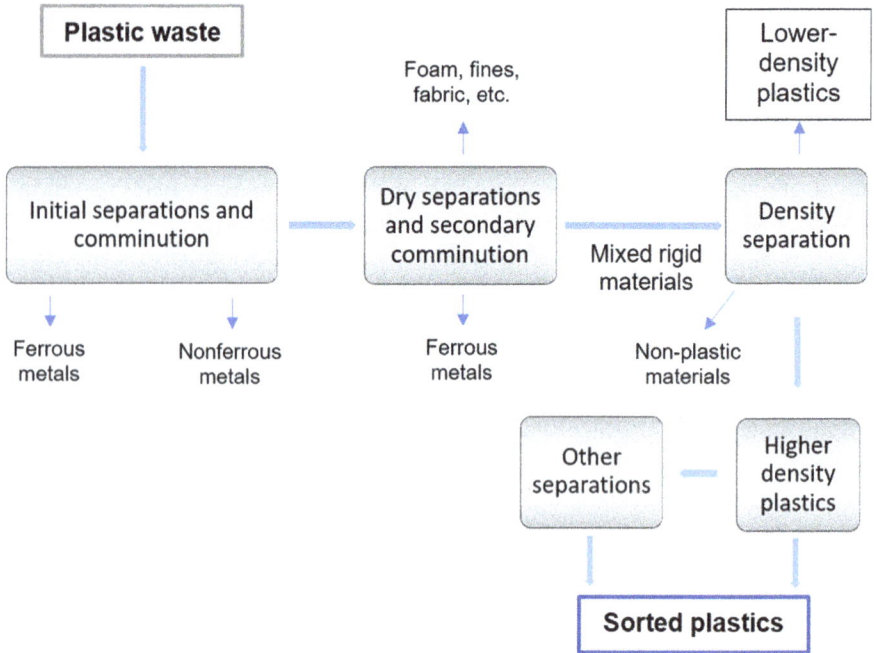

Figure 4.4: Common operations in the pretreatment and sorting of postconsumer plastics as needed to fully exploit mechanical recycling (adapted from [15]).

4.3 Mechanical versus chemical recycling

A natural question that arises is when mechanical recycling is preferred over chemical recycling. Table 4.1 summarizes the main advantages and disadvantages of these two recycling types. As shown in this table, the selection of the most appropriate recycling type depends on the characteristics of the waste stream and the final application of the recycled material.

4.4 Examples of industrial implementations of mechanical recycling

Among the methacrylate plastics family, polymethyl methacrylate (PMMA) is the most well-known polymer. It is obtained by chain-growth polymerization of methyl methacrylate (MMA) [16]. PMMA shows excellent characteristics as high transparency, resistance to weather (ultraviolet light), good impact strength, and tensile properties,

Table 4.1: Advantages and disadvantages of mechanical and chemical recycling.

	Mechanical recycling	Chemical recycling
Advantages	– Presents in principle the lowest environmental impact among all plastic recycling alternatives. – Lower capital expenditure and operational complexity – Primary recycling is simple, inexpensive, and allows closed-loop recycling, substituting virgin plastics – Easy compliance with REACh regulation	– Quality of the recycled materials can be as high as virgin ones, likely with purification – Possibility to obtain high-end products – Suitable to most thermoplastic polymers and their blends – Regenerated material from both post-industrial recycled (PIR) and post-consumer recycled (PCR) sources can be used as a replacement of virgin material – A requirement to be able to recycle most PMMA waste types
Disadvantages	– Limited number of reprocessing cycles, as each cycle leads to a loss of quality due to degradation – Primarily applied to thermoplastic materials – Limitations for the use of recycled plastic in food-contact products – Secondary recycling requires proper washing and sorting steps – Contamination is difficult to fully avoid in secondary recycling – Often leads to downcycling due to degradation or limited application possibilities due to colorants – Cannot be used for plastic wastes containing legacy additives (which are no longer allowed) as they cannot be removed (e.g., phthalates, brominated flame retardants, and heavy metal pigments) – In the case of PMMA: not a suitable option for cast grades and mixed color waste	– Requires high purity of materials for recycling, putting more emphasis on pretreatment steps – May involve complicated chemical agents and/or catalysts – High energy consumption (depends on polymer type) – May result in significant release of volatile organic compounds – Special equipment involved thus high capital expenditure – Each polymer type requires a specialized procedure

among others. Due to these characteristics, PMMA finds applications in automotive lamp covers, lighting equipment, optical fibers, and light guide panels [17].

Concerning its recycling, mechanical recycling of PMMA is normally restricted to scraps and defective parts from extrusion and injection molding of clear grades. However, the transparency of mechanically recycled PMMA products is generally compromised, requiring blending the recycled product with virgin raw material. Specifically, discoloration or spots from contaminants are difficult to avoid by mechanical recycling PMMA, especially after multiple recycling cycles [18]. Chemical recycling is thus much more relevant, and the core of these is given in Chapters 5–11.

However, to make the reader aware of mechanical recycling, the focus of this chapter is on the recycling of polyolefins or polyolefin-rich streams, considering industrial implementation routes. The emphasis is on the recycling of PP and PE (polyethylene), as these two types of polymers account for more than 50% of the plastics produced [19]. PE and PP waste fractions can be obtained from different sources. According to the waste source, the sorting process can present diverse variations.

Gu et al. [20] conducted a life cycle assessment of mechanical recycling of plastic waste in an industrial context and focused on the recycling of PP and PE (both LDPE and HDPE). The types of plastic solid waste (PSW) were very specific, for example, PP components from panels of refrigerators and PP bottle caps; hence, a presorting was already conducted. The recycling flow for this PSW fraction is shown in Figure 4.5. The collected waste is initially washed and sorted. Then, the material is shredded, aiming to increase the bulk density, facilitating both transport and extrusion. Finally, the extruded material is pelletized, stored, and reused in the manufacture of new plastic products.

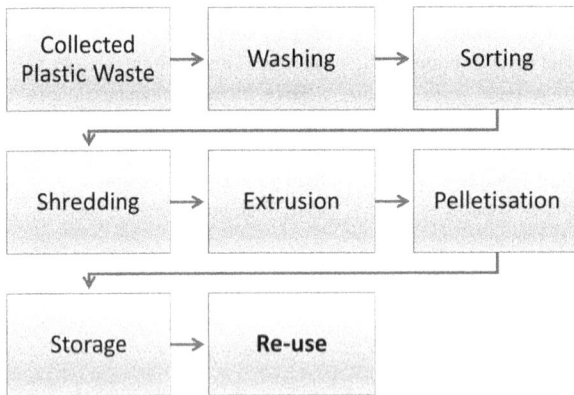

Figure 4.5: Mechanical recycling flow for postconsumer PP and PE, after presorting of PSW (adapted from [20]).

Mass balance flows have also been reported [20], indicating losses of ca. 2 m% after washing and sorting, compared to the initial mass intake of PP and PE. This small loss is due to the presorting step conducted before the material enters the flow presented in Figure 4.5.

For more heterogeneous waste streams, as the one obtained from a household collection system employed in Belgium called PMD (*Plastics, Metalen en Drinkkarton*),

another recycling scheme is necessary. In the PMD system, plastic bottles, metal cans, and carton drink packaging are put together in a collection bag by the final consumer. These bags are collected and then delivered to sorting facilities. The initial sorting flow of PMD is summarized as follows [13]:
1. The sorting process starts by passing the mixed waste through a progressive rotating sieve, sorting the waste by size. Here, both very small and very large objects are removed from the process.
2. A wind shifter is used to remove light objects such as loose paper and plastic labels from the remaining waste fraction.
3. Next, a magnetic device is used to remove ferrous metals, while an optical system removes the carton, and the nonferrous metals (mostly aluminum) are separated by an eddy current device.
4. A ballistic separator removes the "soft" plastics (foils) from the "hard" plastics.
5. Near-infrared spectroscopy is used to separate PET and PE. PET is further sorted by an optical system into clear, blue, and green PET.
6. Finally, all the sorted waste streams are submitted to human inspection so that they can be further separated.

The plastic waste prevenient from the PMD system contains mainly PET, PE, and PP, as well as a fair amount of PS, and small quantities of PVC, poly(acrylonitrile-co-butadiene-co-styrene) polymer, polyamide, and polycarbonate. Also organic contaminants (food residues) are present [13]. The recycling flow is summarized as:
1. The plastic waste is put on a belt and fed into a crude shredder.
2. The shredded material is submitted to a first washing in a rotating drum, gravitationally separating contaminants (e.g., rocks, metals, and glass).
3. Friction washers remove the organic contaminants.
4. The washed fraction is again shredded, resulting in plastic particles of 1–12 mm.
5. Once more, friction washers are employed.
6. A float–sink separation system, consisting in a water bath, separates PP and PE (which float in water) from the other polymers that will sink.
7. An overhead magnet is used for the sink fraction to remove ferrous contaminants, and the plastic fraction is then dried.
8. The float fraction is also dried and then put into a wind sifter, where foils are removed.
9. The heavier ("hard") particles consist in mixing technical plastics and can be further sorted or reprocessed as a blend.
10. The float fraction (mostly PP and PE, called mixed polyolefin – MPO) is then extruded (with melt filtration) and pelletized, becoming ready as a secondary raw material.

As one can observe from item 10, the MPO fraction consists of a mix of PE and PP and can be extruded without further sorting. This fraction is less expensive than the pure recycled fractions of PE or PP. However, MPO shows lower mechanical properties since PP and PE present poor adhesion in the blend. The use of compatibilizers added during the extrusion process can improve the quality of the MPO [21].

Another important fact regarding the recycling of PE and PP is related to the degradation mechanisms during mechanical recycling. While PP tends to present chain scission, reducing its (average) molar mass, PE can eventually become prone to cross-linking, increasing the (average) molar mass [21]. Therefore, degradation reactions of both PE and PP will affect the rheological, morphological, and mechanical characteristics of MPO, making the prediction of MPO quality even more complex. The use of certain additives such as antioxidants can reduce the degradation of polyolefins during the extrusion step.

References

[1]   The future of plastic. Nature Communications 2018.
[2]   Klemeš JJ, Fan YV, Jiang P. Plastics: Friends or foes? The circularity and plastic waste footprint. Energy Sources, Part A: Recovery, Utilization and Environmental Effects 2021, 43(13), 1549–1565. 10.1080/15567036.2020.1801906.
[3]   Geyer R, Jambeck JR, Law KL. Production, use, and fate of all plastics ever made. Science Advances 2017, 3(7), 1–5. 10.1126/sciadv.1700782.
[4]   Simul Bhuyan M, Venkatramanan S, Selvam S, Szabo S, Maruf Hossain M, Rashed-Un-Nabi M, Paramasivam CR, Jonathan MP, Shafiqul Islam M. Plastics in marine ecosystem: A review of their sources and pollution conduits. Regional Studies in Marine Science 2021, 41, 21. 10.1016/j.rsma.2020.101539.
[5]   Plastics Europe – Association of Plastic, M. *Plastics – the facts 2020 – An analysis of European plastics production, demand and waste data*; 2020.
[6]   Jambeck J, Geyer R, Wilcox C, Siegler TR, Perryman M, Andrady A, Narayan R, Law KL. Plastic waste inputs from land into the ocean. Science 2015, 347(6223), 768–771.
[7]   Raut SR, Makarande S, Kalmegh V. Re-use of polyethylene plastic waste in concrete with plasticizer. Science and Technology 2007, 27, 1870–1876.
[8]   Jubinville D, Esmizadeh E, Saikrishnan S, Tzoganakis C, Mekonnen T. A comprehensive review of global production and recycling methods of polyolefin (PO) based products and their post-recycling applications. Sustainable Materials and Technologies 2020, 25, 1–34. 10.1016/j.susmat.2020.e00188.
[9]   Faraca G, Astrup T. Plastic waste from recycling centres: Characterisation and evaluation of plastic recyclability. Waste Management 2019, 95, 388–398. 10.1016/j.wasman.2019.06.038.
[10]   Vollmer I, Jenks MJF, Roelands MCP, White RJ, van Harmelen T, de Wild P, van der Laan GP, Meirer F, Keurentjes JTF, Weckhuysen BM. Beyond mechanical recycling: Giving new life to plastic waste. Angewandte Chemie – International Edition 2020, 59(36), 15402–15423. 10.1002/anie.201915651.
[11]   Thiounn T, Smith RC. Advances and approaches for chemical recycling of plastic waste. Journal of Polymer Science 2020, 58(10), 1347–1364. 10.1002/pol.20190261.
[12]   ASTM, ASTM D7611. Guiding the Plastics Recycling Value Chain, ASTM International, USA, 2013.
[13]   Ragaert K, Delva L, Van Geem K. Mechanical and chemical recycling of solid plastic waste. Waste Management 2017, 69, 24–58. 10.1016/j.wasman.2017.07.044.

[14] Oksana H, Cabanes A, Domene-López D, Fullana A. Applications and future of recycling and recycled plastics. In Parameswaranpillai J, Rangappa SM, Rajkumar AG, Siengchin S, eds. Recent Developments in Plastic Recycling. Springer, e-Book, 2021, 354–358.

[15] Fisher MM. Plastics recycling. In Andrady AL, ed. Plastics and the Environment. John Wiley & Sons, USA, 2003, 598–608.

[16] Pomes B, Richaud E, Nguyen JF. Polymethacrylates. In Grumezescu V, Grumezescu AM, eds. Materials for Biomedical Engineering. Elsevier, 2019, 217–271.

[17] Kikuchi Y, Hirao M, Ookubo T, Sasaki A. Design of recycling system for poly(methyl methacrylate) (PMMA). Part 1: Recycling scenario analysis. International Journal of Life Cycle Assessment 2014, 19(1), 120–129. 10.1007/s11367-013-0624-y.

[18] Moens EKC, De Smit K, Marien YW, Trigilio AD, Van Steenberge PHM, Van Geem KM, Dubois JL, D'Hooge DR. Progress in reaction mechanisms and reactor technologies for thermochemical recycling of poly(methyl methacrylate). Polymers 2020, 12(8). 10.3390/POLYM12081667.

[19] Vogt BD, Stokes KK, Kumar SK. Why is recycling of postconsumer plastics so challenging? ACS Applied Polymer Materials 2021, 3(9), 4325–4346. 10.1021/acsapm.1c00648.

[20] Gu F, Guo J, Zhang W, Summers PA, Hall P. From waste plastics to industrial raw materials: A life cycle assessment of mechanical plastic recycling practice based on a real-world case study. Science of the Total Environment 2017, 601–602, 1192–1207. 10.1016/j.scitotenv.2017.05.278.

[21] Schyns ZOG, Shaver MP. Mechanical recycling of packaging plastics: A review. Macromolecular Rapid Communications 2021, 42(3), 1–27. 10.1002/marc.202000415.

Yoshi W. Marien

5 Kinetic concepts of PMMA thermochemical recycling

5.1 Introduction

Chemical recycling, which is also referred to as tertiary or feedstock recycling, encompasses all chemical processes that convert polymers or polymeric materials into smaller molecules that can be used as feedstock for the production of fuels, chemicals, and new polymers [1, 2]. Important processes include chemolysis, pyrolysis, gasification, catalytic cracking, hydrocracking, and reactive extrusion [3–9].

Much attention has been given to thermally driven processes. For example, pyrolysis of polyurethane has been demonstrated to allow the recovery of polyols and toluene diisocyanate [10]. For vinyl polymers, significant focus has been on pyrolysis of polyolefins, which leads to multicomponent mixtures that can be used as fuel or petrochemical feedstock [11]. Compared to the previous two examples, much higher monomer yields can be obtained via thermochemical recycling of polyacrylics, with core polymer poly(methyl methacrylate) (PMMA) [12, 13]. In order to be able to design chemical recycling or controlled degradation processes that maximize the monomer yield, an understanding of the interplay between various chemical reactions and their effect on the product spectrum is needed.

In this chapter, an overview is given of the reactions involved in thermochemical recycling processes, focusing on a radical degradation/depolymerization mechanism and selecting PMMA as the reference polymer. It is highlighted that the relevant reactions differ for PMMA synthesized via free radical polymerization (FRP) and anionic polymerization (AP), explaining why also the PMMA synthesis mechanisms are discussed. Finally, several kinetic modeling tools are introduced and applied to PMMA thermochemical recycling under ideal reactor conditions. Main emphasis is thus on the kinetic level or on the so-called microscale as commonly defined in the polymerization synthesis field [14].

Yoshi W. Marien, Laboratory for Chemical Technology (LCT), Ghent University, Technologiepark 125, Zwijnaarde 9052, Belgium; Intelligence in Processes, Advanced Catalysts and Solvents (iPRACS), Faculty of Applied Engineering, University of Antwerp, Groenenborgerlaan 171, 2020 Antwerp, Belgium

https://doi.org/10.1515/9783111076997-005

5.2 Radical PMMA thermochemical recycling reactions

In this section, an overview is given of the main (elementary) reactions occurring during thermochemical recycling processes via a radical degradation/depolymerization mechanism. A distinction can be made between initiation, depropagation, and termination reactions.

Head-tail fission

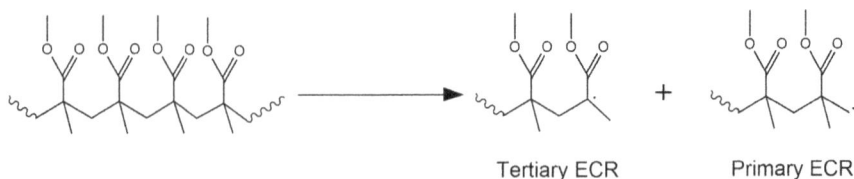

Tertiary ECR   Primary ECR

Head-head fission

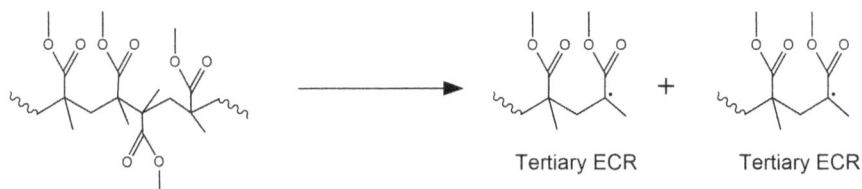

Tertiary ECR   Tertiary ECR

Unsaturated chain-end fission

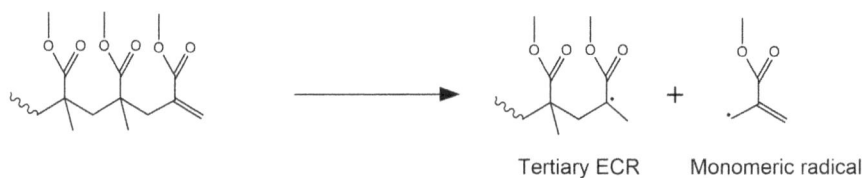

Tertiary ECR   Monomeric radical

Side-group fission

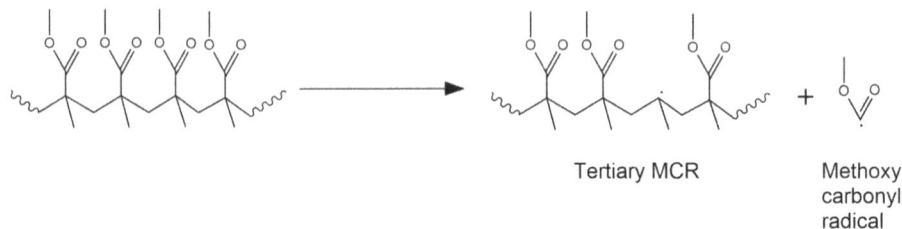

Tertiary MCR   Methoxy carbonyl radical

Figure 5.1: Examples of initiation reactions in thermochemical recycling of PMMA.

The initiation reactions correspond to homolytic bond cleavage, thus fission reactions, leading to the formation of (macro)radicals. In the case of PMMA, these fission reactions can occur at head-tail or head-head linkages, unsaturated chain ends, and side-group linkages. As illustrated in Figure 5.1, head-tail fission results in a primary and tertiary end-chain radical (ECR), while head-head fission results in two tertiary ECRs. For unsaturated chain-end fission, a tertiary ECR and an allylic radical are formed. Side-group fission leads to the formation of a tertiary mid-chain radical (MCR) and a methoxy carbonyl radical. It should be noted that depending on the temperature also other fission reactions (e.g., at methyl groups) might take place. Moreover, in case an "external" radical source (e.g., an impurity) is present, also radical addition reactions to unsaturated chain ends and hydrogen abstraction (i.e., chain transfer) reactions could initiate degradation/depolymerization.

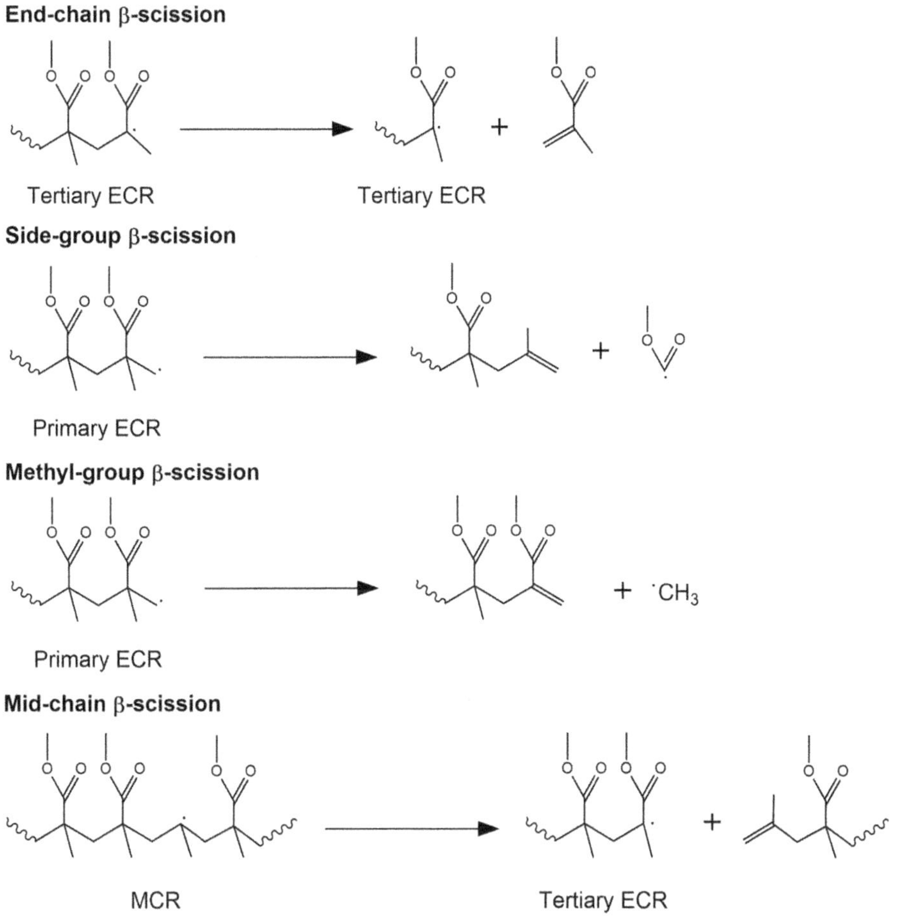

End-chain β-scission

Tertiary ECR   Tertiary ECR

Side-group β-scission

Primary ECR

Methyl-group β-scission

Primary ECR

Mid-chain β-scission

MCR   Tertiary ECR

Figure 5.2: Examples of depropagation reactions in thermochemical recycling of PMMA.

Following the creation of a macroradical via a fission reaction, depropagation can occur via β-scission reactions. As illustrated in Figure 5.2, a distinction can be made between β-scission of an ECR and an MCR. In the case of an ECR, a further distinction can be made between β-scission involving a tertiary and primary ECR. In the former case, β-scission at the chain end results in the formation of an MMA molecule, and a new tertiary ECR with its chain length reduced by one is formed. This ECR can undergo further end-chain β-scission reactions, thereby further reducing its chain length and releasing MMA molecules, which is referred to as unzipping. In the case of a primary ECR, in addition to end-chain β-scission, it has been suggested that side-group β-scission can occur [15], leading to an isobutylene-capped PMMA chain and a methoxy carbonyl radical. Alternatively, the primary ECR can undergo methyl group β-scission, leading to a PMMA chain with an unsaturated chain end and a methyl radical. In the case of an MCR, β-scission can occur to the "left" and "right," leading to a tertiary ECR and an isobutylene-capped PMMA chain.

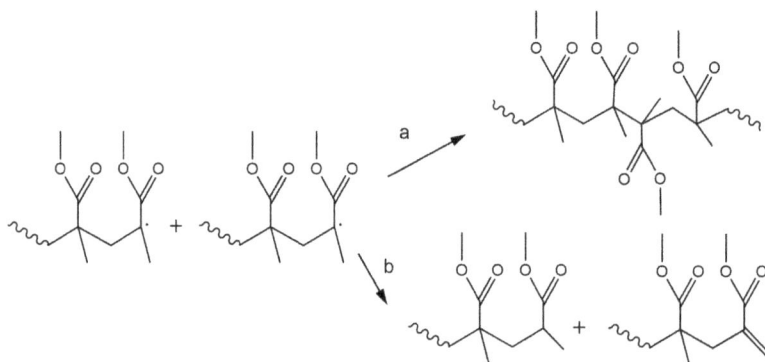

Figure 5.3: Examples of termination in thermochemical recycling of PMMA. For illustration, only recombination (a) and disproportionation (b) of two tertiary end-chain radicals are shown.

The degradation/depolymerization cycle can be stopped via termination reactions, which can occur via either recombination or disproportionation, as illustrated in Figure 5.3 for two tertiary ECRs.

Certain side reactions could lead to more complex product spectra and lower MMA yields. For example, at higher temperatures, MMA could decompose, as illustrated by the consorted reactions (a) and (b) in Figure 5.4, with reaction (a) proposed to be the dominant pathway [16]. However, MMA decomposition can be minimized if the gaseous MMA is removed sufficiently fast from the reaction environment. Additional side reactions include the conversion of methoxy carbonyl radicals to CO_2 (Figure 5.4(c)) or CO (Figure 5.4(d)), also forming a CH_3 or OCH_3 radical, respectively. These radicals could participate in hydrogen abstraction reactions, thereby forming methane or methanol.

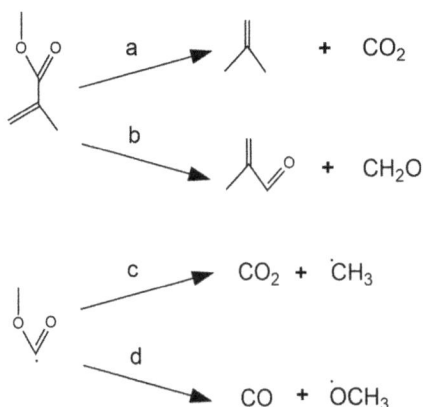

Figure 5.4: Examples of side reactions in thermochemical recycling of PMMA: (a) and (b) consorted decomposition reactions of MMA forming CO_2 and formaldehyde; (c) and (d) conversion of methoxy carbonyl radicals to CO_2 and CO.

5.3 Extending the reaction scheme to thermochemical recycling of MMA-rich copolymers

If one wants to thermochemically recycle PMMA waste, it should be clear that MMA-rich copolymers that contain small amounts of comonomers can be present, which could affect the degradation/depolymerization behavior [17–20].

Depending on the comonomer type, there are two general possibilities. The first possibility is that the copolymer chain can unzip in an unhindered manner. This is, for example, the case if the comonomer is also a methacrylate, as illustrated in Figure 5.5 (a), where R represents an alkyl group. The second possibility is that the comonomer delays or blocks the unzipping. This is the case if a comonomer leads to a less stable radical. For example, as illustrated in Figure 5.5(b), the formation of an acrylate terminal unit after unzipping of an MMA unit leads to a secondary radical, whereas a tertiary radical remains in case no such acrylate unit is present in the penultimate position. Hence, the formation of an acrylate terminal unit during unzipping is delayed and even potentially blocked so that the unzipping can even be reduced to only the units in the MMA segment. It should, however, be noted that the relevance of this delay/blocking depends on the temperature.

In addition to the previous possibilities, cyclization involving an MMA and adjacent comonomer unit could take place [19, 21, 22]. This is, for example, the case for vinyl chloride, (meth)acrylic acid, and vinyl acetate. For vinyl chloride, this is illustrated in Figure 5.5(c).

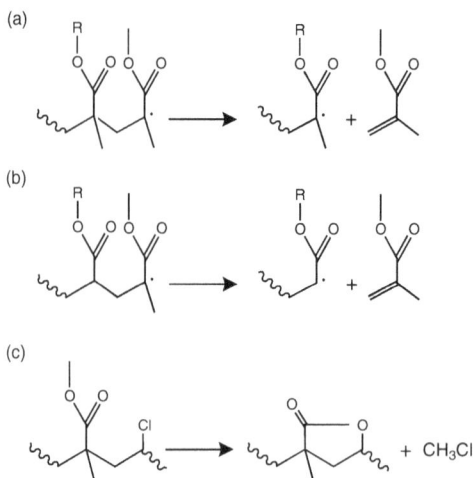

Figure 5.5: Possible effects of comonomers on thermochemical recycling of MMA-rich copolymers: (a) the MMA-rich chain can unzip unhindered, which is the case for methacrylate comonomers (R = alkyl group); (b) the comonomer blocks/hinders the unzipping, which is the case for acrylate comonomers; (c) cyclization involving a comonomer, illustrated here for vinyl chloride.

5.4 The relevance of thermogravimetric analysis to identify dominant reaction sequences

A frequently used experimental technique to study thermal polymer degradation/de-polymerization at lab scale is thermogravimetric analysis (TGA) [23–30]. In TGA, a small sample is placed in an oven. There, the sample is subjected to a predefined temperature profile, leading to degradation/depolymerization via the reactions outlined in Section 5.2, and thus the formation of gaseous species, which are removed from the oven via an inert purge gas. This leads to a decrease of the sample's mass w, which is detected by the balance on which the sample is put and is generally plotted relative to the initial mass w_0 and is referred to as the thermogravimetric (TG) curve. By taking the derivative of w/w_0 with respect to temperature (or time), one obtains the so-called differential thermogravimetric (DTG) curve, which generally allows the identification of multiple peaks that can be linked to the different initiation mechanisms discussed in Section 5.2 [23, 24, 26, 31–33].

This is illustrated in Figure 5.6, which shows the TG (right axis) and DTG (left axis) curves for a PMMA sample synthesized via conventional FRP (full line) and AP (dashed line). For the FRP-made PMMA sample, three peaks can be identified, while for the AP-made sample only one peak is distinguishable. This can be explained by comparing the FRP and AP reaction mechanisms which involve the formation of different structural defects and functional groups, as shown in Figure 5.7.

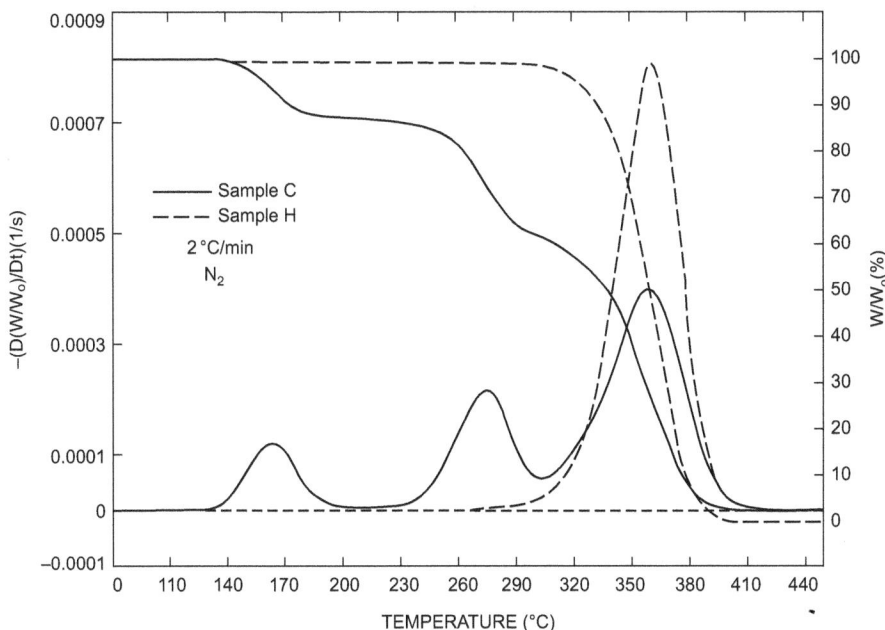

Figure 5.6: Thermogravimetric analysis (TGA) of an FRP-made (full line) and AP-made (dashed line) PMMA sample, highlighting the relevance of the polymerization mechanism, which determines the structural defects and functionalities and thus the relevant initiation mechanisms and the number of peaks in the TGA spectrum (reprinted with permission from [23]). FRP, free radical polymerization; AP, anionic polymerization (Copyright 1986 American Chemical Society).

In FRP, typically a thermal initiator (I_2) is used, which dissociates into two I radicals (Figure 5.7(a)). These radicals can add to a monomer molecule, leading to oligoradicals of length 1 (Figure 5.7(b)). These oligoradicals can propagate further; in the case of MMA, this dominantly occurs via head-to-tail propagation (Figure 5.7(c)), until either a side reaction such as chain transfer to monomer (Figure 5.7(d)), or termination with another radical via recombination (Figure 5.7(f)) or disproportionation (Figure 5.7(g)), occurs. As chain transfer to monomer leads to monomeric radicals, chain initiation with these radicals can take place (Figure 5.7(e)).

Compared to FRP, AP is characterized by a more reduced reaction scheme. In AP, chain growth is initiated by the addition of a nucleophile (A^- in Figure 5.7(h)) to a monomer molecule. This leads to an active chain of length 1, which can propagate further almost exclusively via head-to-tail propagation in view of the attractive force between the opposite (partial) charges (Figure 5.7(i)). In contrast to FRP, in the absence of impurities such as traces of water, the active chains remain living in AP, unless a chain transfer or terminating agent is deliberately added. Hence, contrary to FRP-made PMMA, AP-made PMMA does not contain head-to-head linkages or unsaturated chain ends. This explains why in the DTG curve of FRP-made PMMA samples gener-

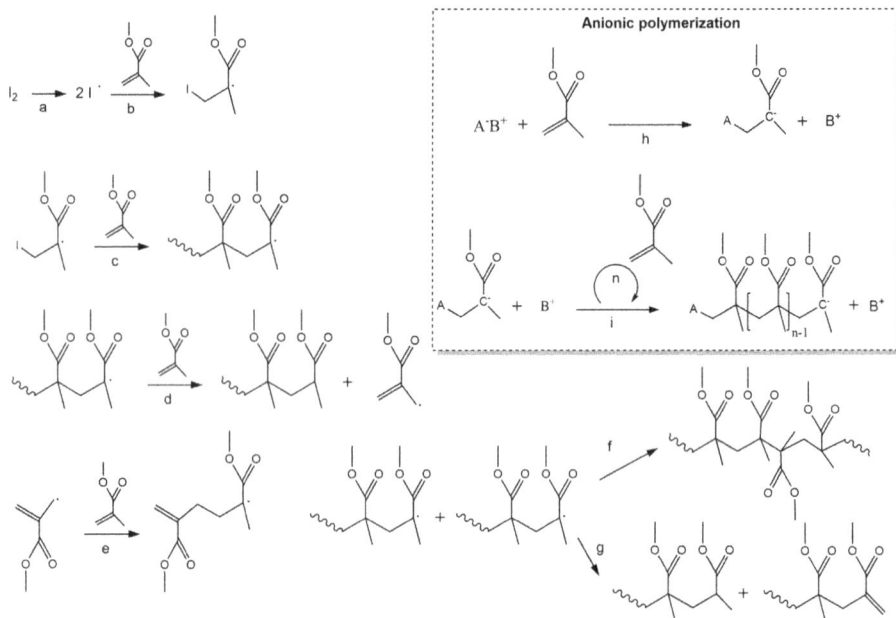

Figure 5.7: Reaction mechanism of free radical polymerization (FRP) and anionic polymerization (AP) of MMA. The occurrence of termination by recombination and disproportionation in FRP leads to head-to-head defects and unsaturated chain ends, resulting in additional initiation mechanisms during thermochemical recycling of PMMA.

ally multiple peaks are distinguishable (cf. the full line in Figure 5.6), while for AP-made PMMA samples generally only one dominant peak is distinguishable (cf. the dashed full line in Figure 5.6).

While several efforts have been devoted to assign the peaks in the DTG curves of different FRP-based PMMA samples to specific initiation reactions [15, 24, 31–34], a dedicated analysis using an elementary reaction-based model that accounts for diffusional limitations, as recently presented by Moens et al. [26], is recommended to quantify the contribution of individual fission reactions to the overall mass loss (rate) in different temperature regions. Moreover, it should be stressed that a peak in a DTG curve does not correspond to an individual elementary reaction but to a sequence of elementary reactions. Hence, activation energies or rate coefficients derived from TGA spectra should be treated as overall apparent parameters only [12], unless a dedicated analysis is performed, as presented in Chapter 6.

5.5 Kinetic models for thermochemical recycling

Kinetic models are a key tool to design or improve thermochemical recycling processes as they allow screening the effect of varying process/reaction conditions on the degradation/depolymerization behavior. As illustrated in Sections 5.2 and 5.3, many reactions can take place during thermochemical recycling with their importance strongly determined by the process conditions and polymer feedstock (cf. Section 5.4). This leads to a complex dynamic product spectrum, which further highlights the relevance of kinetic modeling.

In this section, focus is on the kinetics under ideal lab-scale conditions, assuming perfect mixing and temperature control. For thermochemical recycling, such conditions are often associated with a TGA setup (see Section 5.4), explaining why a large number of kinetic modeling studies aiming at the estimation of overall kinetic parameters have been conducted [12, 23, 24, 27, 28, 33–36]. The majority of these studies thus utilize lumped elementary reactions related to reaction sequences that lead to degradation (cf. the assignment of different peaks in the DTG curve; see Section 5.4). Hence, these global (conversion-based) models are less fundamental than elementary reaction-based models that have been put forward in the MMAtwo project, benefiting from elementary reaction-driven kinetic modeling experience within the consortium [26, 37–39].

In what follows, global conversion-based models and deterministic and stochastic elementary reaction-based models are discussed.

5.5.1 Global conversion-based models

Global conversion-based models are one of the most frequently used models for thermal polymer degradation [12, 27, 28]. Typically, these models are linked to TGA experiments in which mass losses are associated with reaction sequences (cf. the peaks in Figure 5.6). These mass losses are expressed by the conversion α, with $\alpha = (w_0 - w)/w_0$ or alternatively $\alpha = (w_0 - w)/(w_0 - w_\infty)$, where w_∞ is the mass of the solid residue (e.g., char) which does not degrade. The time evolution of α is pragmatically expressed as follows:

$$\frac{d\alpha}{dt} = kf(\alpha) \tag{5.1}$$

where $f(\alpha)$ reflects the overall reaction model (e.g., $f(\alpha) = 1 - \alpha$ in the case of first-order kinetics) and k is the overall (apparent) rate coefficient. The temperature dependence of k follows from the Arrhenius law:

$$k = A \exp\left(-\frac{E_a}{RT}\right) \tag{5.2}$$

where A is the (overall/apparent) pre-exponential factor, E_a is the (overall/apparent) activation energy, R is the universal gas constant, and T is the temperature.

If a linear temperature profile with heating rate β is enforced, the time t can be linked to T via $\beta = dT/dt$. Introducing β in equation (5.1) leads to

$$\frac{da}{dT} = \frac{A}{\beta} \exp\left(-\frac{E_a}{RT}\right) f(a) \tag{5.3}$$

By taking the natural logarithm of equation (5.3) and differentiating it at a certain a over $1/T$, one obtains equations (5.4) and (5.5), respectively:

$$\ln\left(\beta \frac{da}{dT}\right) = \ln(A\,f(a)) - \frac{E_a}{RT} \tag{5.4}$$

$$\left[\frac{d\left(\ln\left(\beta\frac{da}{dt}\right)\right)}{d\left(\frac{1}{T}\right)}\right]_a = -\left[\frac{E_a}{R}\right]_a + \left[\frac{d\ln(f(a))}{d\left(\frac{1}{T}\right)}\right]_a \tag{5.5}$$

Often the so-called isoconversional method [26, 27] is used, for which the second term in the right-hand side of equation (5.5) is formally zero, allowing a direct determination of E_a. By using this simple method, it is often found that E_a depends on a as sequences with a different dominant initiation reaction (see Section 5.2) are triggered at different temperatures (cf. the peaks in Figure 5.6) [12, 23, 24].

Alternatively to the isoconversional method, the overall reaction model $f(a)$ can be explicitly taken into account, as, for example, done by Ferriol et al. [24]. These authors have put forward a four-stage degradation mechanism and assumed for each stage i an overall rate law given by

$$\left(\frac{da}{dt}\right)_i = A_i \exp\left(-\frac{E_{a,i}}{RT}\right)(1-a)^{n_i} \tag{5.6}$$

where n_i, A_i, and $E_{a,i}$ are the apparent reaction order, pre-exponential factor, and activation energy of the ith stage. By applying Lyon's approximation upon integration and assuming that each sample possesses the same relative number of structural defects, equation (5.6) becomes

$$\frac{da}{dt} = \sum_{i=1}^{4} r_i \left[A_i e^{-\frac{E_{a,i}}{RT}}\left[1 - (1-n_i)\cdot\frac{A_i.R}{\beta}\cdot\frac{T^2}{E_i+2RT}\cdot e^{-\frac{E_{a,i}}{RT}}\right]^{\frac{n_i}{1-n_i}}\right] \tag{5.7}$$

where r_i is the relative contribution of the ith weight loss stage, thus with $r_i < 1$ and $\sum_{i=1}^{4} r_i = 1$. By fitting equation (5.7) to experimental DTG curves, overall apparent values for A_i, $E_{a,i}$, n_i, and r_i could be determined [24].

5.5.2 Elementary reaction-based models

In elementary reaction-based models, rate laws are constructed in principle for every elementary reaction. These rate laws can then be solved deterministically (e.g., via numerical integration) or can be implemented in a stochastic framework such as the stochastic simulation algorithm used in kinetic Monte Carlo (kMC) simulations [14, 26, 37, 40–42]. While elementary reaction-based models are most fundamental, their use to describe PMMA degradation/depolymerization is rather limited, and in most cases, additional assumptions have been made lowering their fundamental (elementary) character [12, 23, 34, 43–47]. For example, in the previous decades, deterministic models with formal first-order termination [23, 34] or simplified rate laws [44, 45] have been considered.

A very promising alternative to these deterministic models are matrix-based kMC simulations [37], which allow to track the degradation/depolymerization on the level of individual polymer chains, as recently illustrated by De Smit et al. [39]. In kMC simulations, (elementary) reactions and the time step between them are stochastically sampled based on (elementary) reaction rates [37, 41]. By using matrix-based data structures, the position of structural defects (e.g., head-to-head linkages) and functionalities (e.g., comonomers or unsaturated chain ends) can be stored for an ensemble of polymer chains. As illustrated by De Smit et al. [39], for MMA-rich copolymers, the use of a primary matrix storing the position of the MMA segments and the structural defects and functionalities, and a secondary matrix storing the length of the MMA segments is computationally advantageous.

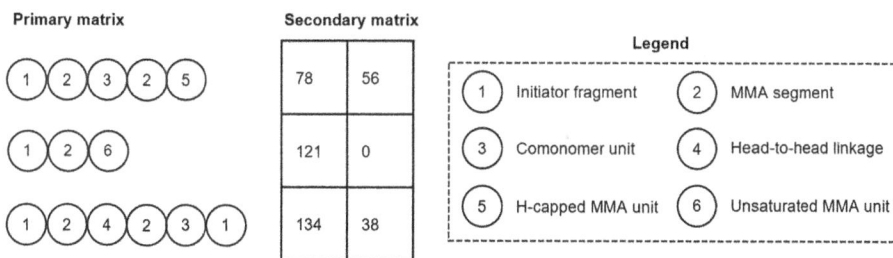

Figure 5.8: Principle of matrix-based kinetic Monte Carlo simulations as highlighted by De Smit et al. [39] to model thermochemical recycling of MMA-rich copolymers, starting from the Ghent University expertise in such modeling tool [37, 38]. In the primary matrix, the location of the MMA segments, the structural defects (head-to-head linkages), and functionalities (comonomers and different chain-end types) along the copolymer backbone are stored, here illustrated for three chains. In the secondary matrix, the length of the MMA segments is stored.

This concept is illustrated in Figure 5.8, where the rows of the matrices correspond to different polymer chains (here only three chains are shown for illustration) and the columns of the primary matrix follow the polymer backbone (short-chain lengths are drawn here for illustration). By assigning a label to MMA segments consisting of conventional head-tail connected monomer units and to the structural defects and functionalities, one is able to simulate the time evolution of these defects and functionalities, which allows to accurately define the elementary reaction rates, ensuring reliable and fundamentally sound model predictions.

A true exploitation of the matrix-based kMC modeling approach requires reliable elementary reaction Arrhenius parameters. The determination of such parameters from lab-scale depolymerization experiments is discussed in the next chapter.

References

[1]   Ragaert K, Delva L, Van Geem K. Mechanical and chemical recycling of solid plastic waste. Waste Management 2017, 69, 24–58.
[2]   Al-Salem SM, Lettieri P, Baeyens J. Recycling and recovery routes of plastic solid waste (PSW): A review. Waste Management 2009, 29, 2625–2643.
[3]   Al-Salem S, Lettieri P, Baeyens J. The valorization of plastic solid waste (PSW) by primary to quaternary routes: From re-use to energy and chemicals. Progress in Energy and Combustion Science 2010, 36, 103–129.
[4]   Dogu O, Pelucchi M, Van de Vijver R, Van Steenberge PHM, D'Hooge DR, Cuoci A, et al. The chemistry of chemical recycling of solid plastic waste via pyrolysis and gasification: State-of-the-art, challenges, and future directions. Progress in Energy and Combustion Science 2021, 84, 100901.
[5]   Dogu O, Plehiers PP, Van de Vijver R, D'hooge DR, Van Steenberge PHM, Van Geem KM. Distribution changes during thermal degradation of poly(styrene peroxide) by pairing tree-based kinetic Monte Carlo and artificial intelligence tools. Industrial & Engineering Chemistry Research 2021, 60, 3334–3353.
[6]   Broadbelt LJ, Stark SM, Klein MT. Computer generated pyrolysis modeling: On-the-fly generation of species, reactions, and rates. Industrial & Engineering Chemistry Research 1994, 33, 790–799.
[7]   López G, Olazar M, Aguado R, Bilbao J. Continuous pyrolysis of waste tyres in a conical spouted bed reactor. Fuel 2010, 89, 1946–1952.
[8]   Kiliaris P, Papaspyrides CD, Pfaendner R. Reactive-extrusion route for the closed-loop recycling of poly(ethylene terephthalate). Journal of Applied Polymer Science 2007, 104, 1671–1678.
[9]   Hernández-Ortiz JC, Van Steenberge PH, Duchateau JN, Toloza C, Schreurs F, Reyniers M-F, et al. A two-phase stochastic model to describe mass transport and kinetics during reactive processing of polyolefins. Chemical Engineering Journal 2019, 377, 119980.
[10] Deng Y, Dewil R, Appels L, Ansart R, Baeyens J, Kang Q. Reviewing the thermo-chemical recycling of waste polyurethane foam. Journal of Environmental Management 2021, 278, 111527.
[11] Donaj PJ, Kaminsky W, Buzeto F, Yang W. Pyrolysis of polyolefins for increasing the yield of monomers' recovery. Waste Management 2012, 32, 840–846.
[12] Moens EKC, De Smit K, Marien YW, Trigilio AD, Van Steenberge PHM, Van Geem KM, et al. Progress in reaction mechanisms and reactor technologies for thermochemical recycling of poly(methyl methacrylate). Polymers 2020, 12.
[13] Lopez G, Artetxe M, Amutio M, Elordi G, Aguado R, Olazar M, et al. Recycling poly-(methyl methacrylate) by pyrolysis in a conical spouted bed reactor. Chemical Engineering and Processing: Process Intensification 2010, 49, 1089–1094.
[14] D'hooge DR, Van Steenberge PH, Reyniers M-F, Marin GB. The strength of multi-scale modeling to unveil the complexity of radical polymerization. Progress in Polymer Science 2016, 58, 59–89.

[15] Kashiwagi T, Inabi A, Hamins A. Behavior of primary radicals during thermal degradation of poly (methyl methacrylate). Polymer Degradation and Stability 1989, 26, 161–184.

[16] Forman R, Mackinnon H, Ritchie P. Studies in pyrolysis. Part XXV. Acrylic, methacrylic, and crotonic acid, and some derivatives: Novel decarbonylation of αβ-unsaturated carboxylic acids. Journal of the Chemical Society C: Organic 1968, 2013–2016.

[17] Özlem-Gundogdu S, Gurel EA, Hacaloglu J. Pyrolysis of poly (methy methacrylate) copolymers. Journal of Analytical and Applied Pyrolysis 2015, 113, 529–538.

[18] McNeill IC. Products of thermal degradation in polymer systems based on Methyl Methacrylate. International Journal of Polymeric Materials 1994, 24, 31–45.

[19] McNeill I. A study of the thermal degradation of methyl methacrylate polymers and copolymers by thermal volatilization analysis. European Polymer Journal 1968, 4, 21–30.

[20] Szabo E, Olah M, Ronkay F, Miskolczi N, Blazso M. Characterization of the liquid product recovered through pyrolysis of PMMA–ABS waste. Journal of Analytical and Applied Pyrolysis 2011, 92, 19–24.

[21] Jamieson A, McNeill I. The thermal degradation of copolymers of methyl methacrylate with methacrylic acid. European Polymer Journal 1974, 10, 217–225.

[22] McNeill I, Straiton T. Thermal degradation of copolymers of 36Cl-vinyl chloride and methyl methacrylate. European Polymer Journal 1979, 15, 1043–1049.

[23] Kashiwagi T, Inaba A, Brown JE, Hatada K, Kitayama T, Masuda E. Effects of weak linkages on the thermal and oxidative degradation of poly (methyl methacrylates). Macromolecules 1986, 19, 2160–2168.

[24] Ferriol M, Gentilhomme A, Cochez M, Oget N, Mieloszynski J. Thermal degradation of poly (methyl methacrylate)(PMMA): Modelling of DTG and TG curves. Polymer Degradation and Stability 2003, 79, 271–281.

[25] Godiya CB, Gabrielli S, Materazzi S, Pianesi MS, Stefanini N, Marcantoni E. Depolymerization of waste poly(methyl methacrylate) scraps and purification of depolymerized products. Journal of Environmental Management 2019, 231, 1012–1020.

[26] Moens EKC, Marien YW, Figueira FL, Trigilio AD, De Smit K, Van Steenberge PHM, d'hooge DR. Coupled matrix-based Monte Carlo modelling for a mechanistic understanding of poly(methyl methacrylate) thermochemical recycling kinetics. Chemical Engineering Journal 2023, 475, 146105.

[27] Das P, Tiwari P. Thermal degradation kinetics of plastics and model selection. Thermochimica Acta 2017, 654, 191–202.

[28] Pérez-Maqueda LA, Sánchez-Jiménez PE, Perejón A, García-Garrido C, Criado JM, Benítez-Guerrero M. Scission kinetic model for the prediction of polymer pyrolysis curves from chain structure. Polymer Testing 2014, 37, 1–5.

[29] Rimez B, Rahier H, Van Assche G, Artoos T, Biesemans M, Van Mele B. The thermal degradation of poly (vinyl acetate) and poly (ethylene-co-vinyl acetate), Part I: Experimental study of the degradation mechanism. Polymer Degradation and Stability 2008, 93, 800–810.

[30] Dubdub I, Al-Yaari M. Pyrolysis of low density polyethylene: Kinetic study using TGA data and ANN prediction. Polymers 2020, 12.

[31] Manring LE, Sogah DY, Cohen GM. Thermal degradation of poly (methyl methacrylate). 3. Polymer with head-to-head linkages. Macromolecules 1989, 22, 4652–4654.

[32] Manring LE. Thermal degradation of poly (methyl methacrylate). 2. Vinyl-terminated polymer. Macromolecules 1989, 22, 2673–2677.

[33] Manring LE. Thermal degradation of poly (methyl methacrylate). 4. Random side-group scission. Macromolecules 1991, 24, 3304–3309.

[34] Holland B, Hay JN. The kinetics and mechanisms of the thermal degradation of poly (methyl methacrylate) studied by thermal analysis-Fourier transform infrared spectroscopy. Polymer 2001, 42, 4825–4835.

[35] Korobeinichev O, Paletsky A, Gonchikzhapov M, Glaznev R, Gerasimov I, Naganovsky Y, et al. Kinetics of thermal decomposition of PMMA at different heating rates and in a wide temperature range. Thermochimica Acta 2019, 671, 17–25.

[36] Fateh T, Richard F, Rogaume T, Joseph P. Experimental and modelling studies on the kinetics and mechanisms of thermal degradation of polymethyl methacrylate in nitrogen and air. Journal of Analytical and Applied Pyrolysis 2016, 120, 423–433.

[37] Trigilio AD, Marien YW, Van Steenberge PHM, D'hooge DR. Gillespie-driven kinetic Monte Carlo algorithms to model events for bulk or solution (bio)chemical systems containing elemental and distributed species. Industrial & Engineering Chemistry Research 2020, 59, 18357–18386.

[38] De Keer L, Kilic KI, Van Steenberge PHM, Daelemans L, Kodura D, Frisch H, et al. Computational prediction of the molecular configuration of three-dimensional network polymers. Nature Materials 2021, 20, 1422–1430.

[39] De Smit K, Marien YW, Van Geem KM, Van Steenberge PHM, D'Hooge DR. Connecting polymer synthesis and chemical recycling on a chain-by-chain basis: A unified matrix-based kinetic Monte Carlo strategy. Reaction Chemistry & Engineering 2020, 5, 1909–1928.

[40] Saldívar-Guerra E. Polymer reaction engineering. Encyclopedia of Polymer Science and Technology 2019, 1–45.

[41] Brandão ALT, Soares JBP, Pinto JC, Alberton AL. When polymer reaction engineers play dice: Applications of Monte Carlo models in PRE. Macromolecular Reaction Engineering 2015, 9, 141–185.

[42] Kruse TM, Woo OS, Broadbelt LJ. Detailed mechanistic modeling of polymer degradation: Application to polystyrene. Chemical Engineering Science 2001, 56, 971–979.

[43] Da Ros S, Braido RS, E Castro N, Brandão AL, Schwaab M, Pinto JC. Modelling the chemical recycling of crosslinked poly (methyl methacrylate): Kinetics of depolymerisation. Journal of Analytical and Applied Pyrolysis 2019, 144, 104706.

[44] Barlow A, Lehrle R, Robb J, Sunderland D. Polymethylmethacrylate degradation – Kinetics and mechanisms in the temperature range 340 to 460 °C. Polymer 1967, 8, 537–545.

[45] Smolders K, Baeyens J. Thermal degradation of PMMA in fluidised beds. Waste Management 2004, 24, 849–857.

[46] Staggs J. Modelling end-chain scission and recombination of linear polymers. Polymer Degradation and Stability 2004, 85, 759–767.

[47] Staggs J. Modelling random scission of linear polymers. Polymer Degradation and Stability 2002, 76, 37–44.

Eli K.C. Moens and Yoshi W. Marien

6 Improving the sensitivity of lab-scale PMMA depolymerization experiments

6.1 Introduction

In Chapter 5, the main reactions and kinetic concepts behind thermochemical degradation or depolymerization of polyacrylics in general and of poly(methyl methacrylate) (PMMA) in particular have been discussed. It has been highlighted that knowledge of kinetic (Arrhenius) parameters for polymer degradation/depolymerization at the elementary reaction level is currently limited [1–4], especially compared to the field of polymer synthesis [5–9]. Such parameters can nevertheless strongly benefit the chemical design of (thermo)chemical recycling processes, provided they are reliably determined [1, 10–12]. In this chapter, the design of lab-scale degradation/depolymerization experiments aiming at the reliable determination of degradation kinetic parameters will therefore be covered.

First, the design of polymer feedstocks for more sensitive degradation/depolymerization kinetic experiments is discussed. Next, it is demonstrated that thermogravimetric analysis (TGA) combined with an elementary reaction-based kinetic model and judiciously designed polymer feedstocks allows tuning of the sought Arrhenius parameters. Here, the importance of diffusional limitations and thus viscosity effects during degradation/depolymerization are illustrated. Finally, the relevance of complementary lab-scale depolymerization experiments targeting isothermicity and considering the measurement of the evolution of the molar mass distribution (MMD) is highlighted. This chapter focuses on PMMA depolymerization, although the introduced concepts are also applicable to other (chain-growth-made) polymers.

6.2 Design of PMMA feedstocks for depolymerization experiments

As illustrated in Chapter 5, PMMA can contain several structural defects, in particular head–head defects, and functionalities, specifically unsaturated chain ends. Depending on the temperature, fission is likely to be initiated at these structural defects and

Eli K. C. Moens, Laboratory for Chemical Technology (LCT), Ghent University, Technologiepark 125, Zwijnaarde B-9052, Belgium
Yoshi W. Marien, Laboratory for Chemical Technology (LCT), Ghent University, Technologiepark 125, Zwijnaarde B-9052, Belgium; Intelligence in Processes, Advanced Catalysts and Solvents (iPRACS), Faculty of Applied Engineering, University of Antwerp, Groenenborgerlaan 171, 2020 Antwerp, Belgium

https://doi.org/10.1515/9783111076997-006

functionalities [1, 2, 13–16]. Hence, the prevalence of structural defects and functionalities can be expected to significantly affect the depolymerization sensitivity in different temperature ranges, especially at the level of individual chains [2].

The prevalence of these structural defects and functionalities is determined by the polymerization mechanism and reaction conditions relevant for the PMMA synthesis [1, 2, 17]. For the dominant PMMA synthesis route via free radical polymerization (FRP), head–head defects can result from termination by recombination, while unsaturated chain ends can result from termination by disproportionation, as illustrated in Figure 5.7.

Knowledge of macromolecular information at the level of individual chains is thus important, for which a combination of experiments and modeling is recommended [18, 19]. The most advanced modeling tool is the so-called coupled matrix-based kinetic Monte Carlo (kMC) simulation, of which the principles have been outlined in previous work considering polymerization and polymer modification [20–22]. In the context of linking polymerization with depolymerization, De Smit et al. [2] have applied this tool (see Figure 5.8) to track the amount and location of structural defects and functionalities in a representative (microscale) ensemble of well-mixed PMMA chains, both during polymer synthesis and degradation.

Essential in the work of De Smit et al. [2] is the validation to experimental data. A polymerization kMC model considering all elementary reactions in Figure 5.7 (FRP part) and accounting for diffusional limitations via the cage, gel, and glass effect has been successfully benchmarked to leading experimental data from Balke and Hamielec [23] on monomer conversion, and number average chain length (x_n) and mass average chain length (x_m). This is illustrated in Figure 6.1, considering three polymerization temperatures (50, 70, and 90 °C) and two initial concentrations of the initiator azobisisobutyronitrile ([AIBN]$_0$ = 0.016 and 0.027 mol/L). It is observed that the model (lines) follows the experimentally observed trends (symbols) well, also taking into account the uncertainties on the experimentally recorded (mass) average chain lengths, due to a less reliable measurement of the MMD tail in the work of Balke and Hamielec [23].

Benefiting from the benchmarked kMC model for isothermal bulk FRP of MMA, three FRP-based PMMA feedstocks are designed, aiming at varying (average) chain lengths and amounts of structural defects and functionalities, which facilitates the tuning of the elementary reaction Arrhenius parameters, as will be demonstrated in Section 6.3. The first feedstock (feedstock A) is synthesized considering an initial ratio of the AIBN to MMA concentration ([AIBN]$_0$/[MMA]$_0$) equal to 0.0015 and a polymerization temperature (T_{pol}) equal to 343 K. The second feedstock (feedstock B) is synthesized considering [AIBN]$_0$/[MMA]$_0$ equal to 0.003 and T_{pol} equal to 343 K. The third feedstock (feedstock C) is synthesized considering [AIBN]$_0$/[MMA]$_0$ equal to 0.003 and T_{pol} equal to 323 K.

As shown in Figure 6.2(a), the highest x_n is obtained for feedstock C (blue line), which can be explained by (i) the lower polymerization temperature leading to lower AIBN dissociation rates (despite the higher initial AIBN concentration compared to feedstock A) and hence lower radical concentrations and termination rates, and (ii) a lower impact of chain transfer to monomer at this lower temperature. As a result,

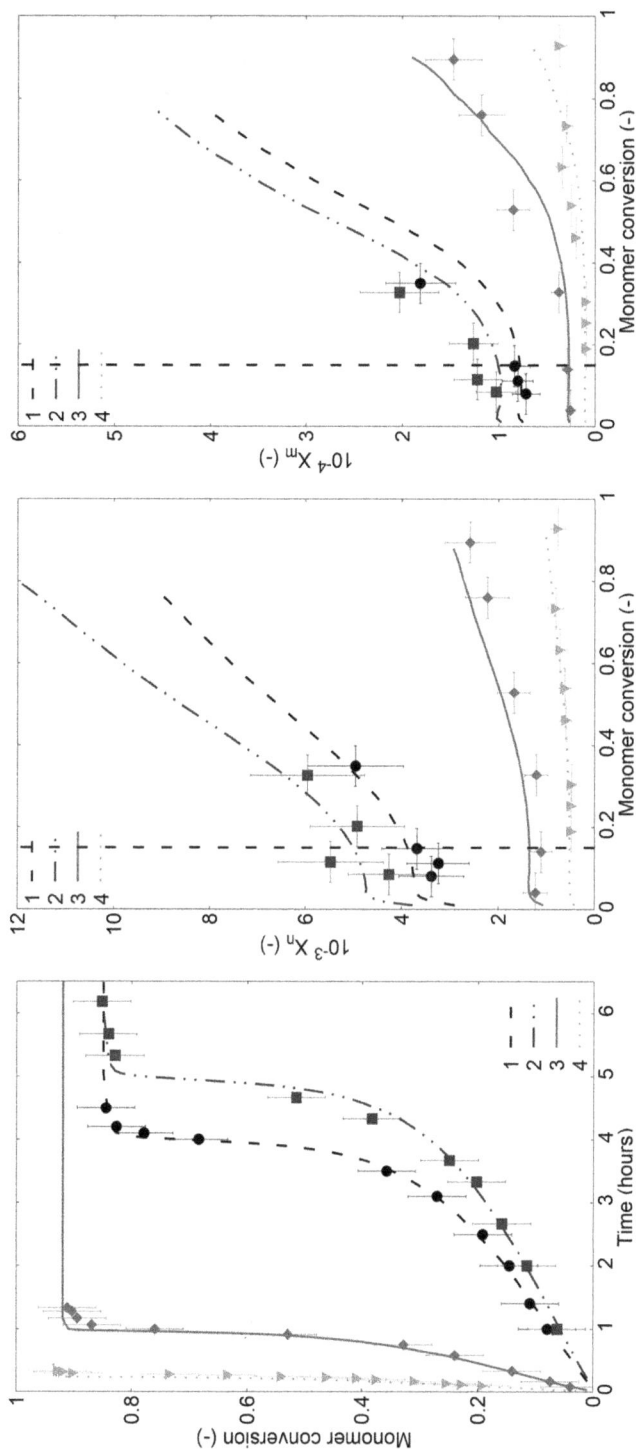

Figure 6.1: Benchmark of the matrix-based kinetic Monte Carlo model of De Smit et al. [2] (lines) to experimental data recorded by Balke and Hamielec [23] (symbols) on monomer conversion (left), number average chain length (middle), and mass average chain length (right) for isothermal bulk FRP of MMA. The benchmarked model of De Smit et al. is used in this work to design FRP-based PMMA feedstocks for depolymerization experiments. Blue dashed line/circles: $[AIBN]_0/[MMA]_0 = 0.003$, polymerization temperature (T_{pol}) = 50 °C; black dashed–doted–doted line/squares: $[AIBN]_0/[MMA]_0 = 0.002$, $T_{pol} = 50$ °C; green full line/diamonds: $[AIBN]_0/[MMA]_0 = 0.003$, $T_{pol} = 70$ °C; red dotted line/triangles: $[AIBN]_0/[MMA]_0 = 0.003$, $T_{pol} = 90$ °C (figure adopted from De Smit et al. [2] with permission from the Royal Society of Chemistry).

feedstock C possesses the highest fraction of head–tail linkages (Figure 6.2(b)), and the lowest fraction of unsaturated chain ends (Figure 6.2(c)) and head–head linkages (Figure 6.2(d)). In contrast, feedstock B (red line) has the lowest x_n (Figure 6.2(a)) as this feedstock is synthesized considering the highest polymerization temperature and initial AIBN concentration, which leads to the highest radical concentrations and hence termination rates, and a higher importance of chain transfer to monomer compared to feedstock C. Consequently, feedstock B has the lowest fraction of head–tail linkages (Figure 6.2(b)) and the highest fraction of unsaturated chain ends (Figure 6.2(c)) and head–head linkages (Figure 6.2(d)).

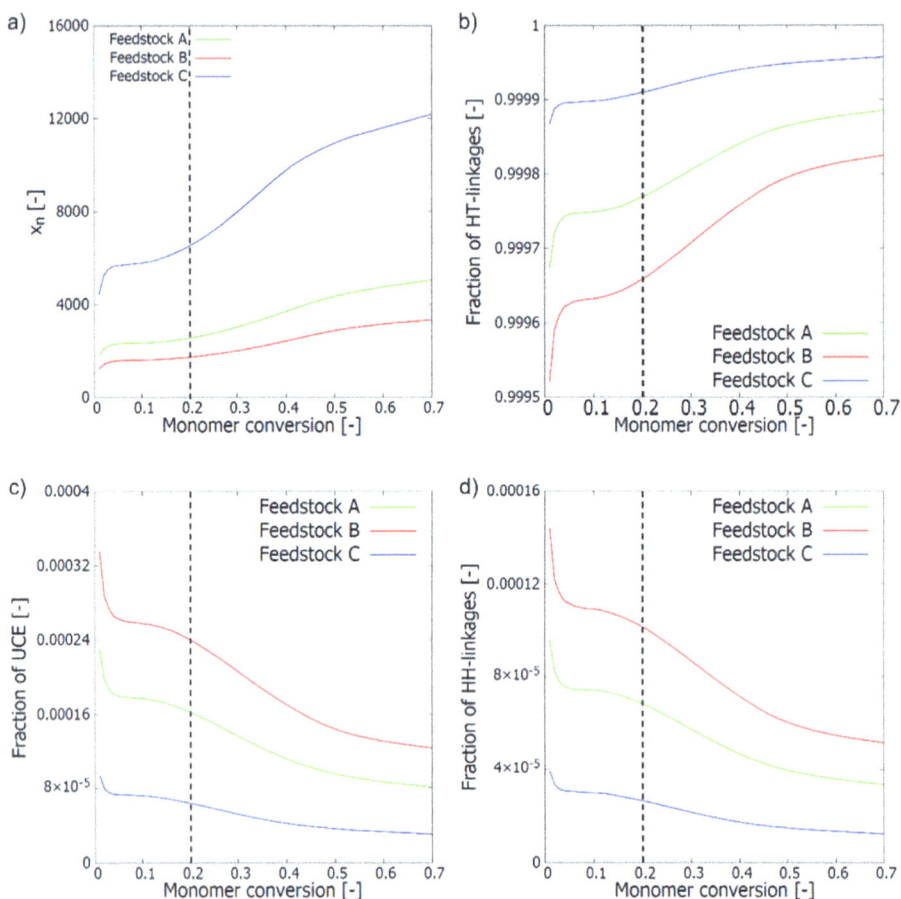

Figure 6.2: Model-based design of three FRP-based PMMA feedstocks displaying variations in (a) the number average chain length (x_n), (b) the fraction of unsaturated chain ends (UCE), and (c) the fraction of head-head (HH) linkages. The actual synthesis is stopped at 20% monomer conversion (indicated by the vertical dashed line) to allow further treatment toward depolymerization. Figure reproduced from [16] with permission from Elsevier.

Practically, in order to avoid a very viscous reaction mixture during the lab-scale synthesis of the feedstocks, which would complicate further treatment toward depolymerization, polymerization is stopped already at 20% monomer conversion, as indicated in Figure 6.2 by the vertical dashed lines. At such a monomer conversion, the sufficient difference in both (average) chain lengths and fraction of structural defects and functionalities between the feedstocks facilitates the tuning of Arrhenius parameters. Here, focus is on an elementary depolymerization reaction model using TGA data and conventional MMD data measured under conditions targeting isothermicity, as illustrated in the next two sections.

6.3 Kinetic modeling using data recorded via thermogravimetric analysis

As highlighted in Section 5.4, TGA has mostly been used for the determination of apparent Arrhenius parameters of dominant reaction sequences [13, 14, 24–29]. In this section, it is demonstrated that upon consideration of the judiciously designed polymer feedstocks from Section 6.2 and a sufficiently detailed elementary reaction-based kinetic model, Arrhenius parameters of individual elementary reactions can be assessed.

In what follows, first, the kinetic model will be described, focusing on the implementation of diffusional limitations, which are crucial to describe polymer degradation kinetics taking into account the highly viscous reaction medium during the main phase of the degradation. For simplicity, perfect macroscale mixing is assumed, and only MMA and small radical species (methyl, methoxy carbonyl, MMA, and isobutylene radicals) are seen as volatile compounds. Next, the tuning of intrinsic Arrhenius parameters of individual elementary reactions will be illustrated for the feedstocks designed in Section 6.2.

6.3.1 Elementary reactions and viscosity effects

In order to tune intrinsic Arrhenius parameters, the reactions shown in Figure 6.3 are included in the matrix-based kinetic Monte Carlo model of Moens et al. [16]. All main degradation reactions described in Chapter 5 are considered, with, in particular, an explicit consideration of fission at head–tail, head–head, and side-group linkages, as well as at unsaturated chain ends and for completeness even at tail–tail linkages formed during degradation via recombination of two primary ECRs. Note that if a chain contains for example a head–head linkage, it can still undergo for example head–tail fission. For completeness it is mentioned that further reactions involving methyl, methoxy carbonyl, MMA, and isobutylene radicals are not considered.

As indicated earlier, since during the main phase of the degradation a highly viscous phase is present, diffusional limitations on the reactions in Figure 6.3 can be expected due to limited macromolecular mobility. Hence, apparent rate coefficients (k_{app} values) accounting for diffusional limitations can be introduced for all reactions in Figure 6.3. To calculate the k_{app} values, a distinction is made between fission, scission, and termination reactions, following the general principles as outlined in D'hooge et al. [30], considering both irreversible and reversible model reaction systems.

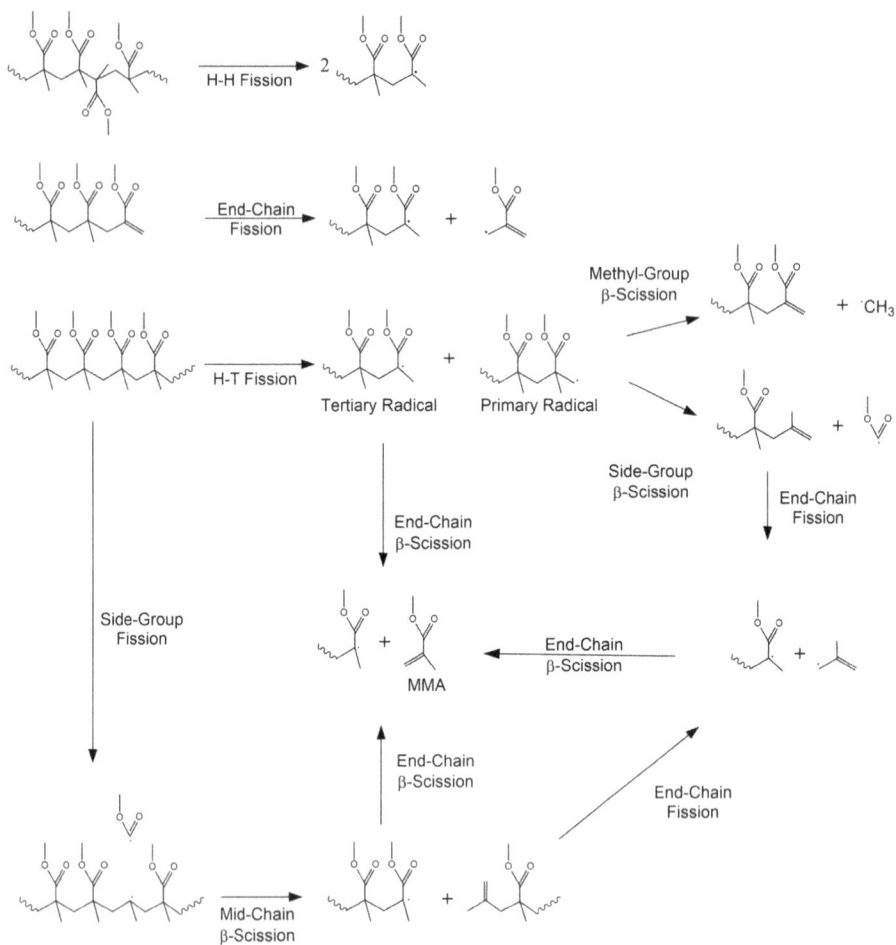

Figure 6.3: Elementary reactions included in the matrix-based kinetic Monte Carlo model of Moens et al. [16] used to simulate the TGA experiments in Figure 6.4. Tail–tail fission, end-chain β-scission of primary ECRs, and termination reactions are included in the model but are not shown to not overload the figure.

For termination, the composite k_t model [31–34] is used, and is also considered in Chapter 10 on repolymerization, which expresses the chain length and polymer mass fraction dependence of the apparent rate coefficient $k_{t,app}$ for termination between two radicals of chain length i at a certain polymer mass fraction w_p:

$$if \ i < i_{gel} \ and \ i < i_{SL} : k_{(t, app, ii)} = k_{t,11} i^{-\alpha_S} \tag{6.1}$$

$$if \ i < i_{gel} \ and \ i \geq i_{SL} : k_{t,app,ii} = k_{t,11} i_{SL}^{\alpha_L - \alpha_S} i^{-\alpha_L} \tag{6.2}$$

$$if \ i \geq i_{gel} \ and \ i < i_{SL} : k_{t,app,ii} = k_{t,11} i_{gel}^{\alpha_{gel} - \alpha_S} i^{-\alpha_{gel}} \tag{6.3}$$

$$if \ i \geq i_{gel} \ and \ i \geq i_{SL} : k_{t,app,ii} = k_{t,11} i_{SL}^{\alpha_L - \alpha_S} i_{gel}^{\alpha_{gel} - \alpha_L} i^{-\alpha_{gel}} \tag{6.4}$$

in which α_S and α_L are the chain length dependence of short and long radicals under dilute conditions, α_{gel} is the chain length dependence of radicals in the gel regime at a certain w_p, i_{SL} is the crossover chain length between short and long radicals under dilute conditions, and i_{gel} is the chain length at the onset of the gel effect, which depends on w_p. Apparent rate coefficients for termination between two radicals with a different chain length i and j ($k_{t,app,ij}$ values) are calculated via the geometric mean of $k_{t,app,ii}$ and $k_{t,app,jj}$.

The practical implementation of the composite k_t model in the polymer degradation kMC model is done considering the shooting method as described by De Smit et al. [33]. The MMA-specific composite k_t parameters are listed in Table 6.1. They are, for simplicity, assumed to be temperature-independent. Moreover, w_p is for simplicity calculated based on the assumption that 5% of the formed MMA is present in the melt phase. For completeness, it is mentioned that to model the TGA experiments in Figure 6.4, Moens et al. [16] considered an extra regime in the composite k_t model to account for the extremely viscous conditions in the melt phase.

Table 6.1: Literature-based MMA-specific composite k_t model parameters to account for diffusional limitations on termination [31].

α	Value (-)
α_S	0.65
α_L	0.15
α_{gel}	$1.66 w_p - 0.06$
i_{SL}	100
gel	$0.53 w_p^{-2.5}$

For fission and scission reactions, diffusional limitations are assumed in a cage configuration as for initiation in radical polymerization, that is, the products obtained by fission and scission can be thought to be present at least in a small time frame in a small theoretical environment. Due to restricted mobility of the macromolecular fission/scission products, only a fraction of the products will diffuse sufficiently fast out of the cage, while the other fraction will undergo the reverse (e.g., termination) reaction, lowering the apparent fission/scission rate.

This effect is accounted for via the (coupled) encounter pair model [30], which states that if fission/scission of a reactant A occurs, the products B and C are formed within a cage with a chemical or intrinsic rate coefficient k_{-chem} (unit s^{-1} for a unimolecular reaction). This can be followed by B and C diffusing out of the cage with rate coefficient k_{-diff}, or alternatively B and C (still as an encounter pair at very close distance) undergoing the reverse reaction with intrinsic rate coefficient $k_{+chem}{}^*$, utilizing a unit of s^{-1} instead of the conventional unit of L/mol/s, hence, the superscript*. Once B and C have diffused out of the cage, they can only react with each other again if they first diffuse toward each other with a rate coefficient k_{+diff} to form an encounter pair. This concept can be expressed as follows:

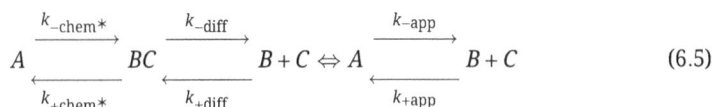

$$ A \underset{k_{+chem*}}{\overset{k_{-chem*}}{\rightleftharpoons}} BC \underset{k_{+diff}}{\overset{k_{-diff}}{\rightleftharpoons}} B + C \Leftrightarrow A \underset{k_{+app}}{\overset{k_{-app}}{\rightleftharpoons}} B + C \tag{6.5} $$

Upon applying the quasi-steady-state approximation, the apparent fission/scission rate coefficient k_{-app} can then be written as follows:

$$ \frac{1}{k_{-app}} = \frac{1}{k_{-chem}} + \frac{K_{eq}}{k_{+diff}} \tag{6.6} $$

in which K_{eq} is the equilibrium coefficient, which can be written as

$$ K_{eq} = \frac{k_{+chem*}}{k_{-chem*}} \frac{k_{+diff}}{k_{-diff}} = \frac{k_{+chem}}{k_{-chem}} \tag{6.7} $$

thus introducing the conventional k_{+chem}.

6.3.2 Tuning of intrinsic Arrhenius parameters of elementary reactions

The three FRP-based PMMA feedstocks defined in Section 6.2 have been subjected to thermogravimetric analysis at three different heating rates (2, 4, and 8 K/min). The experimentally-recorded thermogravimetric (TG) curves are depicted in Figure 6.4 (a), (c) and (e) by the full lines. As expected, the TG curves are shifted to higher temperatures if a higher heating rate is used. Moreover, as anticipated based on the different (average) chain lengths and fraction of head–head linkages and unsaturated chain ends (Figure 6.2), different TG curves are obtained for the three feedstocks. For example, feedstock A and B display a more pronounced shoulder, indicating a clearer separation between the mass loss due to head–head fission at lower temperatures and the mass loss due to other fission reactions at higher temperatures. Such sensitivity has proven to be crucial to be able to tune the intrinsic Arrhenius parameters of all reactions in Figure 6.3, leading to a single set of intrinsic elementary Arrhenius

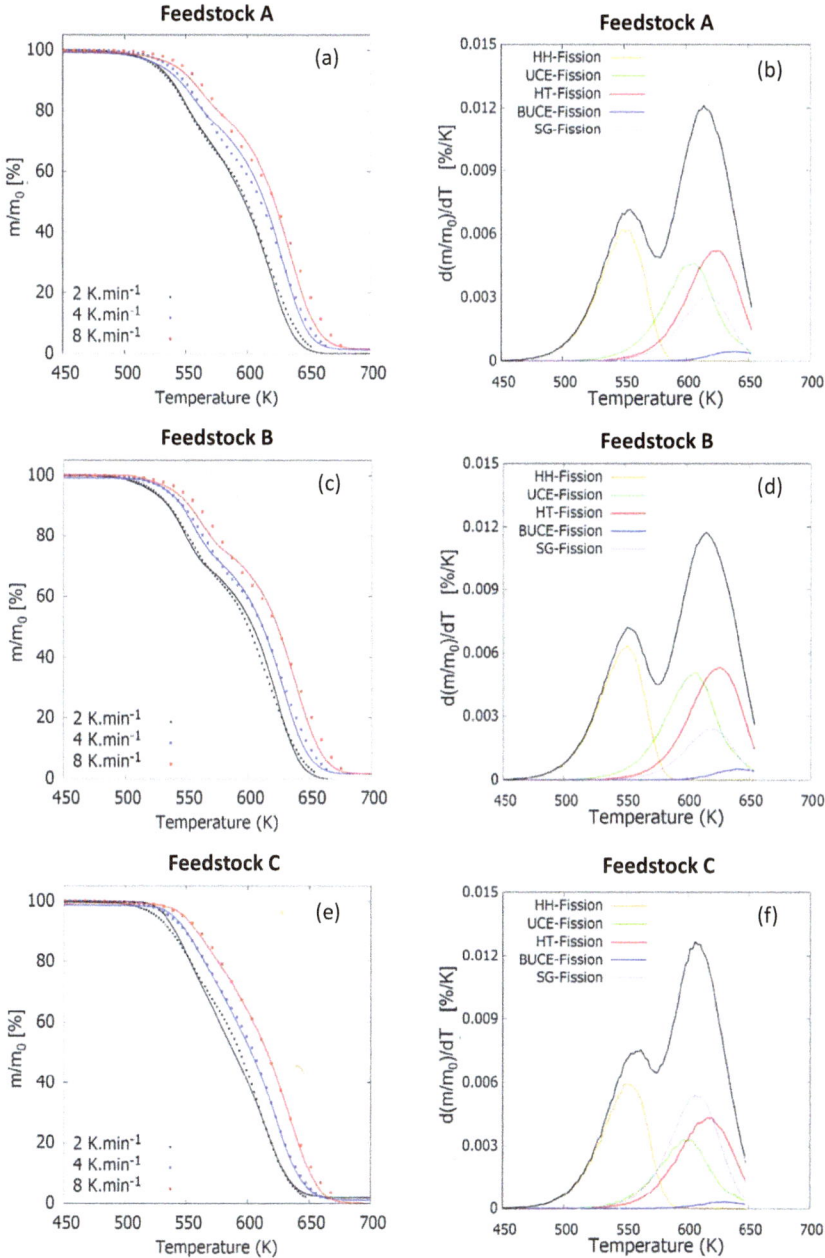

Figure 6.4: (a, c, e) Experimental (lines) and simulated (symbols) thermogravimetric (TG) curves for the three feedstock defined in Figure 6.2 and a heating rate of 2 (black), 4 (blue), and 8 (red) K/min. (b, d, f) Simulated contribution of individual fission reactions to the overall differential thermogravimetric curve (and hence mass loss rate, since dT ~ dt) for the feedstocks defined in Figure 6.2 and a heating rate of 2 K/min. Reproduced from [16] with permission from Elsevier.

parameters Table 6.2 that provide a good agreement between experiment (lines) and model (symbols) for all nine TG curves in Figure 6.4 (a), (c), and (e).

Interestingly, with the intrinsic elementary Arrhenius parameters available, one can analyze the contribution of individual fission reactions to the overall mass loss rate. This is illustrated in Figure 6.4(b), (d), and (f) for a heating rate of 2 K/min. Head–head fission is observed to be the main contributor at low temperatures for all three feedstocks. At higher temperatures, first (MMA-capped) unsaturated chain end fission starts to contribute, followed by head–tail fission, side-group fission, and a minor contribution from fission at isobutylene-capped chains ends at high mass loss stages (since these functionalities are only formed during degradation). Notably, upon integrating the peaks corresponding to the different fission reactions, their overall contribution to the mass loss can be quantified. Upon doing so one sees that, for ex-

Table 6.2: Overview of the intrinsic Arrhenius parameters for elementary fission and β-scission reactions as tuned by Moens et al. [16] using the thermogravimetric data in Figure 6.4(a), (c), and (e); for termination parameters the reader is referred to the work of Moens et al. [16]. A, pre-exponential factor; E_a, activation energy on the level of the elementary reaction; P, "dead" polymer chain; R, macroradical; subscripts i, j indicate chain length; subscripts p, t mean primary/tertiary end-chain radical; subscript MCR means mid-chain radical; subscript = and b mean MMA-capped and isobutylene-capped unsaturated chain; M^*, MC^*, IB^*, and Me^* mean monomer, methoxy carbonyl, isobutylene, and methyl radicals.

Fission reactions	Reactions	$A(s^{-1})$	E_a(kJ/mol)
Head–head fission	$P_{HH,i} \xrightarrow{k_{HHF}} R_{t,i-j}+R_{t,j}$	9×10^{14}	137
Unsaturated chain end fission MMA-capped	$P_{=,i} \xrightarrow{k_{CEF}} R_{t,i-1}+M^*$	3×10^{13}	160
Isobutylene-capped	$P_{b,i} \xrightarrow{k_{CEF}} R_{t,i-1}+IB^*$	3×10^{13}	160
Head–tail fission	$P_i \xrightarrow{k_{HTF}} R_{t,i-j}+R_{p,j}$	5×10^{14}	197
Side-group fission	$P_i \xrightarrow{k_{SGF}} R_{MCR,i}+MC^*$	2×10^{14}	205
Tail–tail fission	$P_{TT,i} \xrightarrow{k_{TTF}} R_{p,i-j}+R_{p,j}$	5×10^{14}	215
β-Scissions	**Reactions**	$A(s^{-1})$	E_a(kJ/mol)
End-chain β-scission of tertiary ECR	$R_{t,i} \xrightarrow{k_{T,ECS}} R_{t,i-1}+M$	8×10^5	27.5
End-chain β-scission of primary ECR	$R_{p,i} \xrightarrow{k_{P,ECS}} R_{p,i-1}+M$	5×10^5	27.5
Mid-chain β-scission	$R_{MCR} \xrightarrow{k_{MCS}} R_t+P_b$	5×10^8	55
Methoxy carbonyl side group β-scission	$R_p \xrightarrow{k_{MF,SGS}} P_b+MC^*$	5×10^7	45
Methyl side group β-scission	$R_p \xrightarrow{k_{M,SGS}} P_=+CH_3^*$	5×10^8	60

ample, side-group fission has a higher contribution for feedstock C, which can be explained by the higher chain lengths (and hence more pronounced diffusional limitations) and the lower fraction of, for example, (MMA-capped) unsaturated chain ends.

6.4 Kinetic modeling of isothermal degradation variations of molar mass distributions

While TGA is useful to identify dominant reaction sequences as a function of temperature, or even to tune Arrhenius parameters of individual elementary reactions as illustrated in Section 6.3, also conventional lab-scale degradation experiments targeting isothermicity and measuring the evolution of the MMD are very relevant. MMD, which is also often referred to as the molecular weight distribution, can specifically reveal information on how chains unzip.

This is theoretically illustrated in Figure 6.5, which shows the MMD evolution during degradation at 352 °C of two PMMA feedstocks obtained by isothermal bulk FRP at a monomer conversion of 85%, as simulated by De Smit et al. [2] considering a simplified degradation reaction scheme. The first feedstock (Figure 6.5(a)) was synthesized at 70 °C with $[AIBN]_0 = 0.027$ mol/L, while the second feedstock (Figure 6.5(b)) was synthesized at 90 °C with $[AIBN]_0 = 0.17$ mol/L. It follows that the second feedstock (curve indicated with "1" in Figure 6.5(b)) has significantly lower chain lengths than the first feedstock (curve indicated with "1" in Figure 6.5(a)), due to the higher initial initiator concentration, less pronounced diffusional limitations on termination, and a more elevated chain transfer to monomer.

For the first feedstock (Figure 6.5(a)), a significant variation of the MMD shape is observed as the weaker chains with head–head defects are in the high molar mass region and heavily attacked. For the second feedstock (Figure 6.5(b)), the high molar mass region is still significantly attacked but due to the shape of the initial MMD a more uniform degradation pattern with chains more likely to unzip completely is obtained so that the shape variation is limited. One should although realize that also here the MMD still shifts to the left as certain chains terminate after a certain extent of unzipping.

In practice, isothermal lab-scale data are thus needed. An example of a lab-scale setup is shown in Figure 6.6 (left). Such a setup allows targeting isothermicity and, in the worst case, obtain the temperature variation. It also allows taking samples during degradation for MMD measurement via gel permeation chromatography.

a)

b)

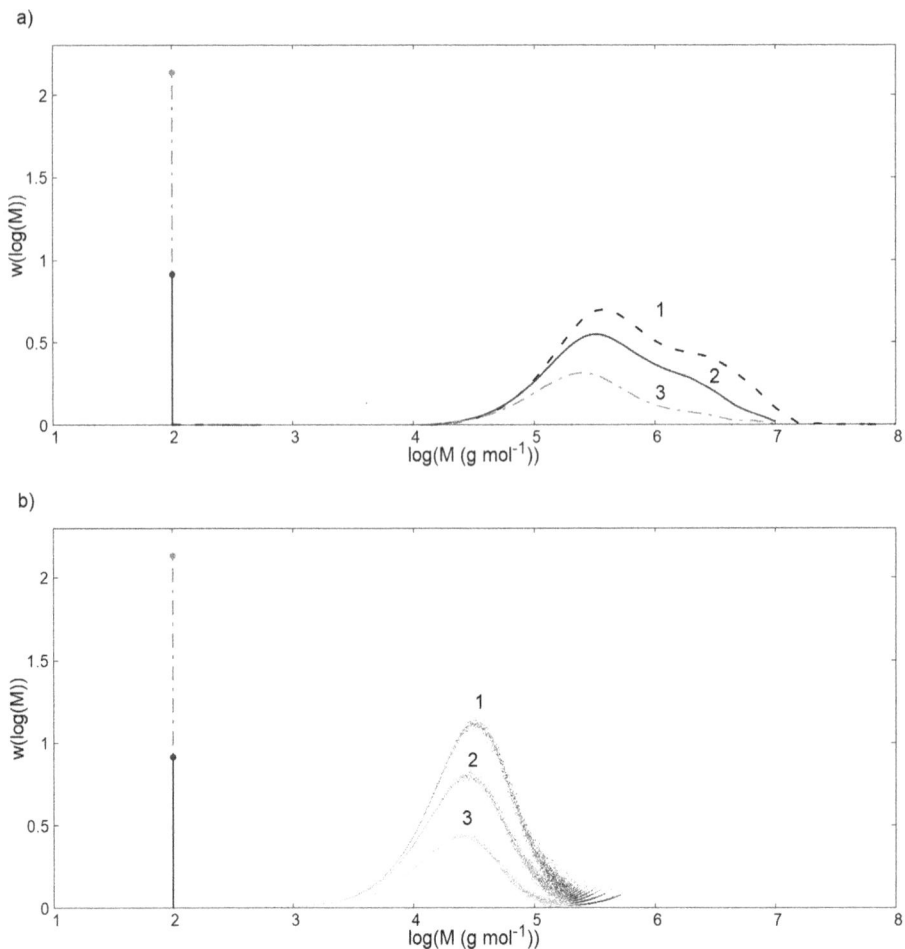

Figure 6.5: Evolution of the molar mass distribution (MMD) during degradation at 352 °C of two PMMA feedstocks obtained by isothermal bulk FRP with a monomer conversion of 85%, as simulated by De Smit et al. [2]. The feedstock in (a) corresponds to T_{pol} (polymerization temperature) = 70 °C and $[AIBN]_0$ = 0.027 mol/L; the feedstock in (b) corresponds to T_{pol} = 90 °C and $[AIBN]_0$ = 0.17 mol/L. Line 1: MMD of feedstock; line 2: MMD at 35% mass loss; line 3: MMD at 70% mass loss. A pronounced effect of the feedstock on the MMD evolution and thus degradation pattern is observed (figure adopted from De Smit et al. [2] with permission from The Royal Society of Chemistry).

An example of the evolution of the MMD during degradation of an FRP-based feedstock obtained from Sigma-Aldrich (M_m (mass average molar mass) = $7.5 \times 10_4$ g/mol, dispersity = 2.4) at a rather low temperature of 200 °C measured using this setup is shown in Figure 6.6 (right). A clear overall shift of the initial MMD (t = 0 min) to lower molar masses is observed, indicating the formation of termination products.

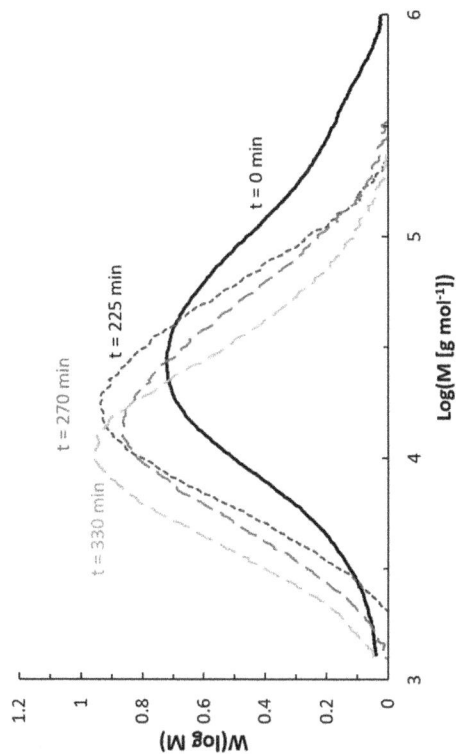

Figure 6.6: Left: Experimental setup at LCT allowing targeting isothermal experiments with measurement of the molar mass distribution (MMD) evolution. Right: Illustration of the MMD evolution during degradation of a FRP-based feedstock (Sigma-Aldrich; M_m (mass average molar mass) = 7.5×10^4 g/mol, dispersity = 2.4) at a low targeted temperature of 200 °C; FRP, free radical polymerization.

6.5 Conclusion

It can be concluded that the combination of TGA data and lab-scale experiments targeting isothermicity and the measurement of the MMD evolution can provide detailed mechanistic and kinetic information for polymer degradation and depolymerization. This is especially the case upon combination with judiciously designed polymer feedstocks and detailed elementary reaction-based kinetic models that account for diffusional limitations and thus viscosity effects.

References

[1] Moens EKC, De Smit K, Marien YW, Trigilio AD, Van Steenberge PHM, Van Geem KM, et al. Progress in reaction mechanisms and reactor technologies for thermochemical recycling of poly(methyl methacrylate). Polymers 2020, 12.

[2] De Smit K, Marien YW, Van Geem KM, Van Steenberge PHM, D'Hooge DR. Connecting polymer synthesis and chemical recycling on a chain-by-chain basis: A unified matrix-based kinetic Monte Carlo strategy. Reaction Chemistry & Engineering 2020, 5, 1909–1928.

[3] Kruse TM, Woo OS, Broadbelt LJ. Detailed mechanistic modeling of polymer degradation: Application to polystyrene. Chemical Engineering Science 2001, 56, 971–979.

[4] Kruse TM, Wong H-W, Broadbelt LJ. Mechanistic modeling of polymer pyrolysis: Polypropylene. Macromolecules 2003, 36, 9594–9607.

[5] Brandrup J, Immergut EH, Grulke EA, Abe A, Bloch DR. Polymer Handbook, Wiley, New York, 1999.

[6] Van Herck J, Harrisson S, Hutchinson RA, Russell GT, Junkers T. A machine-readable online database for rate coefficients in radical polymerization. Polymer Chemistry 2021, 12, 3688–3692.

[7] Marien YW, Edeleva M, Van Steenberge PHM, D'Hooge DR. Chapter Three – Exploiting the pulsed laser polymerization-size exclusion chromatography technique to retrieve kinetic parameters in radical polymerization: State-of-the-art and future challenges. In Moscatelli D, Sponchioni M, eds. Advances in Chemical Engineering. Academic Press, 2020, 59–95.

[8] Beuermann S, Buback M. Rate coefficients of free-radical polymerization deduced from pulsed laser experiments. Progress in Polymer Science 2002, 27, 191–254.

[9] Vir AB, Marien YW, Van Steenberge PHM, Barner-Kowollik C, Reyniers M-F, Marin GB, et al. From n-butyl acrylate Arrhenius parameters for backbiting and tertiary propagation to β-scission via stepwise pulsed laser polymerization. Polymer Chemistry 2019, 10, 4116–4125.

[10] Dogu O, Pelucchi M, Van de Vijver R, Van Steenberge PHM, D'Hooge DR, Cuoci A, et al. The chemistry of chemical recycling of solid plastic waste via pyrolysis and gasification: State-of-the-art, challenges, and future directions. Progress in Energy and Combustion Science 2021, 84, 100901.

[11] Shi C, Reilly LT, Phani Kumar VS, Coile MW, Nicholson SR, Broadbelt LJ, et al. Design principles for intrinsically circular polymers with tunable properties. Chem 2021, 7, 2896–2912.

[12] Iedema PD, Remerie K, van der Ham M, Biemond E, Tacx J. Controlled peroxide-induced degradation of polypropylene in a twin-screw extruder: Change of molecular weight distribution under conditions controlled by micromixing. Chemical Engineering Science 2011, 66, 5474–5486.

[13] Kashiwagi T, Inaba A, Brown JE, Hatada K, Kitayama T, Masuda E. Effects of weak linkages on the thermal and oxidative degradation of poly (methyl methacrylates). Macromolecules 1986, 19, 2160–2168.

[14] Manring LE, Sogah DY, Cohen GM. Thermal degradation of poly (methyl methacrylate). 3. Polymer with head-to-head linkages. Macromolecules 1989, 22, 4652–4654.

[15] Da Ros S, Braido RS, E Castro N, Brandão AL, Schwaab M, Pinto JC. Modelling the chemical recycling of crosslinked poly (methyl methacrylate): Kinetics of depolymerisation. Journal of Analytical and Applied Pyrolysis 2019, 144, 104706.

[16] Moens EKC, Marien YW, Figueira FL, Trigilio AD, De Smit K, Van Steenberge PHM, D'hooge DR. Coupled matrix-based Monte Carlo modelling for a mechanistic understanding of poly(methyl methacrylate) thermochemical recycling kinetics. Chemical Engineering Journal 2023, 475, 146105.

[17] Holland B, Hay J. The effect of polymerisation conditions on the kinetics and mechanisms of thermal degradation of PMMA. Polymer Degradation and Stability 2002, 77, 435–439.

[18] D'hooge DR, Van Steenberge PHM, Reyniers M-F, Marin GB. The strength of multi-scale modeling to unveil the complexity of radical polymerization. Progress in Polymer Science 2016, 58, 59–89.

[19] D'hooge DR, Van Steenberge PH, Derboven P, Reyniers M-F, Marin GB. Model-based design of the polymer microstructure: Bridging the gap between polymer chemistry and engineering. Polymer Chemistry 2015, 6, 7081–7096.

[20] Trigilio AD, Marien YW, Van Steenberge PHM, D'hooge DR. Gillespie-driven kinetic Monte Carlo algorithms to model events for bulk or solution (bio)chemical systems containing elemental and distributed species. Industrial & Engineering Chemistry Research 2020, 59, 18357–18386.

[21] De Keer L, Kilic KI, Van Steenberge PHM, Daelemans L, Kodura D, Frisch H, et al. Computational prediction of the molecular configuration of three-dimensional network polymers. Nature Materials 2021, 20, 1422–1430.

[22] Figueira FL, Wu -Y-Y, Zhou Y-N, Luo Z-H, Van Steenberge PHM, D'Hooge DR. Coupled matrix kinetic Monte Carlo simulations applied for advanced understanding of polymer grafting kinetics. Reaction Chemistry & Engineering 2021, 6, 640–661.

[23] Balke ST, Hamielec AE. Bulk polymerization of methyl methacrylate. Journal of Applied Polymer Science 1973, 17, 905–949.

[24] Kashiwagi T, Inabi A, Hamins A. Behavior of primary radicals during thermal degradation of poly (methyl methacrylate). Polymer Degradation and Stability 1989, 26, 161–184.

[25] Manring LE. Thermal degradation of poly (methyl methacrylate). 4. Random side-group scission. Macromolecules 1991, 24, 3304–3309.

[26] Ferriol M, Gentilhomme A, Cochez M, Oget N, Mieloszynski J. Thermal degradation of poly (methyl methacrylate)(PMMA): Modelling of DTG and TG curves. Polymer Degradation and Stability 2003, 79, 271–281.

[27] Yang J, Miranda R, Roy C. Using the DTG curve fitting method to determine the apparent kinetic parameters of thermal decomposition of polymers. Polymer Degradation and Stability 2001, 73, 455–461.

[28] Briceno J, Lemos MA, Lemos F. Kinetic analysis of the degradation of HDPE+PP polymer mixtures. International Journal of Chemical Kinetics 2021, 53, 660–674.

[29] Das P, Tiwari P. Thermal degradation kinetics of plastics and model selection. Thermochimica Acta 2017, 654, 191–202.

[30] D'hooge DR, Reyniers MF, Marin GB. The crucial role of diffusional limitations in controlled radical polymerization. Macromolecular Reaction Engineering 2013, 7, 362–379.

[31] Johnston-Hall G, Monteiro MJ. Bimolecular radical termination: New perspectives and insights. Journal of Polymer Science Part A: Polymer Chemistry 2008, 46, 3155–3173.

[32] Barner-Kowollik C, Russell GT. Chain-length-dependent termination in radical polymerization: Subtle revolution in tackling a long-standing challenge. Progress in Polymer Science 2009, 34, 1211–1259.

[33] De Smit K, Marien YW, Edeleva M, Van Steenberge PHM, D'hooge DR. Roadmap for monomer conversion and chain length-dependent termination reactivity algorithms in kinetic Monte Carlo modeling of bulk radical polymerization. Industrial & Engineering Chemistry Research 2020, 59, 22422–22439.

[34] Derboven P, D'hooge DR, Reyniers M-F, Marin GB, Barner-Kowollik C. The long and the short of radical polymerization. Macromolecules 2015, 48, 492–501.

Jean-Luc Dubois

7 PMMA chemical recycling reactor technologies

Poly(methyl methacrylate) (PMMA) is one of the few polymers which can be depoly-merized back into its monomer, methyl methacrylate (MMA), through a heat treat-ment. In a pyrolysis process, the polymer is heated in the absence of oxygen at a high temperature to unzip the polymer chain and regenerate the monomer. PMMA has been depolymerized at multiple sites throughout the world for several decades al-ready. There are multiple processes that have been tested and are in operation for MMA regeneration. Continuous processes lead to the best purity of the regenerated MMA, but are not the most common. However, the depolymerization reaction is only one of the multiple steps needed to produce a high-quality rMMA.

7.1 Process steps needed in a recycling plant

The review of the PMMA recycling plants conducted for the MMAtwo project identi-fied the technologies that may be needed to generate a high-quality regenerated MMA (rMMA). Among these, we identified multiple steps: sorting, crushing, wind sifting, washing, drying, pyrolysis, condensation, filtration, washing, rinsing, evaporation, dis-tillation, and water removal. Not all of them are implemented on the same site and not necessarily in that order, and they have a specific role depending on the type of scraps which are to be processed and the quality of rMMA which is to be produced. Some of the operators are selling the rMMA produced, but most of them are using it for internal/captive consumption. So, they are used to tolerating some impurities and have adapted their polymerization process and conditions. In such a case, they do not need to store the crude MMA (produced after pyrolysis) or the rMMA (purified) for a long time, and they don't need to transport it, thus avoiding the use of excessive amounts of stabilizers.

7.1.1 Sorting

This operation is needed when the operator is getting the scraps from multiple sour-ces. PMMA is produced in three major grades: Cast, Extrusion, and Injection. The Cast PMMA has a high molecular weight and usually no comonomers (depending on the

Jean-Luc Dubois, Trinseo France SAS – Altuglas International SAS Tour CB21,16, place de l'Iris, 92400 Courbevoie, France

https://doi.org/10.1515/9783111076997-007

recipe). The Extrusion and Injection grades, however, are produced with co-monomers (acrylates and others) and have lower molecular weights. In addition, the products are sold in multiple colors or with inorganic fillers and additives. Post-production or post-consumer scraps can also be contaminated with other polymers. For example, the PMMA sheets can be covered with paper or polyethylene (PE) film, and injection parts, such as car tail lights, can be soldered onto other polymers. So, the first step in the recycling process is to sort the scraps, and the sorting can be done manually or with various automatic sorting equipment as described in the previous chapters 2 and 3.

7.1.2 Crushing

To complement or facilitate the sorting, the scraps are crushed. It will also facilitate the feeding of the depolymerization reactor. The size of the crushed material is important to reduce the mass and heat transfer limitations during the depolymerization process. Particles between 5 and 15 mm seem to be the optimal size. If the particles are too big, the mass and heat transfer are limiting the depolymerization; if the particles are too small, it requires a lot of energy for crushing the particles. Since PMMA is a thermoplastic polymer, the high energy of crushing can lead to polymer fusion and becomes an intrinsic limitation of the size reduction.

Cast sheets are produced by pouring a liquid syrup (made of MMA oligomers, or eventually a solution of low molecular weight PMMA in MMA) between glass plates, surrounded by a PVC gasket. So, when the sheets are demolded, the gasket is removed together with good-quality PMMA by sawing the sheets a few centimeters away from the gasket. These scraps used to be sent to low labor cost countries to remove the PVC gasket closer to the edge and depolymerize the cast PMMA locally. The crushing step can be combined with other sorting processes to remove the PVC fraction. PVC would depolymerize in the reactor and generate corrosive HCl and a carbonaceous residue, so PVC should be avoided as far as possible. Together with the French start-up X-crusher, the author tested a non-contact crushing process [1]. In this technology, shock waves are used to achieve a selective crushing of the material and help to separate the materials.

PE or paper protective film removal from the PMMA sheets is either done manually in low labor cost countries or a wind sifting step may be necessary. The PE film collected can be sold to other recyclers and can represent several tons per year.

7.1.3 Washing

Sometimes the films are not easily removed. So, a washing step is implemented to separate the PE and paper films more efficiently. PE will also be cracked in the depolymerization reactor and will generate hydrocarbons and waxes, which could contaminate the rMMA because some of them have similar boiling points to MMA, that is, 100 °C (hydrocarbons of six to eight carbons; Table 7.1).

Post-consumer scraps are likely to be contaminated with dirt, which may also require a washing step to remove those impurities. However, any washing step would mean additional equipment (and Capital cost, Capex), water consumption and wastewater disposal (Operating cost, Opex), energy for drying, and would be better avoided.

Table 7.1: Boiling points of hydrocarbons close to MMAs.

Carbon number	6	7	8		6	7	8
2,2-Dimetylalkane		79	106	1-Alkyne		100	126
2-Methylalkane		90	117	Alkylcyclohexane		101	132
2-Methylalkene		92	118	Alkylcyclopentane	72	104	131
3-Methylalkane		92	120	Alkylbenzene	80	111	136
1-Alkene		94	121	Cycloalkene	83	115	
N-alkane		98	126	Cycloalkane	81	119	151

7.1.4 Drying

After a washing step, a drying step would have to be considered. PMMA can contain some water which is dissolved in the polymer. When PMMA granules are used for injection or extrusion, this small water content can generate bubbles in the parts and may require a drying step before use [2]. In the depolymerization process, water can lead to hydrolysis of MMA and generates methanol and organic acids, and when it does not react, it contaminates the crude MMA. Water solubility in PMMA is below 2 wt%, and is then below the water solubility limit in MMA. When methanol is produced during the pyrolysis, it also increases the water solubility in MMA and increases the complexity of the purification process.

7.1.5 Pyrolysis, with or without pre-melting

Pyrolysis is a heat treatment at high temperature in the absence of oxygen. Several technologies have been investigated and are in operation for the regeneration of MMA. They will be detailed in the next section. In the pyrolysis reactor, PMMA scraps have to be heated, so that PMMA becomes liquid (melted); then at a high enough temperature, PMMA depolymerizes, and the MMA evaporates. The heat needed for the depolymerization can then be calculated from the thermodynamic data (specific heat capacities, latent heat of fusion (if any), latent heat of depolymerization, latent heat of evaporation). The heat needed for the depolymerization will then slightly depend on the depolymerization temperature, but much more on the type of technology selected.

The pyrolysis takes place at a temperature close to the MMA self-ignition temperature (435 °C) [3, 4]. So, there is an intrinsic safety risk in case of leakage of hot gases outside the reactor or where air would enter the reactor. The amount of PMMA scraps immobilized in the reactor is, in that sense, a very important parameter to consider. Continuous processes can minimize the amount of hot PMMA/MMA in the reactor and should be preferred. The flash point of MMA is 10 °C (open cup) and 2 °C (closed cup), so even at low temperature special care should be taken when handling this product.

The other safety risk linked with pyrolysis is the formation of pyrophoric carbon residue in the reactor. So special care should be taken when the carbon residues are unloaded hot from the reactor.

The heat required for PMMA depolymerization can be calculated as follows:

a) heat to bring PMMA to depolymerization temperature: $M_{(PMMA)} {}^* Cp_{(PMMA)} {}^* (T_{depoly} - T°)$ where $M_{(PMMA)}$ is the mass of PMMA, $Cp_{(PMMA)}$ is the specific heat of PMMA which varies with the temperature, T_{depoly} is the depolymerization temperature, and $T°$ is the initial temperature;

b) melting energy, usually very small as PMMA is not a crystalline polymer;

c) heat of depolymerization, which can be assumed to be the opposite of the heat of polymerization, although both reactions are conducted at completely different temperatures; and

d) heat of evaporation of MMA, to complete the sequence of elementary steps. In this case, it takes place at the depolymerization temperature, so at a much higher temperature than what has been reported in databases for MMA's boiling point. And to simplify the calculations, we can consider that the residence time of the vapors in the hot reactor is short enough to avoid additional heat transfer.

If the depolymerization takes place at 425 °C instead of 475 °C, then the energy saving corresponds to the specific heat difference mostly (less than 10%), but one also has to take into account the impacts on productivity and product quality. The energy needed is then in the range of 1.7 to 1.8 MJ/kg (without the heat losses on the equipment), or about 0.5 kWh/kg. However, it depends also on the formulation of the PMMA (content of fillers, comonomers, etc.) since they affect the specific heat capacity and also the heat of polymerization.

Example: At 100 kg/h PMMA, 100% conversion, and 425 °C (as depolymerization temperature). Heat to be consumed in the pyrolizer reactor:

∫Cp(T)dT – Heat of polymerization + Heat of evaporation =

2.374/2*((425 + 273)² – (25 + 273)²) + 1129.5(425–25) + 300 000 + 578 000) = 1.8 MJ/kg

Where – 578 kJ/kg is assumed to be the MMA polymerization energy, and 300 kJ/kg is assumed to be the MMA evaporation energy. The heat capacity of PMMA increases with the temperature (data from [5]). Heat of evaporation is extrapolated from data reported in the NIST database [6] and in Chemeo [7], and the heat of polymerization from [8].

Reported heat of evaporation tends to decrease when increasing the temperature, heat capacity is polymer dependent, and the heat of polymerization depends on the acrylate content. Also, the energy consumed depends on the conversion rate. When conversion will start to decrease, the Net specific Depolymerization energy will also start to decrease. If depolymerization takes place at 50 °C lower temperature, the energy demand decreases by 0.13 MJ/kg (or less than 10%).

These values are in line with what has been reported [9]. On top of the theoretical energy consumption, one should also add the energy losses, which are dependent on the pyrolysis technology and equipment used. The heat loss is likely to be constant in a given reactor, or at least less depending on the feeding rate.

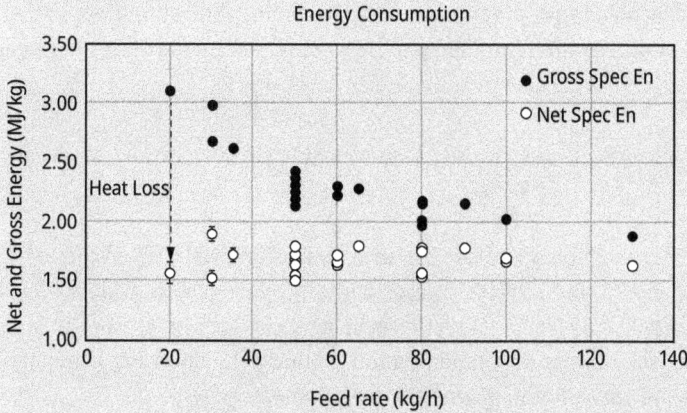

Figure 7.1: Net and gross energy consumption in for PMMA depolymerization, based on trials on a twin-screw extruder, at different feeding rates and for different PMMA grades.

7.1.6 Condensation

The MMA vapors produced in the depolymerization reactor have to be condensed into liquid crude MMA. Two main technologies are used in recycling plants. The tube and shell heat exchangers are quite common in developing countries; otherwise, a "Spray Condenser" can be used, as well as an adiabatic condenser, such as the technology implemented by Pyrovac [10] and Chapter 15 of this book. Hot MMA could repolymerize easily, so it has to be cooled to a low enough temperature. The Self-Accelerating Polymerization Temperature (SAPT) is above 55 °C [4]. A spray condenser or a venturi condenser requires a loop of liquid condensed MMA which has to be cooled, while the tube and shell heat exchanger could be plugged by dust and polymers that would accumulate in the tubes and require frequent cleaning.

7.1.7 Filtration

The depolymerization of PMMA scraps can generate carbon residues and dusts which are carried away with the MMA vapours. For example, 1 kg of PMMA would generate about 600 L of gases at 450 °C, which would generate a significant linear velocity in pipes. In addition, some PMMA grades contain some inorganic pigments such as titanium dioxide or barium sulfate, for example. The dusts generated can plug the condensers or would contaminate the crude MMA collected. In addition, some contamination polymers can also be present. The solid separation can be achieved by decantation, but it may also require a filtration step. Hopefully, the amount in suspension is usually very small, so the filters have to be cleaned only a few times per year. However, cheaper scraps would also include dirtier wastes.

7.1.8 Washing

More often than a filtration, the operators are using a washing step. When the PMMA scraps are containing PVC, the pyrolysis gases contain hydrogen chloride which is highly corrosive. Organic acids such as methacrylic acid can also be present in some formulations and/or freed in the depolymerization. Both can be neutralized by an alkaline solution to avoid corrosion in downstream equipment.

7.1.9 Rinsing

After washing with an alkaline solution, the crude MMA is sometimes (but not always) rinsed; and to save water, the rinsing water is used to make the alkaline water of the washing step. In both steps (washing and rinsing), the MMA organic layer is separated from the aqueous phase by decantation. Although these steps can eliminate acids and water soluble contaminants from the crude MMA, they also contribute to add water in the crude MMA. Water forms multiple azeotropes with MMA and the other organic contaminants present in the crude MMA such as the acrylate comonomers. In addition, when the crude MMA contains methanol, or other alcohols, they contribute to an increased solubility of water in the crude MMA [11]. Methanol also forms azeotropes with MMA and other organics. The presence of water and methanol in the crude MMA is then prone to generate difficulties at the distillation steps. The washing and rinsing steps are then also generating wastewaters which have to be treated appropriately.

7.1.10 Evaporation

After those steps, the crude MMA contains light boilers and heavy boilers, i.e. molecules with lower and higher boiling points than MMA. The heavies include polymers and oligomers soluble in MMA. If the distillation units are directly fed with the crude MMA, the column reboilers could be quickly coated and lose their efficiency, leading to higher energy consumption and/or frequent cleaning. A solution implemented sometimes is to add before the distillation a wiped thin film evaporator, in which the very heavy molecules are withdrawn at the bottom. The evaporated fraction then feeds the distillation units. The energy consumption linked with this step is only a fraction of the heat needed for the whole purification. The residues generated have to be properly disposed as they may accumulate substances of concern.

7.1.11 Distillation

The distillation can be done batch wise or in a continuous process. The distillation can be done at atmospheric pressure or under reduced pressure in a single path or in two steps. A wide diversity of conditions have been seen in the plants treating crude MMA. Preferably the distillation would be done under vacuum to reduce the operating temperature and the risks to trigger MMA polymerization in the column. However, in these conditions the separation of the impurities is more challenging since the boiling points get closer under reduced pressure.

The light and heavy residues from the distillation can be used internally for energy production and fed to a boiler and/or a combined heat and power generator. In other cases, the residues are sold to other companies for energy valorization or proper disposal. Again, the heavy residues might accumulate substances of concern and have to be properly disposed.

7.1.12 Water removal

The regenerated MMA produced by distillation has purities varying between 90 and 99.8 wt%, depending on the site and process combinations. The lowest purities are often associated with the presence of higher amounts of water remaining in rMMA. Indeed, some operators do not separate the water azeotropes, so water remains in the regenerated MMA. When the rMMA is used to produce cast PMMA, it is first polymerized to oligomers forming a syrup. The water can then be easily separated from these oligomers, before the cast PMMA is produced, provided that the plant is equipped with such a unit, which is not common when using only virgin MMA.

In other cases, the traces of water can be removed when contacting the crude MMA with an adsorbent or reagent like calcium chloride or magnesium sulphate.

7.1.13 Preliminary conclusions on the processing steps

The operators have to minimize the capital and operating costs. So, it is unlikely that they would use all these steps on the same site. The choice of the PMMA scraps to be processed also has a strong impact on the processing steps that would be needed. For example, a site that processes only post production cast PMMA would generate a better-quality crude MMA, most likely that would not contain acrylate co-monomer. Therefore, the purification section would be lighter and would lead not only to higher purity rMMA, but also to better yields.

7.2 Depolymerization technologies

As mentioned previously, there is a wide diversity of technologies which have been used for PMMA depolymerization.

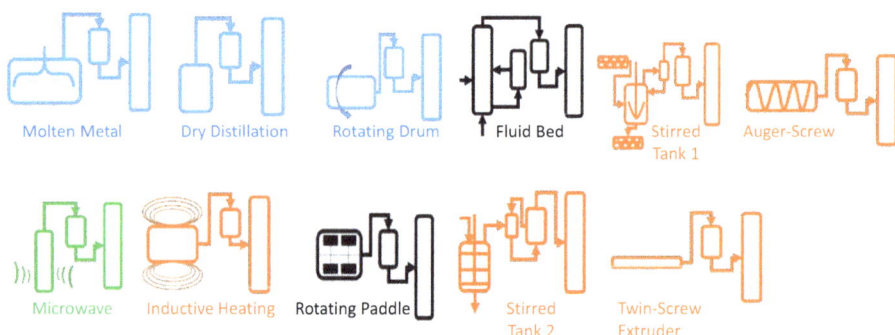

Figure 7.2: Illustration of the diversity of PMMA depolymerization technologies.

7.2.1 Dry distillation

In the Dry-Distillation process [12, 13], the oldest process, a pile of PMMA scraps is loaded in a tank, and fire is lighted underneath the tank, which is directly connected to a condenser. The operation is stopped when no more product is generated from the reactor. Although the process is probably the most common, it has some intrinsic limitations. The heat transfer to the material is not efficient: initially only the material near the walls is heated, and it takes time to transfer heat to the central part of the pile. In addition, as the depolymerization progresses, char is formed and accumulates, further reducing the heat transfer. As char is being formed, hydrogen transfer reactions are taking place. This is translated into the formation of Methyl Isobutyrate, which is the hydrogenated equivalent of MMA. This impurity cannot polymerize unlike MMA, so it would contribute to Volatile Organic Compounds (VOC) and odors. Methyl isobutyrate's boiling point is rather close to MMA's boiling point, so this impurity would be difficult to remove during purification. The accumulation of char in the reactor, like in other processes described below, also restricts the type of PMMA scraps that can be processed. For example, PMMA containing large amounts of inorganic fractions, or that would otherwise produce a large amount of solid residue, would require frequent cleaning operations of the reactor. For these operations, an operator has to enter the reactor and scratch the surface to unload the residues which have to be properly disposed of, and of course, this is a health and safety issue for the exposed personnel.

The process is operated in a batch mode, which means that the reactor is loaded and heated daily in most case. It also means that the reactor has to sustain daily heating and cooling cycles to/from high temperature. These cycles compromise the life of the reactors, which may have to be replaced periodically. Consequently, those reactors and process plants are rather cheap.

7.2.2 Rotating-drum

The Rotating-Drum process is a variation of the Dry Distillation process. To improve the mass and heat transfer, the PMMA scraps are loaded in a rotating cylinder which is heated externally. The reactor is often operated in a batch mode. The heat transfer is made through the wall of the cylinder. This also means that there is an intrinsic limitation to the scale-up of the reactor. To depolymerize more product, it is wiser to increase the length of the cylinder rather than the diameter to keep a high surface to volume ratio of the reactor. But the consequence is a large variation of dimensions when the reactor is heated due to thermal expansion. This is also raising a safety concern since the reactor would have daily heating and cooling cycles and significant dimensions variations that may lead to leakages (and fires). So, the life of the equipment is limited and the unit would require frequent replacements. At the end of the cycle, the reactor can be unloaded to remove the solid residue, by a rotation in the opposite direction. In case of power failure, the rotation of the equipment would be halted as well as the heating. It might be difficult to put the cylinder in rotation again, if the mass of molten PMMA solidifies.

The technology is common in Asia and used for recycling other plastics like used tires. Several companies are selling this type of equipment at rather low cost.

7.2.3 Molten-lead bath process

The heat transfer has been improved by using a molten metal (lead) above 400 °C [14, 15]. Crushed PMMA scraps are introduced in a liquid bath of molten metal and depolymerize in a few minutes. Several metals have been considered for the process such as lead, tin, and zinc. Tin is about 10 times more expensive than lead, and zinc raises serious safety concerns as it is known that it can corrode steel, and that its vapor pressure is higher than lead partial pressure. Lead is still used in several plants. One of the key advantages of this process is linked to the high density of molten lead. So, all the impurities, including metals, would float on the surface of the molten bath. Like in the previous depolymerization processes, char is being formed. The surface is scrapped on regular basis to remove the char and contaminants. But this solid residue is then contaminated with lead. Alloys, especially those forming eutectics are also attractive, as the alloy melting point would be lower than the pure metals themselves. To some extent it can make the process easier to operate. Recently the Italian company Biorennova patented a process based on molten metal, and the company was acquired by Maire Technimont/Nextchem which intends to build a 5000 t/y plant [16] and [17]. The companies called the process CatC and attribute to the molten metal a catalytic effect. An important feature in the patented technology is that there is circulation of molten metal, in a high mass ratio compared to the amount of PMMA being processed. Metal prices can be seen at the London Metal Exchange place for Tin, Lead and Zinc [18].

Consequently, the technology is not appropriate to treat PMMA scraps that would generate a large amount of solid residue. Lead partial pressure is rather low at the depolymerization temperature; but it means that some lead will be anyway carried with the crude MMA and has to be recovered downstream. So, the process generates several streams which are contaminated and have to be properly treated and disposed.

7.2.4 Twin-fluid bed process

In a fluid-bed reactor [19, 20], a solid is kept in suspension by a gas stream and/or mechanical action. The fine solid particles behave like a liquid and can be moved from one reactor to another. In a regenerator fluid bed, the solid (sand) is heated to high temperature and then transported to the reactor where it is contacted with the PMMA which has to be depolymerized. The fluid-bed reactor has a high heat transfer, when the hot solid particles contact the PMMA particles (high heat contact area due the small size of the solid particles). There is also a high mass transfer rate and a short contact time of the gas phase. The fluid bed has then some similarity with the molten lead bath reactor, without the disadvantage of the use of toxic lead. Because of the high mixing rate, the reactor behaves like a stirred tank reactor.

However, the technology has also several disadvantages: when PMMA particles melt in the fluid bed, they may create islands of molten polymer, and the gas cannot circulate in these areas where defluidization occurs, and in the other zones a higher gas linear velocity leads to transportation (carry away) of the solid. The fluid bed becomes unstable when one wants to reach a high productivity. It is an intrinsic limitation to this technology. In addition, solid is continuously withdrawn from the reactor to be sent to the regenerator, where any char and remaining PMMA particles are burned, together with some fuel to reheat the solid and keep the system running. The gas used to fluidize the solid is preferably mainly made of MMA rather than an inert gas, to keep a high MMA partial pressure and increase the performance of the condenser.

This technology had been initially developed by ICI, patented by DuVergier, and later by Mitsubishi Rayon [21] which operated a pilot plant and a demonstration unit. A variation of the fluid bed reactor which was investigated at academic level is the Conical Spouted Bed Reactor [22]. Mitsubishi patented a technology where the fluidization is assisted with a mechanical agitator. A benefit if this process is to reduce the amount of gas needed for the fluidization. Other variations of the mechanical agitation are also possible, but these also require a significant amount of energy.

7.2.5 Stirred-tank process

The stirred-tank process [23] is again a variation of the Dry Distillation process. In this process, there is agitation to improve the heat and mass transfer. To further improve it, the PMMA scraps can be pre-melted in a screw feeder, so that molten polymer reaches the reactor where it is heated to a very high temperature and where depolymerization takes place. This is, for example, the type of technology promoted by Plastic Energy for mixed polymer recycling [24]. Solid residues accumulate in the reactor, from which they can be continuously withdrawn. The challenge in this technology is to bring enough heat to the reactor to maintain high productivity and to keep the agitation in a highly viscous medium (molten polymer). The heat transfer can be facilitated by a solid present in the reactor, and which may have to be reheated periodically [25].

Different stirring technologies are possible. In a continuous process, the solid scraps are dropped in the molten polymer, which improves the heat transfer. The Auger–Screw-type process is a variation of the Stirred-Tank process. In this process, an Archimedes screw is pushes the product to be depolymerized, and the heat is either provided through the screw itself or through the walls. A major disadvantage is that usually the screw is not self-cleaning, and the scale-up capacity is limited by the heat exchange area.

An appropriate stirred tank reactor would then operate in continuous mode, with a high and scalable heat exchange area, a continuous outlet of the solid residue, and a self-cleaning design. Stirred tank reactors operate either with a temperature gradient in a continuous mode or at different temperature in a batch cycle.

The technology patented by Agilyx [26] for polystyrene depolymerization, or mixed plastics pyrolysis, would enter into this family. Two screws with a large internal volume, like in a kneader reactor, bring the material to the temperature.

The technology developed by Pyrovac [34], and Chapter 15 of this book, also falls in this family. It is based on multiple soles heated by a circulating molten salt. The PMMA scrap would fall in the unit from the top of the reactor, and then starts to circulate from one sole to the other in a cascade mode. At the same time, internal agitators are mixing the molten polymer with the PMMA scraps, promoting heat transfer. At the bottom of the unit, the solid residues are collected and evacuated. The pyrolysis is completed by a unique adiabatic condensation/separation step that take benefit of the heat integration [10].

The Korean company R&E [35], and chapter 14 of this book, has also implemented a stirred tank process to recycle solid surface. These products are composites of PMMA and Aluminum TriHydroxyde (ATH), containing about 65% ATH. Because of the nature of the waste to be processed, the rMMA purity cannot be high. However, since in this case the energy demand and CO_2 emissions can be split over 2 marketable products (rMMA and Alumina), the environmental impact is still acceptable.

7.2.6 Twin-screw extruder process

The twin-screw extruder [27–30] is also a variation of the Dry Distillation and is a plug flow stirred reactor, in which the heat transfer has been pushed to the maximum. This is achieved by reducing the diameter of the reactor and increasing its length. The residence time in the reactor can now be a matter of seconds and no longer hours. The process is intrinsically safer than the other alternatives because the amount of PMMA immobilized in the reactor is at any time always very small. In case of emergency shutdown, the unit can be free of polymer very quickly.

JSW and Kosui Chemical (a small Japanese PMMA producer) developed a depolymerization process using a twin-screw extruder technology in 1995. The MMA is produced at high temperature (400–500 °C) from the electric heater and shearing from the mixing screw in the twin-screw extruder. The produced MMA is captured at the end of the twin-screw extruder, cooled in a heat exchanger and purified. The minerals, metal and glass fiber in PMMA are separated at the end of twin-screw extruder. The PMMA depolymerization unit at Kosui Chemical was operated for about 5 years. The PMMA is melted in the first section of the screw to create a plug and to avoid a backflush of the produced MMA [31].

The process can be used for PMMA grades that are generating a lot of solid residues because the reactor continuously withdraws the solid residue which is pushed to the end of the extruder and behaves like a plug-flow reactor.

This is also the technology which has been selected and improved in the MMAtwo Project, and now implemented in Trinseo's Rho plant in Italy [32]. Recently Sumitomo also announced a project using this type of technology and the construction of a pilot plant [33].

7.2.7 Rotating paddle process

The main disadvantage of the Dry Distillation is the inhomogeneous temperature in the reactor, leading to the formation of residues near the walls. British Patent GB460009 [13] describes a process where PMMA is heated with hot sand over several hours. German Patent DE3146194A1 [36] also claims a depolymerization in a pyrolysis reactor provided with a stirrer connected to an extruder to melt the polymer. In this process a solid heat transfer medium (sand) is added to the PMMA and recycled. Solid impurities could accumulate in the reactor, so only high-quality PMMA could be used. According to the process [37], the PMMA scraps are fed into a paddle-reactor (rotating paddles), and a hot solid is placed in the capacity to transfer heat from walls of the reactor to the PMMA. The vapors are continuously removed, and fine particles are trapped. The main disadvantage of this type of technology is that it cannot be operated with products forming a large amount of solid residue. Unlike the lead bath process, it cannot be expected that the impurities would float on the sand's surface. Emig et al. [38] compared the fluid bed technology and the rotating paddle-type technology, such as developed by Roehm, with high sand-to-polymer mass ratios and recommend to investigate the rotating kiln reactor.

The rotating paddle reactor is again a variation of the stirred tank reactor. In this technology, a heat transfer medium, such as sand or metallic balls, is added to the reactor which is heated through the wall. The heat is transferred from the wall to the solid particles, which are agitated by the rotating paddles. It combines the heat transfer of the liquid medium/fluid bed, but also the disadvantages of the Rotating-Drum process. In addition, in such a reactor the solid particles act as a mill and generate very fine dust particles, which can be pyrophoric. The risks associated with the pyrophoric char particles might be exacerbated in this technology.

7.2.8 Microwave process

The heat transfer can be improved with microwaves [39] which should be putting the energy directly into the product. However, PMMA does not absorb the microwaves efficiently. So PMMA scraps have to be mixed with a microwave sensitizer, or PMMA parts already containing such a sensitizer would have to be used. A process was patented by AECI some 20 years ago. The advantage of the technology is clearly evidenced when PMMA is mixed with other materials and the sensitizer is only in the PMMA parts. So in that process, only PMMA should be heated to the depolymerization temperature [40].

Recently, Mitsubishi announced a collaboration with Microwave Chemical to build a pilot plant for PMMA recycling [41, 42]. Microwaves are currently investigated by several companies for the recycling of other polymers, such as Gr3n for PET, Pyrowave for PS, and Enval for mixed plastics [43–45]. The Microwave Chemical concept is that the molten PMMA, and more precisely the liquid residue produced from previously depolymerized PMMA, can be used as a kind of sensitizer to promote the penetration of microwaves in the product.

7.2.9 Inductive heating

Similarly to microwave heating, inductive heating could be used and has been patented by the Turkish company M-D2 [46], and Chapter 16 of this book [47]. This process is then a variation of the rotating paddle reactor. A material able to be heated by induction (metallic balls) is placed in the reactor and transfers the heat to the PMMA. As with microwaves, the advantage is that the heat does not have to go through the wall, but there is still a limitation to the heat transfer due to the penetration depth of the electromagnetic waves.

7.2.10 Other technologies

Other metals and metal salts [48] have been proposed to replace lead in molten bath processes, but lead still offers the best performance. Examples illustrate the process with lithium chloride-potassium chloride salts as heat transfer medium; however, when process scraps generate a lot of solid residues, the technology does not seem the most appropriate.

Depolymerization in the presence of a high boiler solvent [49] has also been described. In a practical example, equal amounts of high boiler solvent and PMMA are heated together under vacuum from 200 to 340 °C. Both solvent and MMA are condensed and MMA is distilled at atmospheric pressure. Solvent and remaining MMA are reused for a subsequent batch. This process has the disadvantage that some impurities will accumulate in the solvent, and that solid residues from the PMMA will be contaminated with solvents, leading to significant losses.

Superheated steam process [50]: ground polymer below centimeter size is used in the process, and superheated steam is used as a heat carrier. The teaching of this patent is that the monomer would be rather stable in the presence of steam and would not hydrolyze easily. This would be in contradiction with other results from the other processes in the absence of steam that report the presence of methanol and water among the products. In addition to the difficulties of separation of MMA and water, the process could generate other impurities through hydrogen transfer, as in the MURA process [51]. That would lead to more methyl isobutyrate which is difficult to separate from MMA.

The challenge with these "new energy form" technologies, whether inductive heating or microwaves, is to identify the clear benefit one can get from them. Often the claim is on energy saving, but current depolymerization technologies, such as the molten lead process or the twin-screw extruder process, are already quite energy efficient. There is little energy to be saved. However, the capital cost of such plants is rather large, and what operators are looking for are technologies that would lead to higher productivities and/or higher purity. Savings could be made either from downstream equipment or from lower quality scraps.

7.3 Intellectual property

The IPScore® tool, developed by the European Patent Office to analyze the value of patents, has been used to compare the IP portfolio of stakeholders. The value of Intellectual Property was analyzed both from the perspective of the technology developers and from the regenerated MMA producer's perspective. Quite often the rMMA producer also produces PMMA sheets from it but does not have multiple products in its portfolio. Concerning the technology developers, it can also be a large PMMA producer, as well as a start-up or small company, or, like Japan Steel Works, a company which sells many other types of equipment used in other fields. The impact for the respective companies is therefore different and is also reflected by the size of the balls and the position in the Risk-Opportunity Matrices (Figure 7.3). The figure also illustrates the multiple technologies which have been patented for PMMA depolymerization, with some patents now in the public domain (or will soon be) and others which have been filed more recently. Based on this analysis, the twin-screw extruder process looks very promising. One of the stirred-tank reactor processes is also attractive in this analysis, but it has not been fully implemented yet for PMMA depolymerization.

In all these technologies, a limitation is the scalability and is linked to the heat transfer capacity. To transfer more heat, most technologies would need more heat transfer area or would use an alternative concept like a fluid bed, microwave, or inductive heating. When increasing the plant capacity, it is necessary to keep a high surface-to-volume ratio. It means that as the capacity increases, the heat transfer area increases proportionally.

The other family of technologies which are relevant are those that allow for minimizing the purification steps. Indeed, each additional step increases the capex needed, but also the energy and utilities consumption.

7.4 MMA regeneration world map

From all the data collected on the PMMA depolymerization plants, a world map of the depolymerization sites has been produced (Figure 7.4). The map includes information on the main technologies in each territory, as well as the cumulative capacities on a regional basis. Altogether, several 10,000 tons of yearly capacity are in operation;

R-MMA producer (left) and Tech Developer (right) perspectives

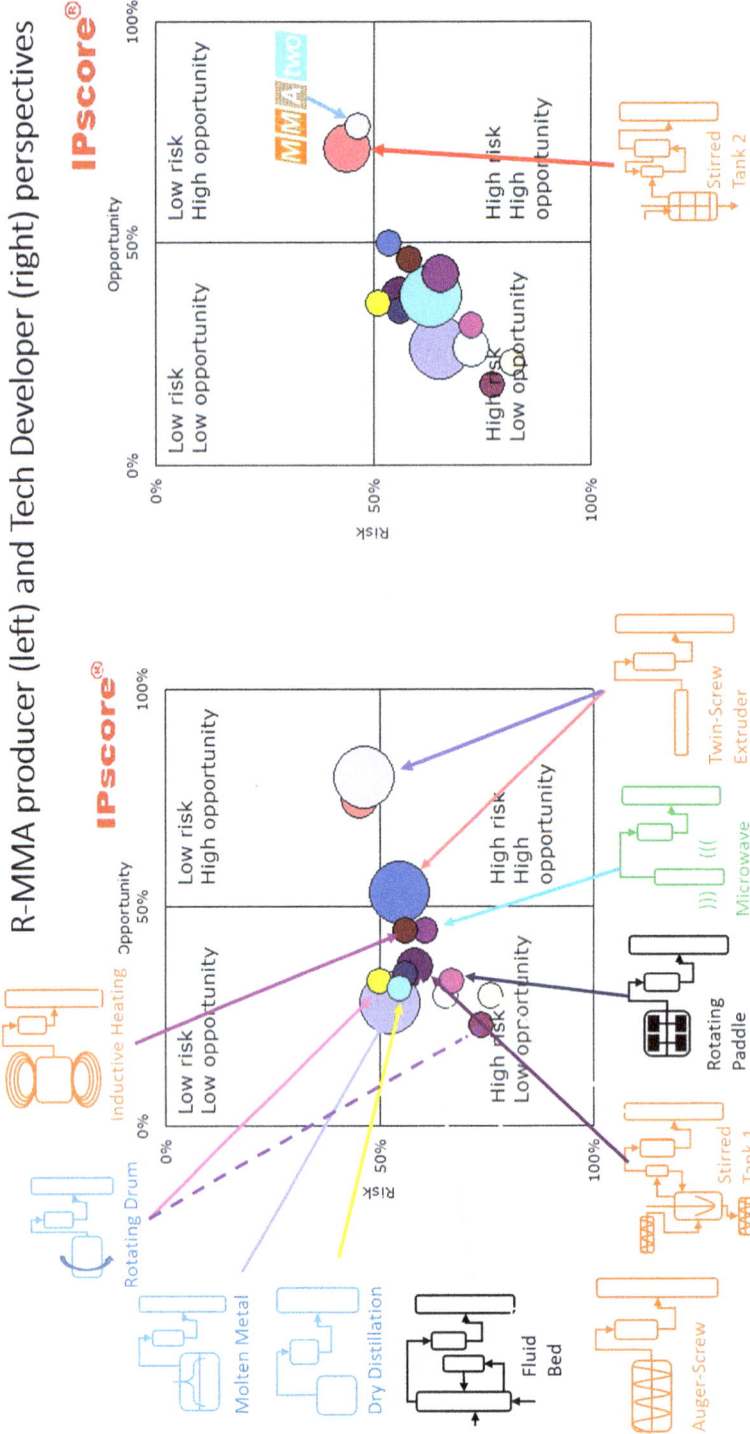

Figure 7.3: IPscore® Risk-Opportunity matrix for the PMMA depolymerization technologies.

The diameter of the circle representing a patent is proportional to the score for assessment factor D6 ("What is the patented technology's contribution to company profits?"

Regenerated Methyl Methacrylate World Map

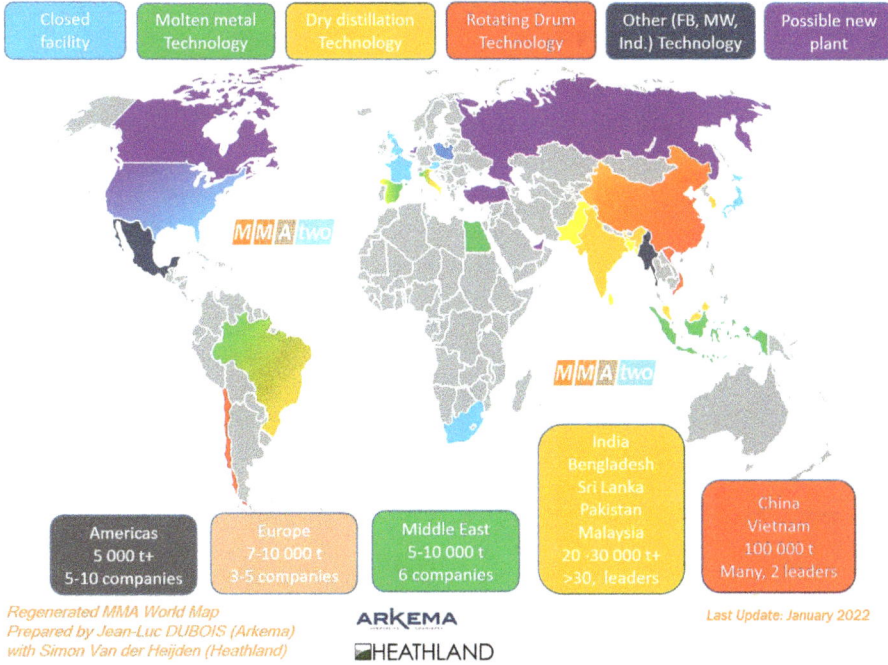

Figure 7.4: MMA regeneration world map.

most of them to produce rMMA for captive use. Since the map was produced several announcements have been made: Trinseo started a demonstration plant in Milano Area (Italy); Sumitomo Chemicals started a pilot plant in Japan; Chimei claims to have recycling activities in Taiwan; BioRenova had a pyrolysis unit in Italy and has sold its technology to NextChem/Maire Technimont which intends to have a demonstration site on PMMA depolymerization, and M-D2 is planning to start a plant in Turkey. Before the infamous invasion of Ukraine by Russia, a Russian start-up was promoting its depolymerization capacity (a few tons per month), but has been rather silent over the last few months.

7.5 Comparison of technologies

The different technologies are further compared in Table 7.2.

The twin-screw extruder process is advantageous over many current methods, as the process is self-cleaning and able to run longer cycles without interruption. It is more flexible than the molten lead-based process and can depolymerize end-of-life acrylics as well as challenging, contaminated acrylic waste (like PMMA with some

Table 7.2: Comparison of MMA regeneration technologies.

Technology	Advantages	Disadvantages
Dry Distillation	Cheap technology, good track record	Batch operation, means frequent heating, and cooling cycles Short equipment life Solid residue accumulation Scale-up limitations Large amount of PMMA immobilized in the unit (safety)
Rotating Drum	Cheap technology Unloading of solid residue Homogeneous mixing	Batch operation, means frequent heating, and cooling cycles Short equipment life Scale-up issues Safety (rotating gaskets, thermal expansion) Limited productivity
Molten Metal	Continuous process High rMMA purity Residence time: a few minutes Fast shutdown of depolymerization Good track record	Generates solid residues contaminated with lead (or other metals) Cannot process scraps that would generate more solid residues Tin is 10 times more expensive than lead
Fluid Bed	Continuous process High heat transfer rate Short residence time (of the gas)	High energy consumption Cannot process scraps that would generate more solid residues Limited productivity Expensive technology
Stirred Tank 1	Continuous process Can be used for a wide diversity of scraps	High equipment cost Difficult agitation of molten polymer Limited heat transfer area
Twin-Screw Extruder	Continuous process Recyclers are familiar with the extruder technology Short residence time – fast shutdown Renewable electricity as energy source	Capacity may be limited to 5,000 tons per line due to the possible heat transfer area
Rotating Paddle	Continuous process	Scale-up limitations (heat transfer) Safety concerns

Table 7.2 (continued)

Technology	Advantages	Disadvantages
Microwave	Input the energy where it is needed No thermal gradient in the product Renewable electricity as an energy source is possible (like for other electrified processes)	Requires to use microwave sensitizer (such as carbon residue, graphite, silicon carbide, etc.) Scale-up limited by the irradiation volume (penetration depth)
Inductive Heating	Renewable electricity as an energy source is possible	Requires the use of a material that absorbs electromagnetic waves Scale-up is limited by the irradiation volume (penetration depth)
Auger-Screw	Continuous process	Not self-cleaning (when single screw) Long residence time
Stirred Tank 2	Continuous process Low temperature operation possible Staged depolymerization (possible temperature gradient) High heat transfer area/volume ratio	High residence time Volume immobilized in the reactor (safety)

glass fiber). Besides, it does not use other substances as heat transfer agent, which is a big advantage to reduce contamination.

The technology developed by JSW [28] has been selected for the MMAtwo project. In this technology, no vacuum vent ports are placed on the extruder. MMA gas generated in the extruder exits at the tip of the extruder, while the patents developed previously by JSW [28] and Breyer [29] have venting ports along the screw in order to remove MMA gas from the extruder.

In German Patent DE3146194A1 assigned to Roehm [36], there are two heating stages. In the first stage, PMMA is melted and then fed to the pyrolysis reactor equipped with a stirrer. Furthermore, a solid heat transfer medium (sand) is added to the PMMA and needs to be recycled. Because of the solid recycle, impurities could accumulate, so only high-quality PMMA could be used. The technology developed by JSW appears much simpler, safer, and easier to operate.

7.6 Economics of recycling

PMMA depolymerization to produce regenerated MMA is done in many countries, with many different combinations of process steps. The quality of the rMMA produced differs significantly, with purities of the rMMA ranging from 90 to 99.8 wt%. The value of the rMMA can range from a discounted price compared to the virgin MMA to a price premium.

To be able to achieve the best rMMA quality, it is necessary to use the best processes and also to use the most appropriate scraps, but at a higher cost. To be competitive, the number of steps has to be minimized, and, for example, the sorting and washing steps should be avoided. In some countries, rMMA is only seen as a cheaper source of raw material which can compete with virgin MMA or imported cast PMMA sheets. Therefore, the cheapest technologies are used, even if some equipment must be replaced on a regular basis. Cheapest PMMA scraps would be used, which means that they are often contaminated with PE films and PVC gaskets, but also that all the grades are mixed together.

Both scenarios, and others, have been compared in a recent publication by our team [52]. As long as there is a market for the products, both scenarios could have similar financial returns.

7.6.1 Capital cost

The cheapest technology (operating units) would have to be replaced more frequently and so would be depreciated over a shorter period. The maintenance cost would be proportionally higher, and because these technologies have more intrinsic safety concerns, the insurance cost would also be higher. These costs are directly linked to the capital cost, and as illustrated in Table 7.3, the impact of the higher capital cost is reduced on the annual production cost.

To reduce the capital cost, which is where the investor is taking risks, it is important to avoid all the pretreatments and washing steps. In a low labor cost country, the scraps could be sorted manually, and the PVC gasket and PE films can be removed manually.

Table 7.3: Impacts on capital costs based on technology choices, for the same plant capacity.

Technology	CAPEX	Depreciation	Maintenance	Insurance	Impact on annual production cost
Cheap	2 M€	5 years	7%	3%	=2 × (0.2 + 0.07 + 0.03) =0.6 M€
Expensive	10 M€	10 years	2%	1%	=10 × (0.1 + 0.03) =1.3 M€

7.6.2 Operating cost

The higher purchasing price of the PMMA scraps can be compensated by a better mass yield and a higher rMMA price. In the example below, Table 7.4, a scrap that is twice as expensive leads to only a 50% higher impact on the production cost. In addition, the energy demand is higher when the yield is low, and cheap scraps require extra purification treatments, increasing the processing costs.

Table 7.4: Impacts on operating costs based on technology choices.

Technology	Scrap price (relative)	Depolymerization mass yield (wt%)	rMMA purity	rMMA price premium/ discount	Impact on production cost (relative)
Cheap	100	67%	<96 wt%	−20%	=100/0.67 =149
Expensive	200	90%	>99 wt%	+20%	=200/0.90 =222

7.6.3 Side products

Some side products can generate value for the recycling plant. For example, sorted PE films and PVC gaskets can be sold to specialized recyclers. Similarly, the side products from distillation and cracking, if not consumed on-site, can be sold for energy valorization to external companies. However, some of these side products, wastewater, and solids from the filtration, might be contaminated with toxic materials such as metallic pigment residues and would have to be disposed of properly, generating a cost for the operator which has to be compensated by a cheaper feedstock price.

7.6.4 Business plans

Most of the operators surveyed prefer to use a constant mix of PMMA scraps to have a stable process. Alternatively, dedicated lines can be built in a plant. This economic model may have the favors of end users who want to have a guarantee on the chain of custody of the waste they generate or raw materials that they use. For example, some companies may provide their wastes to a recycler and want to recover the rMMA derived exclusively from these wastes. So, that would be favored by dedicated recycling lines.

Independently, some scraps are generating much more solid wastes and dust than others and may require specific equipment that does not need to be implemented on all the lines. There might be a market for high-quality rMMA, with a price premium, but that might be limited in volume. So again, some lines could be dedicated to the best products. However, this kind of model has not yet been seen in the operating sites surveyed. To improve the quality of the rMMA produced, the operators preferred to dilute it with virgin MMA.

7.7 Performance of PMMA depolymerization plants

Twenty-four sites or technologies have been surveyed for which mass and energy balances have been made. From the data collected, it was possible to derive the rMMA purity and the conversion from the scrap to the rMMA. In Figure 7.5, the open circles correspond to paper studies and not to physical plants. The blue open circle on the top right corresponds to an over-optimistic paper study, in which a lot of impurities had been omitted. The other blue open circle is a corrected version. The technologies identified to be able to produce the best-quality rMMA are continuous processes and include the Twin-Screw Extruder, a Stirred-Tank process, the Molten Lead bath process, and the Fluid Bed process, but at the expense of the yield.

Some plants are rather inefficient, either by design or by the type of waste they process. Nevertheless, even in those cases, the mass yield is above 50 wt%. Most of the plants have a yield above 75 wt%, and in most cases the rMMA purity is above 95 wt%. Only in a few cases a purity above 99 wt% would be achieved, and these correspond to the continuous processes: Molten-Lead Bath, Stirred-Tank, and Twin-Screw Extruder Processes.

When the technology is developed to have a more sustainable product and not just a cheaper alternative to the virgin MMA, the operator should take care of the overall mass yield of the process. Indeed, a couple of technology-scrap processes which can have a 90% mass yield means that 50% of the carbon remains in the economy, even after six recycling operations. Extending the useful life of the monomer through multiple recycling is driving the sustainability of PMMA.

7.8 Conclusions

All the technologies and sites surveyed have a much better carbon footprint than the virgin MMA, with a reduction which can be higher than 70%. There are three main processes to produce MMA: the C3 processes which use acetone as feedstock and dominate in Europe and the USA; the C4 processes which use isobutene or tertiobutanol and are mostly used in Asia; and the C2 processes which use ethylene and are more recent.

The PMMA recycling processes also consume much less energy to produce MMA than the virgin routes. In addition, all the recycling processes consume much

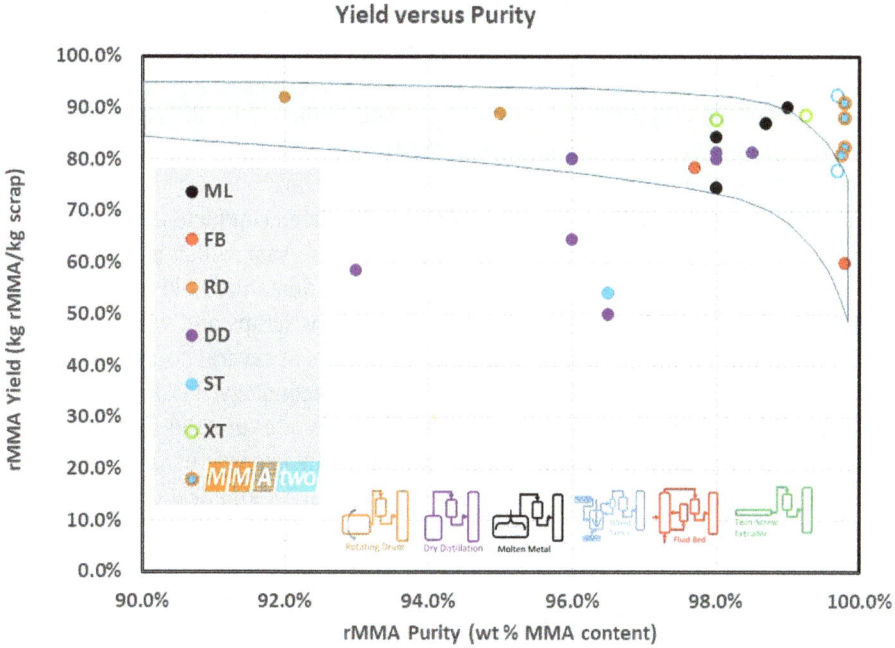

Figure 7.5: rMMA yield versus rMMA purity.

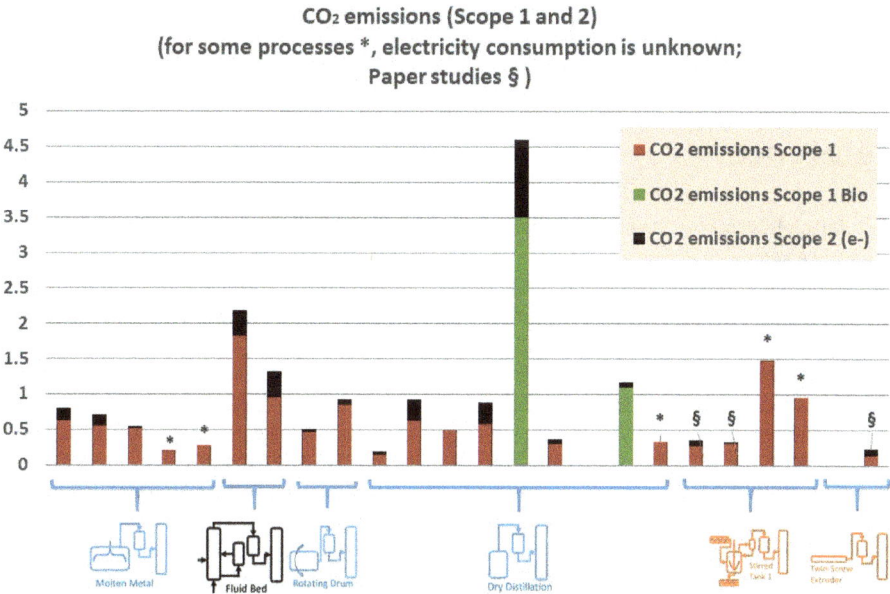

Figure 7.6: CO_2 emissions (scopes 1 and 2) for various technologies.

less water (two orders of magnitude) than the virgin C3 process. In 80% of the cases, the CO_2 emissions from scopes 1 and 2 (on-site non-biogenic CO_2 emissions, and off-site CO_2 emissions linked to electricity production) are below 1 kg CO_2/kg rMMA (Figure 7.6).

The size of the recycling plants can vary greatly, from a few 100 tons per year, to a few 10,000 tons per year, but in Europe the typical plants, either closed or still in operation, are in the range of a few thousand tons per year.

In a separate study [52], an economic analysis has been conducted using a Monte-Carlo simulation, assuming a plant of about 5,000 tonnes/year which is the appropriate size for a new plant located in Europe. European demand is likely to favor the high MMA purity, which can be achieved with the right scraps and with continuous processes. The study analyzed the probability of success of several scenarios, and the best Net Present Value is likely obtained with the best technology/rMMA quality.

Continuous processes, operating in a plug-flow mode, are the most suited to achieve high MMA purities, which is also achieved at the expense of yield or productivity. At low temperature, the units would operate under a kinetically controlled regime, but at higher temperature, the units are most likely limited by heat and mass transfer. So some processes have intrinsic productivity limitations.

The choice of the technology is also governed by safety concerns. The risks related to PMMA depolymerization are related to: (1) the high operating temperature which can be above MMA self-ignition temperature; (2) the formation of solid residues that could be pyrophoric; (3) the cleaning operations; and (4) the amount of material immobilized in the unit. And finally, for the operators, the possibility of retrofitting the equipment easily to process other materials is important to minimize the financial risks.

References

[1]   Dubois JL, Bentaj A, Couchot N, PCT patent application WO2022123176, 2022, assigned to X-crusher and Arkema.

[2]   Anonymous, Altuglas, Injection Moulding Resins, Technical Manual, downloaded from https://www.nevicolor.it/Apps/WebObjects/Nevicolor.woa/wa/viewFile?id=5087&lang=eng on December 31st 2021.

[3]   Methyl Methacrylate, REACH registration dossier, https://echa.europa.eu/fr/registration-dossier/-/registered-dossier/15528/4/13, accessed on January 2nd 2022.

[4]   Anonymous, Methacrylate Esters, Safe Handling Manual, Methacrylate Producers Association, Inc. and Methacrylates Sector Group of the European Chemical Industry Council 2019.

[5]   Gaur U, Lau S-F, Wunderlich BB, Wunderlich B. Heat capacity and other thermodynamic properties of linear macromolecules VI. Acrylic polymers. Journal of Physical and Chemical Reference Data 11(1065), 1982, 1073.

[6]   NIST Chemistry WebBook, https://webbook.nist.gov/chemistry/ accessed on June 19th 2024.

[7]   Chemeo – Chemical & Physical Properties by Chemeo (chemeo.com), https://www.chemeo.com/cid/57-301-4/Methyl-methacrylate, accessed on June 19th 2024.

7 PMMA chemical recycling reactor technologies ━━ 97

[8]   Brandrup J, Immergut EH, eds. Polymer Handbook, 2nd Edition, Wiley, January 29 1975, ISBN 10:
      0471098043 ISBN 13: 9780471098041.
[9]   Dubois JL, Tojo M, Kakizaki J, PCT Patent application WO2023-046942, 2023, assigned to Japan Steel
      Works Europe and Arkema.
[10] Dubois JL, De Caumia B, Blanchette D, Roy C, PCT patent application WO2020079380, 2020, assigned
      to Pyrovac and Arkema.
[11] Gao X, Chen M, Wang T. Design and optimization for the separation of a ternary methyl
      methacrylate-methanol-water mixture to save energy. Energy Sources, Part A: Recovery, Utilization,
      and Environmental Effects 2020. 10.1080/15567036.2020.1829751.
[12] Strain D, US patent US2030901, 1936, assigned to DuPont.
[13] Anonymous, British Patent, British patent GB460009, 1937, and French Patent 809386, 1937,
      assigned to Rohm and Haas.
[14] Domingo Segui E, Cabanero Alarcon B, US Patent US 2 858 255, 1958.
[15] Riedewald F, PCT patent application, WO2014/032843, 2014.
[16] Reverso R, D'abbondanza E, PCT patent application WO2019003253, 2019, assigned to
      Biorenova Spa.
[17] https://www.nextchem.it/en/newsroom/press-releases/detail/maires-nextchem-acquires-control-of-
      catc-an-innovative-plastic-chemical-recycling-technology/, accessed on June 4th 2024.
[18] https://www.lme.com/Metals/Non-ferrous/LME-Tin#Summary;https://www.lme.com/en/Metals/
      Non-ferrous/LME-Lead#Summary;https://www.lme.com/Metals/Non-ferrous/LME-Zinc#Summary
      accessed on June 4th 2024.
[19] Vaughan PW, Highgate DJ, US Patent US5663420, 1997, assigned to Duvergier.
[20] Kaminsky W, Franck J. Monomer recovery by pyrolysis of poly(methyl methacrylate) (PMMA).
      Journal of Analytical and Applied Pyrolysis 1991, 19, 311–318.
[21] Sasaki A, Kikuya N, Ookubo T, Hayashida M, US Patent 830 573, 2012 and European Patent
      EP2157075B1; and Sasaki A, Tsuji T, Proceedings of the 5th ISFR (October 11–14, 2009, Chengdu,
      China), 79–83, http://www.fsrj.org/act/7_nenkai/12-5-IFSR/Proceeding-IFSR05/R-16.pdf.
[22] Lopez G, Artetxe M, Amutio M, Elordi G, Aguado R, Olazar M, Bilbao J. Recycling poly-(methyl
      methacrylate) by pyrolysis in a conical spouted bed reactor. Chemical Engineering and Processing
      2010, 49, 1089–1094.
[23] Muller M, Wenzel F, German Patent DE3146194A1, 1983 assigned to Roehm.
[24] McNamara D, Strivens C, Yabrudy A, Dunphy P, PCT Patent application WO2020/065316 (2020)
      assigned to Plastic Energy.
[25] Weiss HJ, Schmalfeld J, Zenter U, Groschang T, Gropp U, Fuss W, Goedecke R, Schöla E, US Patent
      US6469203, 2002, assigned to Metallgesellschaft Aktiengesellschaft and Rohm GmbH Chemische
      Fabrik.
[26] Cavinaw CS, Lamaze S, Allen D, Christenson E, Rumford M, US Patent US10301235, assigned to
      Agilyx.
[27] Tokushige H, Kosaki A, Sakai T, US Patent US3959357, 1976, assigned to Japan Steel Works.
[28] Heiji S, Tetsuo M, Japanese Patent JP3410343, 1999, assigned to Japan Steel Works and Kosui
      Kagaku Kogyo.
[29] Breyer K, Enders M, Weyers J, German Patent DE19729065, 1999.
[30] Breyer K, Michaeli W. Feedstock recycling of polymethyl methacrylate (PMMA) by depolymerizing in
      a reactive extrusion process, ANTEC. Proceedings of the 56th Annual Technic 1998, 3, 2942–2945.
[31] Koyanagi K. Development of continuous chemical recycling process from waste acrylic resin by using
      twin screw extruder. Proceedings of the 6th Japan International SAMPE Symposium Oct 1999, 26–29.
[32] Trinseo – New Trinseo PMMA Depolymerization Plant Reimagines Plastics Value Chain With
      Sustainability In Mind, and Recycling Today, December 7th 2023 https://investor.trinseo.com/home/

news/news-details/2023/New-Trinseo-PMMA-Depolymerization-Plant-Reimagines-Plastics-Value-Chain-With-Sustainability-In-Mind/default.aspx accessed on June 19th 2024.

[33] Sumitomo Chemical Completes Construction of Pilot Facility for Acrylic Resin Chemical Recycling Accelerating the development of a circular system for plastics through cross-industry collaboration, https://www.sumitomo-chem.co.jp/english/news/detail/20221223e.html, accessed online on June 4th 2024.

[34] de Caumia B, Dubois B, Mbou G J A, Laplante St-Martin D, Mercier G, Pakdel H, and Roy C, Chapter 15 of this book, Pyrovac's waste plastic pyrolysis process, in PMMA recycling Roadmap, DeGruyter, 2025.

[35] de Tommaso J, Patience G, and Dubois J-L, Chapter 14 of this book, Recycling of artificial marble., in PMMA recycling Roadmap, DeGruyter, 2025.

[36] Muller M, Wenzel F, German Patent DE3146194 A1, 1983, assigned to Rohm.

[37] Schola E, Ruzicka M US patent application 2006/205845 assigned to Roehm GmbH.

[38] Sasse F, Emig G. Chemical recycling of polymer materials. Chemical Engineering Technology 1998, 21, 10.

[39] Porée ID, Cameron KP, Bloem JA, Schlosser FD, McGowan A, US6160031, 2000, assigned to AECI.

[40] Dubois J-L, PCT Patent application WO2019207265, 2019.

[41] Mitsubishi – Microwaves Chemical, Press release May 24th 2021, https://mwcc.jp/en/news/detail.php?no=MTk last access on January 3rd 2022.

[42] Yamauchi T, Hagimoto A, Kakuta K, Deguchi Y, European patent application EP4342876, 2024, assigned to Microwave Chemical.

[43] Parravicini M, Crippa M, Bertele MV, PCT Patent application WO2013/014650, 2013.

[44] Doucet J, Chaouki J, Sobhy A, PCT Patent application WO2012/097448 (2012); and Doucet J, Laviolette JP, PCT Patent application WO2018/018153 assigned to Pyrowave Inc.

[45] Ludlow-Palafox C, Chase HA, US Patent US7951270.

[46] Ural AT, Sarioglu R, Uludag Y, Ural MK, PCT Patent application WO2019/226135, (2019).

[47] Ural A T, Güzel E, and Uludag Y, Chapter 16 of this book, A novel method for PMMA recycling: mixed induction reactor, in PMMA recycling Roadmap, DeGruyter, 2025.

[48] Tatsumi T, Yoshihara H, Uesaka G,US Patent US3886202, 1975, assigned to Denshi Kagaku KK.

[49] Miller I, Beiser AL, US2470361.

[50] Mannsfeld SP, Paalsea KJ, Bucbhollz E, US Patent US3494958 (1970) assigned to Degussa.

[51] MURA HydroPRS™ process, http://79.170.40.224/muratechnology.com/hydroprs-process/, last accessed on January 3rd 2022.

[52] De Tommaso J, Dubois JL. Risk analysis on PMMA recycling economics. Polymers 2021, 13, 2724. https://doi.org/10.3390/polym13162724.

Freddy L. Figueira, Kevin M. Van Geem, Marek Blahušiak,
and Juraj Hrstka

8 Process design of chemical recycling

8.1 Introduction

Engineering design is the a creative manner to envision a system or process to best meet a stated objective and encompasses several steps. To accomplish this objective, an engineer must rely not only on basic and engineering sciences but also on social sciences and humanities [1]. For a new product being developed in research laboratories or the an improvement of existing processes due to the changes in specifications dictated by market conditions, the process design team is challenged. Under such circumstances, profitable processes need to be designed that can produce goods in sufficient quantities and qualities to satisfy foreseeable market demands [2].

Design is also an essential part of the product life cycle. This cycle starts with the inception of the inception of product and ends with its recycling or environmental-friendly disposal [3]. Nowadays, recycling/recovery/management of plastic solid waste (PSW) is a growing matter of concern. Plastics have become a crucial part of our lifestyle. Since so many products are being manufactured from plastics, plastic production has increased immensely during the past 50 years [4]. PSW is produced on a massive scale worldwide, and its production exceeds 150 million tons per year [5]. PSW accounts for approximately 30 million tons of municipal solid waste in the United States, of which only less than 10% is recycled [6], while in Eastern and Central Asia, the PSW annual generation is at least 93 million tons [4]. In 2018, a total of 29.1 million tons of PSW was collected in the European Union, United Kingdom, Norway, and Switzerland, of which 61% came from plastic packaging, 32.5% was recycled, and 42.6% was incinerated [7]. Rahimi and García [6] mention four main techniques for plastic recycling: primary recycling being closed loop; secondary or equivalently mechanical recycling; tertiary or equivalently chemical recycling; and energy recovery through incineration. Tertiary recycling deals with processes for the depolymerization of PSW to recover chemicals, preferably monomers as building blocks of the original plastic product. Examples are pyrolysis, gasification, liquid-gas hydrogenation, viscosity breaking, reactive extrusion, and steam or catalytic cracking [8]. Typical reactions for chemical recycling are highlighted in Chapter 5, typical experimental procedures in Chapter 6, and common reactor technology in Chapter 7.

Freddy L. Figueira, Kevin M. Van Geem, Laboratory for Chemical Technology, Ghent University, Technologiepark 125, Zwijnaarde 9052, Belgium
Marek Blahušiak, Process Design Center, Paardeweide 7, 4824EH Breda, The Netherlands; Inprocess Technology and Consulting Group, S.L Carrer de Pedro i Pons 9 08034 Barcelona
Juraj Hrstka, Process Design Center, Paardeweide 7, 4824EH Breda, The Netherlands; RWE Technology International, Altenessener Straße 32, 45141 Essen, Germany, e-mail: hrstka.juraj@gmail.com

https://doi.org/10.1515/9783111076997-008

In this chapter, the principles of chemical process design, applicable also to tertiary recycling processes, are first introduced. The focus is then on the hierarchy decomposition method, with specific illustrations to produce regenerated methyl methacrylate (rMMA) from poly(methacrylate) (PMMA) waste.

8.2 Principles of chemical process design

Design is a creative activity that begins with a specific objective, for example, a customer need, and, by developing and evaluating possible designs, arrives at the best way of achieving that objective. The final design will be constrained by many factors of different natures. There are also constraints that are fixed and invariable, for example, physical laws, government regulations, and standards. These constraints are not in control of the designer; hence, they are classified as external constraints (Figure 8.1, full line). Other constraints are less rigid so that they can be considered depending on the strategy of the design team. These types of constraints are defined as internal constraints (Figure 8.1, dashed line). The boundaries set by external constraints define several possible designs that will be reduced to a number of plausible designs bounded by the internal constraints, for example, the choice of process, process conditions, materials, and equipment [9]. Specifically, economic considerations constitute a major constraint as an industrial process must be profitable.

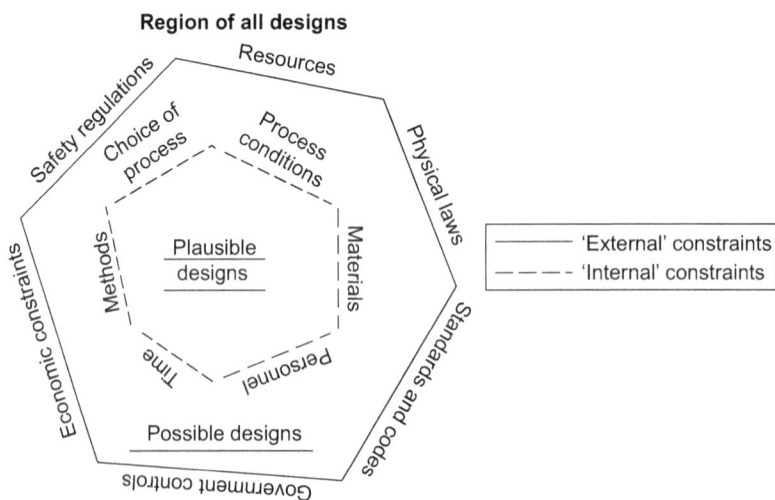

Figure 8.1: Design constraints (reproduced with permission from Sinnott [9]).

In some cases, experienced process designers use heuristics, for example, rules of thumb, to quickly identify the probably workable region of the design space, as defined by the design constraints (Figure 8.1). These rules of thumb do not replace more rigorous analysis when it comes to detailed design, but their compilation of knowledge gained from experience is useful to speed up the design in conceptual stage. Of course, their limitations and validity must be kept in mind [10].

As the number of plausible designs for a problem or need is unknown, several methodologies have been proposed, more in an academic context, to design a chemical process. The three main methodologies are hierarchy decomposition, superstructure optimization, and evolutionary methods [1, 2, 9, 11, 12]. In the next section, the hierarchy decomposition approach is introduced and applied to the PMMA recycling process.

8.3 Hierarchy decomposition

The hierarchy decomposition method was proposed in 1988 [12] and improved later on [11, 13]. This method is originally organized into five core levels: (i) the input information and the evaluation of batch versus continuous operation; (ii) the input/output structure of the flow sheet; (iii) the recycle structure of the flow sheet; (iv) the separation system(s); and (v) the heat exchanger networks. Additional levels are usually added, for example, the handling of process chemical agents, that is, water and solvents, and the conceptual design of the process utilities and process control system. The first four levels belong to the process synthesis phase as the basic flow sheet is developed there. The other phases are grouped in the process integration phase, in which the addition, design, and sizing of units are implemented [13].

Dimian and Bildea [14] improved the hierarchical decomposition method (Figure 8.2) by optimizing the interactions between the process synthesis and process integration phases. The goal of improved method, which consists of eight levels, including a basic one and a final flow sheet, is to obtain a feasible and close-to-optimum efficient flow sheet of the process. The method is empowered using state-of-the-art thermodynamics and is intended to generate design decisions, while steps in the synthesis phase generate several alternatives employing systematic conceptual techniques and heuristics. The evaluation steps are based on economic and technological criteria that eliminate the less attractive alternatives. The definition of updated hierarchical methodology levels is explained below, considering rMMA production.

8.3.1 Level 0: basis of design

The basis of design comprises the definition of products to be manufactured, their quality specifications, and the use of alternative raw materials. At this point, the decision between a continuous or a batch process is made [13]. Material Safety Data Sheets

(MSDSs) are also collected at this point. Such sheets provide information on the physical properties of the components and materials and their potential health, safety, and environment (HSE) hazards. Economic data, for example, prices of products and raw materials, are collected as well, along with the cost of common utilities required for plant operation and the cost of emissions and waste disposal, which are necessary to estimate sustainability indices.

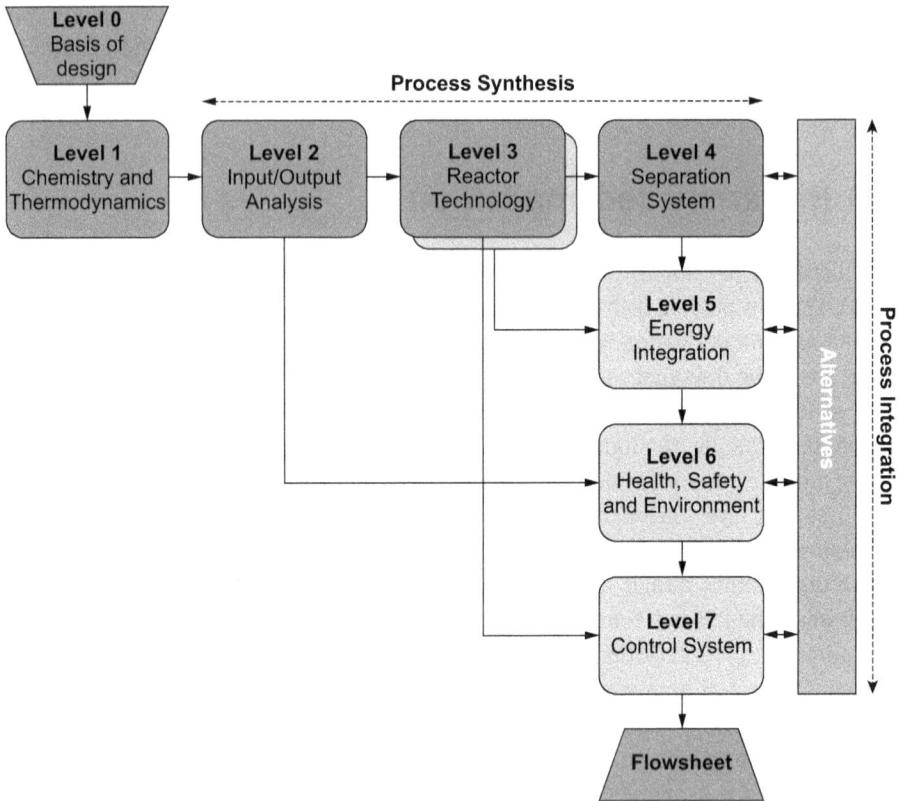

Figure 8.2: Improved hierarchical methodology of conceptual design (adapted from Dimian et al. [13]).

At this level, a review of existing technologies using technological encyclopedias, patents, monographs, and research papers is performed. In the case of PMMA chemical recycling, Moens et al. [8] reviewed different patented and proposed reactor technologies. The most used in the industry is the molten-lead bath technology, patented by Segui and Alarcon [15] in 1958, while several other technologies based on fluidized bed reactors or reactive extrusion are either patented or researched.

8.3.2 Level 1: chemistry and thermodynamics

The knowledge of physical and thermodynamic properties of present compounds and the kinetics and equilibria of the chemical reaction network are crucial in the design of a process. Main and secondary reactions must be identified to understand the complexity of reaction phases in which they occur. Temperature and pressure ranges, changes in aggregation state, technological/safety constraints, for example flammability, explosion limits, and reaction exothermicity, need to be included as well. Attention is further paid to intermediate components that can be separated and recycled, and to by-products and impurities that can be formed in the reactor but also in other unit operations, or present in the raw materials.

More in detail, potential limitations of the reaction kinetics must be investigated, meaning that it should be clear if they are controlled by chemical equilibrium or by mass or heat transfer limitations. In general, kinetic lab-scale data are necessary for sizing the chemical reactor and for acquiring an understanding of the dynamic behavior of the system, although alternative approaches can be followed in the absence of this information [9, 13].

Moens et al. [8] reviewed the reaction mechanism and kinetics of the thermal depolymerization of PMMA. Despite rather simple chemical structure of MMA, an ongoing debate exists regarding the degradation mechanisms of PMMA toward the formation of rMMA. The review shows that an essential link exists between PMMA thermochemical recycling and its synthesis, as certain structural defects in PMMA play an important role in the degradation mechanisms, which can be assessed by kinetic modeling tools.

8.3.3 Level 2: input/output analysis

In this level, design decisions, regarding feed purification, recycling reactants, and treatment of emissions and waste aim to obtain the most efficient material balance. The key decision relates to the expected performance of reaction systems, although at this level there is no actual selection or design. Also, at this level, a preliminary assessment of health, safety, and environment (HSE) hazards should be performed. The idea is to evaluate alternative design decisions from an HSE viewpoint, for example, safer chemicals or smaller holdups of hazardous chemicals. Hence, a basic, qualitative hazard identification is performed, but not yet a quantitative risk assessment based on the identified hazards. Information regarding HSE regulations and laws is usually available via the official communication channels of regulatory agencies of each country or region [13].

The economics of different alternatives generated at this level can be evaluated by means of an economic potential (EP) [2, 11–13], which represents the difference between revenues obtained by selling products and valuable by-products, and the expenses, including the cost of raw and auxiliary materials, environmental and utilities costs, and on-site costs. At this point, the cost associated with the reaction system, pumps, and compressors for stream recycles and separation has not been considered, so a large positive EP is needed to proceed with further steps toward an optimized flow sheet [13]. The EP is updated at each level by subtracting the utilities, operational, and additional environmental costs of new equipment being added [12]. Design alternatives at each level that do not meet a specified overall minimum EP are discarded [13].

The importance of basic material balances to assess material efficiency and viability of processes can be illustrated in the design of chemical recycling of PMMA to regenerate its building block, MMA. To put the recycling process in perspective, its anticipated material efficiency is compared to the production of virgin MMA by the C_3 route via cyanohydrin, which is the prevalent MMA production process in Europe [16].

The C_3 route production of virgin MMA comprises a cascade of reaction stages, each coupled with purification processes of the intermediate products (Figure 8.3). Similarly to other synthetic routes (e.g., C_2 and C_4), the process feedstock is based on hydrocarbons from petroleum resources. Hydrocarbons are also used as a source of energy to drive the processes, either directly combusted for heat or indirectly in power plants to provide electricity for the processes.

In the case of PMMA recycling process, PMMA is the hydrocarbon feedstock that could otherwise be used as fuel, thus serving as an energy source for incineration. Energy sources are needed for the PMMA recycling process and for pretreatment, depolymerization, and separation processes to purify the final MMA product.

A straightforward comparison of the material efficiency of virgin and regenerated MMA processes can be performed using the heat value of all hydrocarbon resources involved in the production of 1 kg of MMA. This approach, instead of using the mass of input feedstocks, has the advantage that various petroleum-based resources and hydrocarbon intermediate products are interchangeable for the same or similar production technologies; for example, hydrogen used for ammonia production obtained via steam reforming as an intermediate.

Figure 8.3 shows the comparison between virgin MMA production via the C_3 route and PMMA recycling technology. The data for the virgin route is based on the published life cycle assessment of overall value chain [16]. The generation of all MMA process feedstocks, for example, ammonia, acetone, and methanol, from the primary petroleum resources in their respective production processes, is also included in this assessment.

Based on literature data [16], a cumulative amount of petroleum resources equivalent to 101 MJ of combustible energy is needed to produce 1 kg of virgin MMA, of which 19% accounts for energy consumption and 81% is in the process feedstocks. The residual content of the combustible energy in the produced MMA is 27 MJ/kg; that is, a total of 74 MJ/kg of the energy in resources is consumed in the process.

(a)

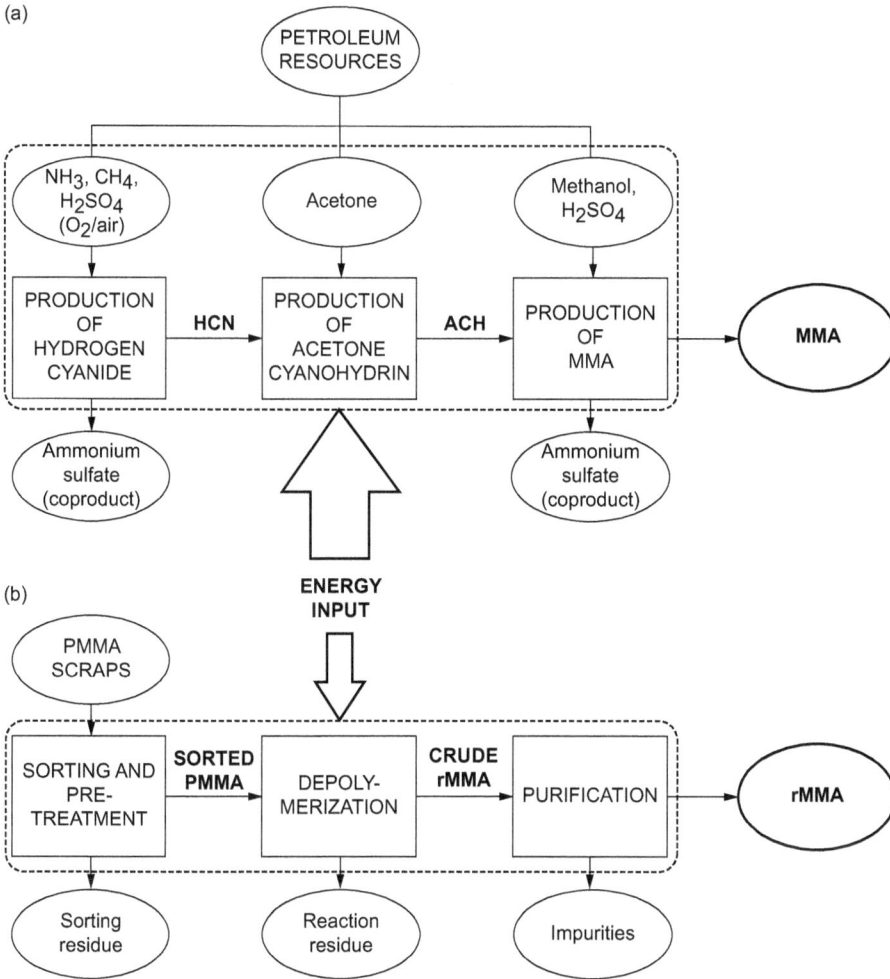

(b)

Figure 8.3: Units involved in the production of MMA monomer (a) via the C_3 acetone-cyanohydrin route [16] and (b) in a general process for the chemical recycling of PMMA.

To estimate the relative viability of PMMA recycling process, this value needs to be compared to its anticipated performance. The exact performance depends on many factors or process boundaries, such as the quality of recycled PMMA stream or the required product purity, as well as more concrete process design decisions, which are addressed in the next levels. However, the following assumptions are used to evaluate the process at this high-level analysis:

- Ninety percent yield of all three recycling unit operations, that is, 73% yield of PMMA scraps to pure rMMA. This implies that (i) the feed is practically pure PMMA and (ii) the PMMA thermal decomposition is highly selective toward rMMA [8], leading to a lower amount of reactor residues and impurities in the crude product.

– Five times the equivalent of PMMA depolymerization as a conservative estimate of the depolymerization reactor energy consumption: 5×0.5 MJ/kg of reactor feed-stock.

– Five times MMA vaporization energy: 5×0.5 MJ/kg of crude rMMA as a conservative energy consumption estimate for product purification by relatively simple distillation process.

Using these numbers, as shown in Table 8.1, the cumulative material consumption of the PMMA recycling process is 40.5 MJ/kg of produced rMMA. Excluding the residual energy content of rMMA, only 13.5 MJ/kg worth of hydrocarbon material was spent in the process. While this result is based only on a high-level assessment with very crude assumptions, the material loss to obtain 1 kg of MMA for the depolymerization process is only ca. 20% of the material lost in virgin MMA production. Conversely, 66% of the depolymerization process inputs are converted in the MMA product energy, while only 27% of the virgin MMA production inputs end in the MMA product.

The benefits of the recycling process in terms of relative process viability can be explained by the higher complexity involved in producing virgin MMA, which requires a larger number of operations. From a thermodynamic point of view, each operation contributes to a loss of useful energy and/or degradation of the useful energy content of the processed material, that is, exergy (see also Level 5). Hence, more synthetic and separation operations can generally be expected to lead to a larger consumption of feedstock and energy in a process, for example, the virgin MMA production via the C_3 route shown in Figure 8.3, with numerous intermediate products and inputs.

Table 8.1: Material (expressed as the heat value of the feedstock) and energy inputs and outputs (expressed as heat value) in the production of 1 kg MMA via the C_3 acetone-cyanohydrin route [16] and in a general process for the chemical recycling of PMMA.

	Material input (MJ)	Energy input (MJ)	Material output (MJ/kg$_{MMA}$)
Virgin production C_3 route	81.9	19.1	27.0
PMMA depolymerization	36.4	3.1 (depolymerization) 2.8 (purification)	27.0

As a result of this analysis, an advantageous prospect for the PMMA recycling process can be expected. A more exact understanding will, however, depend on the specifics of the design decisions, which are described in the following sections.

8.3.4 Level 3: reactor technology

With the knowledge gained in previous levels, the initial process equipment design can be done at this stage. Main tasks include chemical reactor design, possible introduction of process recycles due to incomplete conversion, and interlinking the reactor with the process separation section. The main goal is to select a process architecture that is feasible and appropriate for stable and robust operation within the defined operational window, providing flexibility in the production rate and for the expected variability of raw materials.

The selection of chemical reactor type depends on various aspects of the reactions to be carried out in the process. For (ideal) homogeneous reactions, plug flow reactors generally offer better volumetric productivity and yield, whereas continuous stirred reactors display better mass and heat transfer characteristics. For heterogeneous reactions, the range of reactor-type options is larger, and selection criteria would depend on a multitude of aspects, for example, aggregation phases of the inlet materials, intermediate streams and products, heat and mass transfer, type of catalyst, and hydrodynamic flow regimes. Dimian et al. [13, 17], Douglas [12], and Seider et al. [2] present a series of heuristics for reactor-type selection. The process designer must consider whether design choices on the reactor have other consequential effects on the downstream processing of the product.

The PMMA depolymerization process can be described as a heterogeneous, endothermic, non-catalytic system with the presence of rather complex mass and heat transfer phenomena at elevated temperatures above 370 °C. The heterogeneity aspect must be thoroughly considered as multiple physical phases occur in the process. First, solid PMMA material is fed into the reactor and then, due to heat addition, melts into a highly viscous polymeric melt. The viscosity of this intermediate phase decreases with the gradual decrease of (average) molar mass of the polymer chains. Additionally, side reactions might take place in the MMA-rich product vapor phase or again on the melt/vapor interface. Finally, if the product collection section is embedded within the reactor design, the final product would be present in liquid phase.

Overall, PMMA recycling is a hydrodynamically challenging process, especially when continuous production is selected in the previous design levels. Due to the diffusional limitations, especially in the melt phase, appropriate design choices must be made to maximize heat distribution through the bulk material flow and assert proper mixing. As the process is endothermic and runs at elevated temperatures, the source of energy supply must be considered in the reactor design. Traditionally, a choice is made between gas-fired, electric, thermal oil, fluidized hot sand bed, and other forms of heat supply. Putting those aspects together may lead to a broad variety of potential reactor-type candidates, where often small details in practical considerations lead to a definite answer.

A comprehensive study on different PMMA depolymerization reactor types was conducted by Moens et al. [8]. The review has shown that, at first sight, a straightforward and simple way of processing PMMA scraps is by using molten-lead baths in batch operation. The feed is immersed in a hot molten metallic medium, and process vapors are collected at the top of the tank. Such technology is still used due to its simplicity; however, the disadvantage of this reactor technology is the production of carbonaceous residues from the depolymerization. Such residues must be removed and inevitably contaminate the molten bath and present a problematic waste stream of the process.

Design choices vary when it comes to the continuous processing of PMMA scraps, primarily in terms of the feeding systems and heating media. One of the prospective designs shown in Moens et al. [8] proposes direct heating using high-temperature steam in a counter-current contactor, although there is no information on this design being realized in running production. However, adding steam directly to the process might promote undesired hydrolysis reactions, and water is known to form azeotropes with MMA, which then need to be treated downstream. Another option aims at using a fluidized bed with solid carriers in a hot inert gas. These systems are inherently good for heat transfer, which minimizes the diffusivity problems of the viscous PMMA melt. Numerous process layouts for fluidized reactors have been proposed, including the shapes of the fluidization chamber, the presence of internal product gas recycles and cyclones for solids removal, and the type of solid carriers. The quality of generated product is acceptable [8]. However, these units are mostly prevalent at the lab scale and are less stable and hence more challenging in terms of operability.

Various mechanically agitated reactors are also proposed, in which the viscous PMMA melt is continuously mixed by a rotating element. Such a concept is applied in the externally heated rotating paddle reactor design, where the PMMA is fed under an inert atmosphere into a cylindrical body and is mixed by a propeller equipped with paddles. In addition, metallic beads are introduced to enhance mixing and heat transfer inside the reactor.

Extrusion-based reactors also fall under the mechanically agitated reactor category. These systems, already known in the polymer industry, use mechanical shear forces of rotating screws to deliver heat directly to the solid polymer. By adding extra heat supply from the barrel jacket to reach depolymerization temperatures and provide the endothermic heat of the reaction, this concept can be applied to the PMMA process. Extruders are, in general, easier to operate than fluidized beds; however, they are more difficult to scale up due to the limitation of reactor productivity by heat transfer through the external reactor wall, that is, the decreasing area-to-volume ratio for increasing reactor barrel diameter. In the MMAtwo project, extrusion-based technology was chosen as the most suitable reactor design [18].

8.3.5 Level 4: separation system

In this level, the separation system required to isolate the products at required purity is devised. For this, several task-oriented methodologies based on heuristics and process simulation are available [19–21]. The overall idea is to generate a sequence of operations to increase the purity of the products and separate the unconverted feed material, process impurities, or solvents. A good understanding of different unit operations, for example, distillation, membrane separators, filtration, physical/chemical adsorption, and the advantages, drawbacks, and applicability for characteristic properties of the mixture to be treated is important. Hence, specialized teams should carry out this level of design.

The first step in the synthesis of separation system is the selection of an appropriate sequence of splits of the mixed components, for example, the formation of two immiscible liquid phases, according to the mixture composition and the specifications of products. The second step is to identify suitable separation method for the proposed splits by ranking the mixture components versus a characteristic property. The final step is the detailed design of each piece of equipment involved in the separation architecture [13].

In the chemical recycling of PMMA, some impurities have been found in the rMMA product streams. Achilias [22], who performed pyrolysis experiments in a laboratory-scale fixed-bed reactor, found rather small amounts of organic compounds (mainly methyl esters) accompanying the rMMA liquid product. These impurities acted as non-ideal reaction retarders, altering the rMMA polymerization and thus affecting the average molar mass of the PMMA produced, causing structural defects, and reducing the quality of the product. Godiya et al. [23], studying the thermal depolymerization of PMMA using thermogravimetric analysis coupled with mass spectrometry (TGA-MS), also found that the produced rMMA contained traces of byproducts that negatively impacted the repolymerization of rMMA. Among the impurities they found 2,3-butanedione, which was found to be responsible for the bad smell of rMMA and to be present in the re-polymerized PMMA. This highlights the importance of the separation system for this type of process.

8.3.6 Level 5: energy integration

After Level 4, a close-to-optimum flow sheet with consistent material balance has been obtained, so the objective in Level 5 is the optimization of energy and material resources. Tasks to be performed include: (i) selection of appropriate utilities for motive power supply, for example, steam turbines and electric drives, for rotating equipment, for example, pumps, fans, and compressors, and for heating, for example, electric, hot oil, and steam, for individual parts of the process; (ii) the design of heat exchangers network, aiming to reuse the process heat to save energy in reactors and separators; (iii) consideration of waste heat recovery, for example, via heat-pumping; (iv) minimization of process recycle streams to reduce energy consumption. Some of these optimizations might imply changes in the process flow sheet architecture generated in Levels 3 and 4 [13]. From this perspective, process design can be viewed as an iterative procedure, as suggested in Figure 8.2, to find unit operations that not only perform optimally on their own but also fit well the overall energy-integrated process.

Among the most important tools available to optimize energy resources are thermodynamics-based energy balances and the pinch point analysis (PPA) [13]. The objective is to minimize the generation of entropy, that is, to minimize the loss of potential thermodynamic work (also called "exergy") for each operation as well as the whole plant or its specific sub-parts. PPA is mainly used in the design of the heat exchanger network, where large differences in temperature between the contacted fluids cause losses of valuable temperature potential for heat exchange, which is translated into higher irreversible generation of entropy and manifested by larger consumption of energy utilities by a process. The concept of preserving resources and utility potential has also been extended to other valuable resources used in a chemical plant, for example, steam, water, solvents, and hydrogen [9].

In terms of heat integration, a proper definition of the reference case is crucial, as the viability studies of proposed energy-saving measures must be weighed against a sound base. Figure 8.4 shows a reference design for the PMMA depolymerization process in which, as a standard approach, the downstream processing is carried out in two steps. Initially, the hot reactor effluent is cooled and condensed, and subsequently, the cold liquid is sent to the purification step.

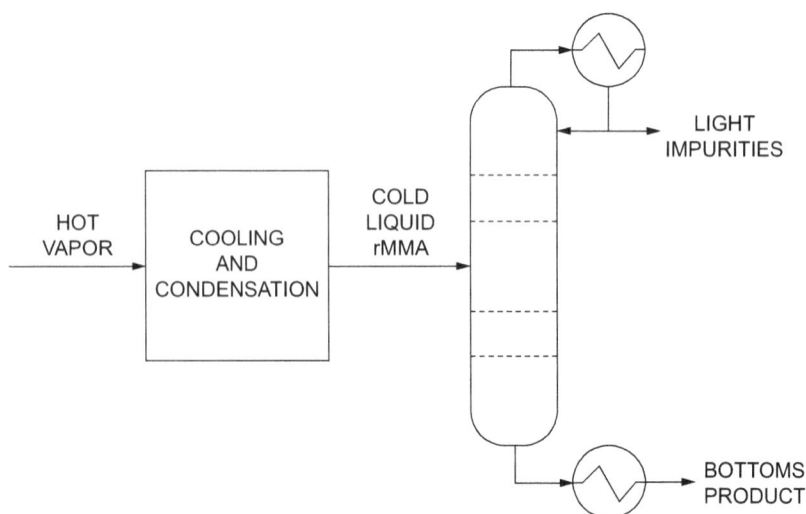

Figure 8.4: Reference concept of PMMA depolymerization downstream processing.

The first distillation column, separating the impurities from crude rMMA, offers an opportunity for heat integration by reducing the external energy use in the reboiler. As the depolymerization reactions occur at high temperatures, the reactor effluent vapors are a rich source of process heat that can be used. Several integration alternatives have been identified:

1. The addition of a column feed preheater (Figure 8.5a) is a simple and straightforward solution that involves using an extra heat exchanger with the reactor effluent on the hot side and the liquid column feed on the cold side, providing a direct reduction of the duty of a utility heating in the reboiler.

2. Use of hot vapors as a heating medium in the reboiler (Figure 8.5b). In this alternative, the reactor effluent can be used as a heating medium for the reboiler and is subsequently led to the column as a vapor feed. No additional heat exchanger is required, as only a different heating medium (process stream instead of a utility) is used in the reboiler. This measure requires further upstream modifications, for example, the omission of the cooling and condensation step.

3. Direct injection of hot vapors into the reboiler stage (Figure 8.5c): In this alternative, the hot product vapors are injected into the bottom stage of the distillation tower as both a material inlet and a direct heating medium, with the prospect of not using an external reboiler at all. As in the previous alternative, the product condensation stage is also omitted. The energy required to reboil the separated mixture is transferred from the hot stream by its cooling and condensation in the bottom stage. It must be checked whether the light impurities can be successfully removed from crude rMMA under given conditions in Level 4, given that the hot vapors need to be injected into the very bottom of the column, while stages above and below the column inlet are generally needed to achieve good separation in a distillation process.

In general, heat integration helps reduce the operating costs associated with energy sources, typically at the expense of increased process complexity, which may translate, for instance, into reduced flexibility, reduced actionability in case of contingencies of the operation parameters, or more challenging process control. The process designer should pay attention to the most energetically significant sections of the process to save the most energy while keeping the amount of interaction between the individual process parts at a reasonable level.

8.3.7 Level 6: health, safety, and environment

Safety considerations must be embedded in every level of process design, as early decisions made purely for process reasons can later lead to problems with SHE that would require complex solutions [11]. Therefore, in Levels 0–5, the key issues affecting SHE are qualitatively identified for each decision made along the design process. At this level, a quantitative evaluation of these effects is performed. For this, the impact factors, for example, on toxicity, safety, and the environment, are evaluated using MSDS documents for components or computer-aided methods [13]. The goal is to achieve an inherently safe design.

(a)

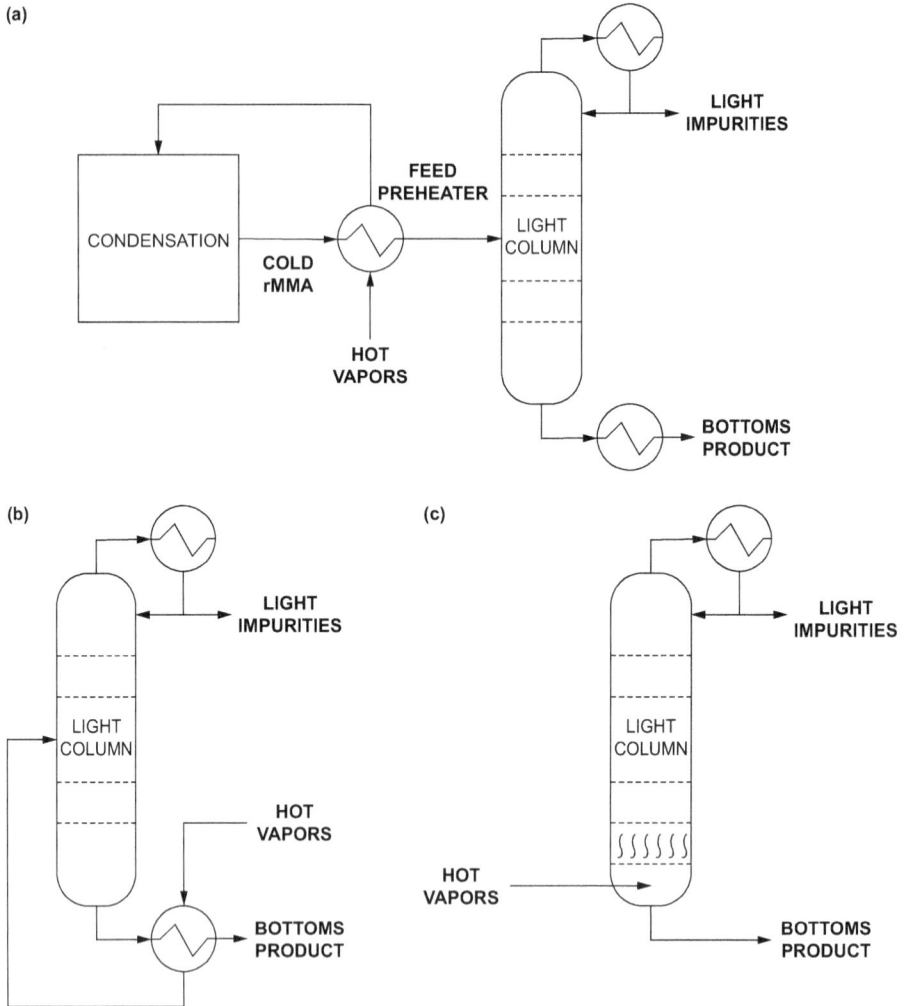

Figure 8.5: Conceptual alternative schemes for PMMA depolymerization downstream processing: (a) feed preheater addition, (b) reactor effluent used as a reboiler heating medium, and (c) direct injection of a hot vapor stream in the reboiler stage.

8.3.8 Level 7: control system

In this level, the plant-wide control strategy is addressed. Although the key issues of process control must have been treated for each production unit in previous levels, for example, stability issues related to reactor operation, the controllability of individual unit operations does not guarantee controllability of the whole plant. The cause is that recycling of raw materials, energy integration, and reduced size of buffer vessels produce strong interactions between units, which adds complexity to the overall controllability of the plant. Luyben et al. [24] and Dimian et al. [13] present different strategies to design a plant-wide control system.

References

[1]   Biegler LT, Grossmann IE, Westerberg AW. Systematic Methods of Chemical Process Design, Prentice Hall PTR, Upper Saddle River, New Jersey, USA, 1997.
[2]   Seider WD, Lewin DR, Seader JD, Soemantri W, Gani R, Ng KM. Product and Process Design Principles: Synthesis, Analysis, and Evaluation, 5th Edition, John Wiley & Sons, New York, 2017.
[3]   Pahl G, Beitz W, Feldhusen J, Grote K-H. Engineering Design: A Systematic Approach, 3rd Edition, Springer, London, 2007.
[4]   Singh N, Hui D, Singh R, Ahuja IPS, Feo L, Fraternali F. Recycling of plastic solid waste: A state of art review and future applications. Composites Part B: Engineering 2017, 115, 409–422.
[5]   Garcia JM, Robertson ML. The future of plastics recycling. Science 2017, 358, 870–872.
[6]   Rahimi A, García J. Chemical recycling of waste plastics for new materials production. Nature Reviews Chemistry 2017, 1, 0046. https://doi.org/10.1038/s41570-017-0046.
[7]   PlasticEurope. Plastics – The Facts 2020 An analysis of European plastics production, demand and waste data, 2020.
[8]   Moens EKC, De Smit K, Marien YW, Trigilio AD, Phm VS, Van Geem KM, et al. Progress in reaction mechanisms and reactor technologies for thermochemical recycling of poly(methyl methacrylate). Polymers 2020, 12, 1667.
[9]   Sinnott R, Towler G. Chemical Engineering Design, Butterworth-Heinemann, Amsterdam, 2019.
[10]   Moran S. An Applied Guide to Process and Plant Design, Elsevier, Amsterdam, 2019.
[11]   Smith R. Chemical Process: Design and Integration, Wiley, New York, 2005.
[12]   Douglas JM. Conceptual Design of Chemical Processes, McGraw-Hill, New York, 1988.
[13]   Dimian AC, Bildea CS, Kiss AA. Integrated Design and Simulation of Chemical Processes, 2nd Edition, Elsevier, Amsterdam, 2014.
[14]   Dimian AC, Bildea CS. Chemical Process Design: Computer-Aided Case Studies, Wiley, Weinheim (Germany), 2008.
[15]   Segui ED, Alarcon BC. Process and Device for the Regeneration of Monomers Starting from Polymethacrylate And, More Especially, Methyl Polymethacrylate. U.S. Patent US2858255A, 28 October 1958.
[16]   CEFIC. Methyl Methacrylate (MMA) Eco-profiles and Environmental Product Declarations of the European Plastics Manufacturers, 2014.
[17]   Dimian AC, Bildea CS, Kiss AA. Applications in Design and Simulation of Sustainable Chemical Processes, Elsevier, Amsterdam, 2019.
[18]   De Tommaso J, Dubois J-L. Risk analysis on PMMA recycling economics. Polymers 2021, 13, 2724.

[19] Barnicki SD, Fair JR. Separation system synthesis: A knowledge-based approach. 1. Liquid mixture separations. Industrial & Engineering Chemistry Research 1990, 29, 421–432.

[20] Barnicki SD, Fair JR. Separation system synthesis: A knowledge-based approach. 2. Gas/ vapor mixtures. Industrial & Engineering Chemistry Research 1992, 31, 1679–1694.

[21] Barnicki SD, Siirola JJ. Separation system synthesis. In Kirk-Othmer Encyclopedia of Chemical Technology, 4th Edition. Wiley-Interscience, Hoboken, NJ, USA, 1997, 923–962.

[22] Achilias DS. Chemical recycling of poly(methyl methacrylate) by pyrolysis. Potential use of the liquid fraction as a raw material for the reproduction of the polymer. European Polymer Journal 2007, 43, 2564–2575.

[23] Godiya CB, Gabrielli S, Materazzi S, Pianesi MS, Stefanini N, Marcantoni E. Depolymerization of waste poly(methyl methacrylate) scraps and purification of depolymerized products. Journal of Environmental Management 2019, 231, 1012–1020.

[24] Luyben WL, Tyreus B, Luyben ML. Plantwide Process Control, McGraw Hill, New York, 1999.

Fady Boutros, Michiel Van Melkebeke, Clément Lemenu,
and Philippe De Groote

9 Purification of regenerated monomers

9.1 Introduction

As discussed in Chapters 5–7, thermal degradation has been extensively investigated for the chemical recycling of various polymers [1–3]. Upon submission to high temperatures in the absence of oxygen, polymer chains break down into lighter hydrocarbons. Typically, this degradation process is less controllable in terms of the degradation selectivity, resulting in a distribution of molar masses and chemical structures in the output stream [4–6]. As such, the aim of the degradation process is to transform polymer chains back into basic chemicals.

Table 9.1 gives an overview of product spectra per polymer time. For instance, pyrolysis of polyolefins produces a product stream that is similar to the petroleum fractions in the refining of crude oil [7–10]. Consequently, this process is also referred to as thermal cracking or a type of plastic-to-fuel conversion. For certain polymers such as poly(methyl methacrylate) (PMMA), polystyrene (PS), and polytetrafluoroethylene (PTFE), thermal degradation/depolymerization can be used to recover monomer (oligomers) in a highly selective way, which can thus be seen as the reverse operation of polymerization, thus depolymerization [2, 7, 10–14]. This thermal depolymerization is generally described as a plastic-to-monomer conversion.

A crucial step is the purification stage once the reactor step has been concluded, which is addressed in this chapter. Emphasis is on the purification of monomer-rich product streams as obtained by plastic-to-monomer conversion processes, focusing in particular on thermal degradation of PMMA. The latter degradation process has been proven to be economically feasible due to its relatively high product value and the possibility to recover its monomer methyl methacrylate (MMA) in high yields [14–21]. For example, it has been reported that at relatively low temperatures (e.g., 450 °C), approximately 97% of the MMA monomer can be recovered [15, 19, 22–24], this in contrast to PS, for which a higher degradation temperature is needed (520–600 °C) and a lower monomer recovery is obtained (e.g., 75%) [12, 25].

In what follows, first an overview of the impurities in crude regenerated MMA (rMMA) as reported in literature is given. Next, key techniques for the purification of crude rMMA and the characterization of crude and purified rMMA are discussed. Fi-

Fady Boutros, Speichim, Allée des Pins, Saint-Vulbas 01150, France
Michiel Van Melkebeke, Ghent University, Laboratory of Environmental Toxicology and Aquatic Ecology, Gent B-9000, Belgium; Department of Green Chemistry and Technology, Kortrijk B-8500, Belgium
Clément Lemenu, Philippe De Groote, Certech, Rue Jules Bordet 45, Seneffe B-7180, Belgium

https://doi.org/10.1515/9783111076997-009

nally, results of a two-step distillation process as investigated in the MMAtwo project are presented. This is supported by a detailed evaluation, including analysis of the purified rMMA composition, the measurement of the odor concentration, and the investigation of the health and safety aspects related to processing purified rMMA.

Table 9.1: Product fractions obtained by thermal degradation of different waste polymer feedstocks. Table based on data in reference [12].

Feedstocks	Degradation temperature (°C)	Gas (wt%)	Oil (wt%)	Residue (wt%)	Other products (wt%)
Polyethylene PE	760	55.8	42.4	1.8 °C	
Polyethylene PE	530	7.6	50.3	0.1	42 waxes
Polypropylene PP	740	49.6	48.8	1.6 °C	
Polystyrene PS	580	9.9	24.6	0.6	64.9 styrene
Mixtures of PE/PP/PS	750	52.0	46.6	1.4	
Polyester	768	50.8	40.0	7.1	2.1 H_2O
Polyurethane	760	37.9	56.3	0.5	5.0 H_2O 0.3 HCN
ABS copolymer	740	6.9	90.8	1.1	1.2 HCN
Polyamide PA	760	39.2	56.8	0.0	3.4 HCN
Polycarbonate	710	26.5	46.4	24.6	2.5 H_2O
Phenolformaldehyde resins	780	14.4	28.1	49.5	8.0 H_2O
Poly(methyl methacrylate) (PMMA)	450	1.25	1.4	0.15 °C	97.2 MMA
Poly(vinyl chloride) (PVC)	740	6.8	28.1	8.8	56.3 HCl
Polytetrafluoroethylene (PTFE)	555	18.8	5.2	0.3	76 TFE
EPDM elastomer	700	32.3	19.2	47.5	1.0 H_2O
SB rubber	740	25.1	31.9	42.8	0.2 H_2S 5.1 H_2O

9.2 Main impurities in crude rMMA: The need for purification

Prior to discussing purification techniques, it is important to consider the nature and quantity of the impurities and byproducts that have been reported to be present in crude streams obtained by thermal degradation or depolymerization. In general, as shown in Table 9.1, thermal degradation products are divided in three fractions, namely gas, liquid (after condensation), and solid residue [3, 26]. To that end, a distinction is ideally made between virgin-grade feedstocks and commercial or postconsumer feedstocks. Aside from the composition of the PMMA feedstock, the residence time (distribution) and temperature play an important role in the type and distribution of the degradation products [22, 27–32]. In what follows, an overview is given of the most relevant results for PMMA thermal degradation.

At an optimal temperature PMMA, thermal degradation research indicates that the liquid product is by far the most dominant one making up 97–99% of the output stream [15, 19,22–24, 33, 34]. The MMA monomer content in this liquid product is particularly high, yet varies considerably between virgin and commercial or post-consumer PMMA as starting material. For example, Kaminsky and Franck [19] showed that the MMA purity of the liquid product for both virgin and postconsumer PMMA is approximately 98%. The MMA yield, however, dropped from 97% for virgin PMMA to 91% for postconsumer PMMA. The latter is largely explained by the increased mass fraction of the gaseous product for postconsumer PMMA amounting to 7% compared to merely 1% for the virgin PMMA [19]. Kang et al. [15] found in turn a comparable MMA purity of 96–97% of the liquid product yet the mass fraction of the gaseous output stream was significantly higher for postconsumer PMMA. This logically resulted in an overall lower MMA yield in the range of 94–96% as opposed to 95–97% for virgin PMMA.

In addition, two studies by Achilias [23, 24] compared virgin PMMA with commercial PMMA. The liquid fraction showed a MMA purity of 99% for virgin PMMA and 97–98% for commercial PMMA. The work done by Popescu et al. [33] reported a MMA purity of 98% for the liquid product derived from virgin PMMA, while a purity of 90% was obtained after degradation of postconsumer PMMA [33]. Furthermore, the research by Kaminsky et al. [22] compared the degradation of virgin-grade PMMA with two different commercial types of filled PMMA, namely one filled with 62% silica and another with 71% granite [22]. The results displayed a MMA purity between 97 and 99% for the virgin PMMA and between 92 and 96% for the commercial types of PMMA. The monomer recovery yielded 97–98% for virgin PMMA and 83–90% for filled PMMA. Later work by Kaminsky et al. [12] included commercial PMMA filled with 67% aluminum trihydroxide (ATH). There, monomer recoveries of 97% and 58% were reported for virgin and filled PMMA, respectively. Subsequent efforts by Grause et al. [13] demonstrated a MMA yield of 86% for the degradation of commercial ATH-filled PMMA materials [13]. In contrast, a study by Lopez et al. [34] reported remarkably lower levels of MMA in the liquid fraction of approximately 87%. However, the PMMA used in this study was a commercial-grade copolymer of MMA and ethyl acrylate (EA). This explains both the relatively low purity of MMA and the high level of EA (i.e., 6%) found in the liquid product.

Next to MMA, a series of other products are known to be present in the liquid fraction of pyrolyzed PMMA. Kaminsky et al. [22] identified methyl acrylate (MA), methyl isobutyrate (MIB), 1,4-cyclohexane dicarboxylic acid dimethyl ester, and methanol as the main byproducts from the degradation of both virgin and postconsumer PMMA [19]. Later work done by Kaminsky et al. [22] confirmed the presence of such components and additionally identified methacrylic acid and dimethyl ethyl cyclohexene as the main liquid byproducts of virgin and commercial PMMA. In a similar context, analysis of the degradation of virgin PMMA and a commercial PMMA filled with ATH revealed methacrylic acid, methanol, cyclopentanone, methyl propionate, and isobutyric acid as the main liquid byproducts [12, 13]. The studies performed by Achilias [23, 24] in turn reported primarily ester compounds as the main liquid degradation byproducts for virgin and commercial PMMA, mostly verifying the results by Kaminsky et al. [12, 23, 24]. Kang et al. [15] found MA and MIB as the main liquid byproducts for both virgin and postconsumer PMMA.

From the postconsumer PMMA pyrolysis, significant amounts of EA and butyl acrylate (BA) have also been found in the liquid fraction. Both EA and BA are commonly known as comonomers of particular types of commercial PMMA [15]. Lopez et al. [34] reported MA, acetaldehyde, 2-propenal, 1,3-cyclapentadiene, and 1,3-pentadiene as the main liquid byproducts for the degradation of a PMMA copolymer next to its comonomers MMA and EA. Szabo et al. [35] investigated the degradation products of PMMA blends, in particular, a commercial PMMA-acrylonitrile butadiene styrene (ABS) polymer blend with approximately 61% PMMA and 39% ABS. Next to MMA and styrene monomers, the results showed methyl styrene, ethyl benzene, dimer derivatives, and trimer derivatives as the most significant components in the liquid fraction. Godiya et al. [36] additionally degraded virgin PMMA and identified MIB as the main byproduct in the liquid fraction with traces of methyl pyruvate and 2,3-butanedione. These authors also claim that 2,3-butanedione rather than MIB is responsible for the unpleasant smell in rMMA and subsequently regenerated PMMA. With respect to the gas fraction of pyrolyzed PMMA materials, the main components are carbon dioxide, carbon monoxide, hydrogen, methane, ethene, and propene [15, 16, 19, 22–24].

Increasing the temperature beyond its optimal value has been shown to significantly increase the gas fraction and reduce the MMA monomer yield; hence, one should not only look at the initial composition but also the degradation temperature. For instance, moderately increasing the temperature from 450 °C to 480 °C has been reported to increase gas fraction from 0.5 to 1.4% and reduce the MMA yield from 98.4% to 93.7% [22].

The above literature results allow to conclude that the liquid fraction of a crude thermally degraded PMMA stream predominantly consists of MMA thus monomer.

For virgin PMMA, the crude MMA purity can be as high as 99%, while for commercial and postconsumer PMMA the crude purity averages around 97% with some exceptions. As a result, it has been suggested that there is little to no need for purification [12, 19, 22]. However, PMMA produced directly from the liquid fraction of depolymerized PMMA waste was found to lead to lower average molar masses [24]. In addition, impurities present in postconsumer PMMA were found to render the recycled polymer opaque rather than optically clear and transparent [33]. A disagreeable smell of recycled PMMA is also known to be associated with insufficient purification [36]. The small fraction of byproducts and impurities present in the liquid fraction can act as a reaction inhibitor or retarder for subsequent repolymerization. This decreases the reaction rate and lowers the glass transition temperature [24, 36].

Consequently, recycled PMMA derived from the liquid MMA stream resulting from thermal degradation will lead to inferior polymer properties in case no further purification is implemented prior to repolymerization. It was argued that even a 0.2% impurity in recycled MMA results in a reduced performance of optical devices based on PMMA [37]. Therefore, purification is an essential part of the recycling process of PMMA in order to avoid (strong) downcycling and to enable a true polymer circularity roadmap.

9.3 Principles of key purification and characterization techniques for rMMA

In this section, focus is on distillation as purification technique and gas chromatography-mass spectrometry/flame ionization detection (GC-MS/FID) and dynamic olfactometry as characterization techniques. More techniques exist but focus is mainly on those relevant for the MMAtwo project. This is also confirmed in the next section in which novel results obtained by the extruder-assisted depolymerization process from the MMAtwo project are described. Virgin MMA is used as a benchmark.

9.3.1 Distillation

Distillation is one of the most universal separation techniques in the chemical industry. It is based on a difference in composition between a boiling liquid and its vapor. Typically, a vapor–liquid mixture with a certain concentration of one or more components is fed to a distillation column. In the reboiler, an input of heat is provided to create a vapor flow which moves upward along the column length, while in the condenser heat is dissipated to create a liquid flow which moves down along the trays from top to bottom.

More in detail, on an arbitrary column tray, the liquid flow from the tray above meets the vapor flow from the tray below. Since these flows are not in equilibrium, a mass exchange takes place. Part of the components with a relatively low boiling point transfer from the liquid to the vapor phase, while part of the components with a relatively high boiling point transfers from the vapor to the liquid phase. As a result, the composition on the trays varies along the length of the column. The highest concentration of light components is located at the top trays, while the highest concentration of heavy components can be found at the bottom trays. Notably the condenser allows part of the rising vapor to be extracted as liquid distillate and circulates the other part back into the column as reflux. In the opposite manner, the reboiler evaporates part of the collected liquid from the bottom and adds it back to the column, while the remaining liquid is removed as bottom product. In this fashion, a separation of the initial mixture is possible. From technical description, it is clear that distillation is essentially a two-phase multistage equilibrium process with a countercurrent flow [38].

At present, distillation is the most widely applied technique for the purification of liquid pyrolysis products [2]. Research regarding the thermal degradation of PMMA thus predominantly uses distillation processes to purify the crude MMA. For instance, Braido et al. [39] utilized a packed bed distillation column with two distilling receivers in a laboratory setup to purify the liquid fraction obtained after pyrolysis of virgin and postconsumer PMMA. The mixture was distilled under atmospheric pressure and hydroquinone was added in order to prevent spontaneous polymerization. It was shown that impurities could be removed efficiently yet small amounts of light components were still present.

It needs to be mentioned for completeness that the purification process is largely dependent on the targeted application. The more stringent the specifications on the purity of the monomer, the more advanced the purification installation should be. Considering the aim to generate recycled PMMA polymers with near identical properties as in their virgin form, large distillation towers or complex distillation setups with multiple columns could be necessary. A study performed by Kikuchi et al. [37] illustrates the application of two distillation columns in series in order to purify a crude MMA monomer stream derived from the thermal degradation of PMMA. The first column removed components with a relatively low boiling point, while the second column allowed for optimal MMA purity. The resulting monomer stream reached a purity of 99.8%. This two-step distillation setup has also been considered in the MMAtwo project at laboratory and pilot scale, as discussed in Section 9.4.

9.3.2 Gas chromatography-mass spectrometry/flame ionization detection

Gas chromatography (GC) coupled to mass spectrometry (MS) is a technique for separation, detection, and identification of compounds in the gaseous state. In mass spectrometry, the response factor is relative to the ionization efficiency of the molecule and is therefore strongly dependent on the chemical groups present. MS mainly aims to identify the nature of the molecules, while additional flame ionization detector (FID) allows to cover a wide range of concentration for quantification.

The combination of FID and MS allows both detectors to be used in a single analysis and under the same conditions. This equipment allows analysis of compounds at trace levels, identification, and quantification of compounds in complex mixtures, and analysis of impurities and contaminants that could be present in crude/purified regenerated monomers.

In the present study, rMMA is introduced into the system by liquid injection, the gas chromatography unit is an Intuvo 9000 and the mass spectrometer is equipped with a specific high efficiency source, enhancing sensitivity.

9.3.3 Dynamic olfactometry

Different techniques have been developed to evaluate the emissions and odors from a wide range of materials. Among these, dynamic olfactometry aims to quantify the odor at detection threshold with human panelists representative of the whole population. The method is described into a specific European standard, EN13725. The primary goal of this standard is to provide a common basis for the regulation of outdoor odor emissions by the member states of the European Union. Therefore, this standard defines a universal odor scale based on a common unit of measurement, being the European odor unit per cubic meter or ouE/m^3.

The European odor unit [ou_E] is the amount of odorant(s) that if evaporated into $1 m^3$ of neutral gas at standard conditions elicits a physiological response from a panel (D_{50} detection threshold) equivalent to that elicited by one European Reference Odor Mass (EROM) under the same conditions. One EROM corresponds to 123 µg of n-butanol (CAS-Nr. 71-36-3). This relationship effectively expresses odor concentrations in terms of mass equivalents of n-butanol.

In practice, the odor concentration is measured by determining the dilution factor required to reach the detection threshold by the panel. It is then expressed in terms of multiples of the detection threshold. The range of measurement is typically from $10^1 ou_E/m^3$ to $10^7 ou_E/m^3$ (including pre-dilution). The dilution of the odorant sample within the olfactometer is done with a neutral gas that must be safe for breathing and perceived as odorless so that it does not interfere with the perception of the odor under investigation. The method starts from an "infinite" dilution so that the panel cannot detect any odor from the sample. The dilution is then decreased allowing the panel members to establish the detection threshold. The dilution factor at the panel threshold corresponds to the dilution factor that was applied to the sample to arrive at the physiological response of the panel. Hence, the odor concentration in ou_E/m^3 is equal to the numerical value of this dilution factor. The dynamic olfactometer station used in this study is a TO-evolution from Olfasense.

9.4 Two-step distillation results for rMMA

In the MMAtwo project, a two-step distillation process for the purification of crude rMMA has been investigated. Prior to distillation, the crude rMMA stream derived from thermal degradation of PMMA is separated by decantation from excess of water (if present) and any suspended matter is filtered or separated by centrifugation. After removal of the excess water and suspended matter, the crude rMMA stream with typical components and properties given in Table 9.2 is sent to a two-step distillation process.

Table 9.2: Physical properties of the main impurities in crude rMMA; MM: molar mass; B_p: boiling point, M_p; melting point.

Product	Formula	MM (g/mol)	Specific gravity g/cm³ at 20 °C	B_p (°C)	M_p (°C)	Latent heat at T_b (cal/g)	Specific heat at 20 °C (cal/g)
Methanol	CH_4O	32	0.794	64.7	−98	263	0.598
Me acrylate	$C_4H_6O_2$	86	0.954	80.2	−77	89	0.444
Me propionate	$C_4H_8O_2$	88	0.914	79.5	−88	87.5	0.460
Me isobutyrate	$C_5H_{10}O_2$	102	0.889	92.3	−85	77.5	0.448
Et acrylate	$C_5H_8O_2$	1,000	0.922	99.5	−71	82.3	0.410
MMA	$C_5H_8O_2$	100	0.944	100.3	−48	80	0.447
Benzene	C_6H_6	78	0.878	80.1	5.5	90.5	0.414
Toluene	C_7H_8	92	0.868	110.6	−95	86.4	0.403
Water	H_2O	18	0.9965	100	0	540	1.001
Methacrylic acid	$C_4H_6O_2$	86	0.014	161	15	100	0.444

As shown in Figure 9.1, the first step of the two-step distillation process is a lights/ water removal at the top of the first column (Figure 9.1(a)). The bottom stream of the first column is fed to the second column (Figure 9.1(b)), in which the heavies are removed at the bottom and an anhydrous MMA stream is recovered at the top. Thanks to the high purity, as further illustrated in Section 9.5, the obtained rMMA can be used to replace virgin MMA in most applications, as discussed in more detail in Chapters 10 and 11.

More detailed analysis revealed that EA cannot be removed from the rMMA as the boiling point is very close to the one of MMA (99.5 °C for EA). Also, it is important to realize that MA, methyl propionate, 1-butanol, MIB, EA, and MMA constitute azeotropes with water. These azeotropes are heterogeneous, and water decants in a heavy layer, as shown in Figure 9.2(b) showing the MMA/water equilibrium curves as an ex-

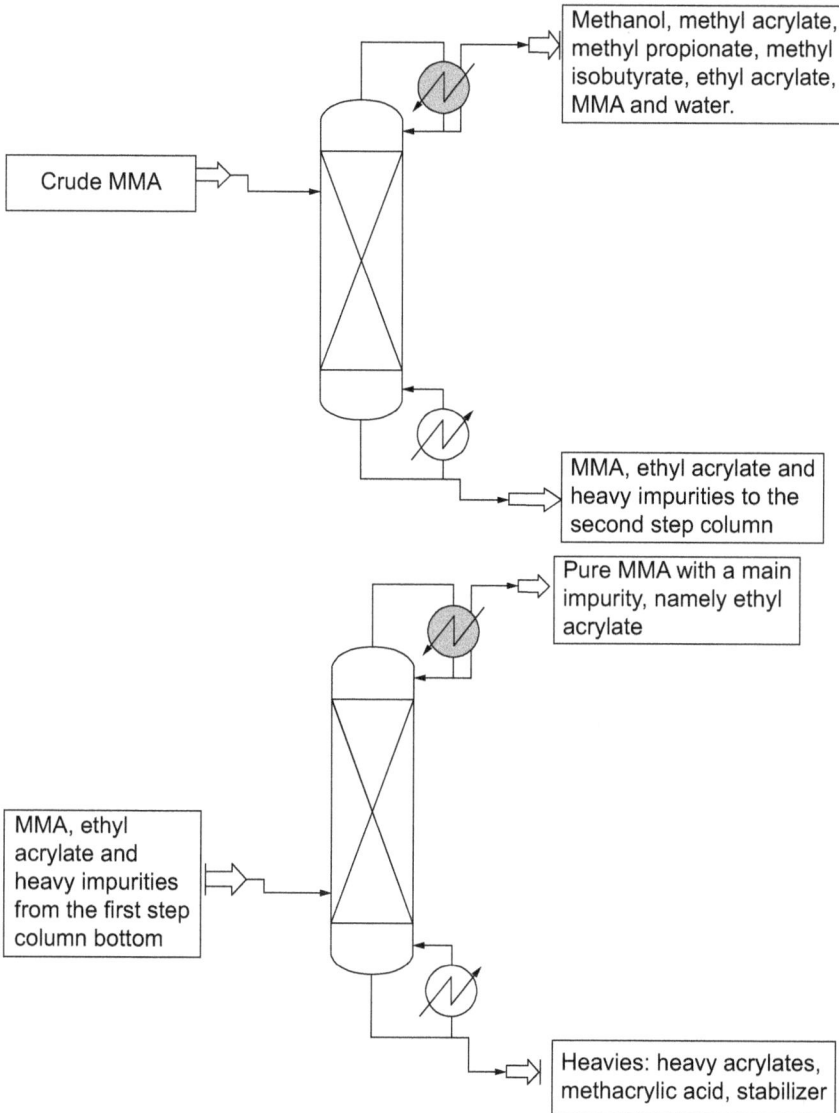

Figure 9.1: Two-step continuous distillation process to purify crude rMMA. Lights/water removal over the top of the first column (a). The bottom stream of the first column is fed to the second column, where heavies are removed via the bottom and a purified anhydrous MMA stream is recovered via the top (b).

ample. In the light fraction (top of the first step), the decanted water is methanol rich, at least if water is present in the crude MMA.

It should be noted that crude rMMA could be purified by batch distillation as well. A continuous process as presented here has although the advantage of minimizing the residence time, thereby also reducing the risk on polymerization taking place.

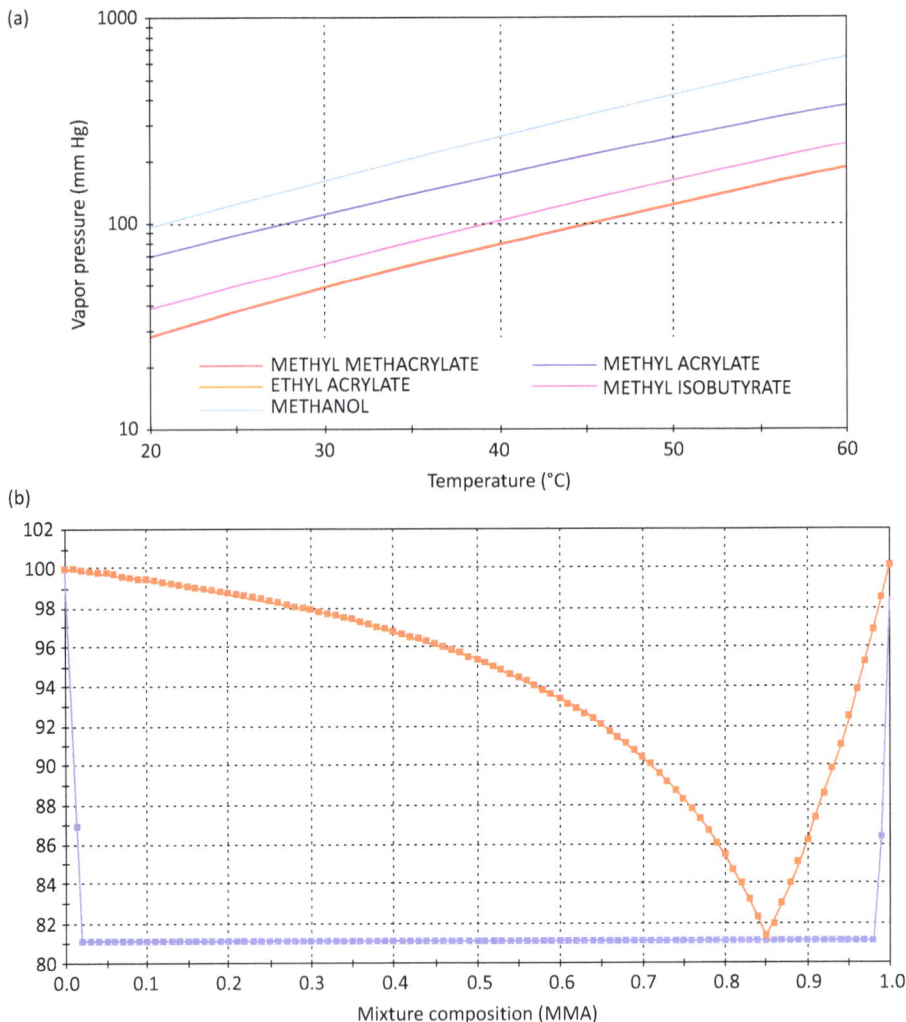

Figure 9.2: (a) Vapor pressure curves of the main components in crude rMMA; (b) MMA/water equilibrium curve.

This risk can be further reduced by working at reduced pressure, as temperatures are then strongly reduced, as illustrated in Figure 9.2(a). Nevertheless, a stabilizer solution is continuously added at the top of both columns in the continuous process.

9.5 Evaluation of the quality of purified rMMA

The composition and odor analyses have been carried out on virgin MMA and rMMA samples. The virgin MMA is a benchmark liquid, purchased from Sigma Aldrich M55909). The following purified rMMA samples have been considered: (i) PIL1 being regenerated monomer obtained by depolymerizing a stream composed of postproduction clear transparent cast PMMA sheets; (ii) MIX1 being regenerated monomer obtained from a mix of four streams composed of postproduction mixed color injection PMMA grades from car-lights, end-of-life (EoL) PMMA from LCD scrap, postproduction white extrusion PMMA sheets and, postproduction PMMA dust with PE contamination; (iii) MIX2 being regenerated monomer obtained from a mix of four streams composed of two postproduction mixed cast and extrusion PMMA (50/50), originating from the Netherlands, the same originating from Germany, EoL PMMA from LCD scrap, and postproduction PMMA dust with PE contamination coming from cutting of extrusion sheets; and (iv) MIX3 being regenerated monomer obtained from a mix of several streams composed of postproduction clear extrusion PMMA sheets, postproduction smokey transparent cast PMMA, and post-production white cast PMMA from sanitary sheets.

9.5.1 Composition of purified rMMA

The composition of virgin MMA, PIL1, MIX1, MIX2, and MIX3 as obtained by GCMD/FID is shown in Table 9.3, with all percentages calculated as the percentual peak area in the chromatograms obtained with FID.

Table 9.3: Composition of various purified rMMAs compared to virgin MMA. Components are given per increasing retention time (RT), with percentages calculated as FID % peak area.

		Purity (%)				
C	RT (min.)	Virgin MMA	PIL 1	MIX 1	MIX 2	MIX 3
Methyl isobutyrate	4.62		0.02		0.02	0.02
Ethyl acrylate	4.84		0.12	0.05	0.04	
Methyl methacrylate	5.25	99.96	99.80	99.88	99.68	99.84
Methacrylic acid	6.38	0.02				
Other compounds (sum)		0.02	0.06	0.07	0.27*	0.14*

*No single compound > 0.1%.

From Table 9.3, it can be observed that the purified rMMAs have a purity ranging from 99.68 to 99.88%, which is close to the purity of the benchmark virgin MMA (99.96%). This indicates that these purified regenerated monomers are worth considering for high-end applications (see Chapter 11). While virgin MMA sample contains 0.02% methacrylic acid, the rMMAs main impurities are methyl isobutyrate (MIB) and ethyl acrylate (EA). All these impurities are below 0.1%, except for rMMA PIL1, which contains 0.12% EA. This last observation is quite unexpected as the cast material stream from which this specific rMMA is generated should not contain any acrylate comonomer. This suggests that some contamination may have occurred at some point in the chain.

9.5.2 Health and safety aspects of rMMA

The presence of impurities in the purified rMMA could generate health and safety concerns in applications such as the manufacturing of large composite parts in which much more significant exposure involving workers takes place, as opposed to other techniques in confined equipments, for example, extruders. In order to make correct conclusions, exposure calculations in scenarios representative of composite applications have been carried out.

These calculations involved purified rMMA, and impurities having tabulated exposure limit (EL) values. Two scenarios in which virgin MMA is used under safe conditions have been considered: a continuous diffuse leak and an accidental spill, in a representative workplace volume with imposed air renewal and an exposure duration of 8 h.

Notably, calculations based on a worst case purified rMMA (maximum concentration in impurities) showed that the impurities mentioned above do not generate additional risks, that is, none of the EL values are reached throughout the total exposure duration of 8 h. Hence, purified rMMA can be used like virgin MMA. In other words, no additional safety risks are generated.

9.5.3 Odor of purified rMMA

Odor assessment on purified rMMA has been carried out in order to see whether the presence of the impurities discussed before have an impact on the rMMA odor and therefore jeopardize its acceptance as substitute for virgin MMA.

Direct odor evaluation of the samples by smelling like practiced in the field of odors methodology (Champ des Odeurs© = Field of Odeurs) was ruled out in this work because of the high acrid odor intensity of MMA, likely to cause discomfort for human panelists. With this in mind, dynamic olfactometry was used for odour evaluation. As explained above, this technique starts with "infinite" dilution (pure air) and as the gas flow originating from the sample of interest is increased, the measurement will stop once 50% of the panel has detected something. A difference between olfactometric results is considered as significant if at least a factor 2 is observed between two evaluations.

The dynamic olfactometry results for virgin MMA, MIX1, MIX2, and MIX3 are shown in Figure 9.3. It follows that while MIX3 is practically identical to virgin MMA, MIX1, and MIX2 are significantly higher in odor concentration, more than a twofold

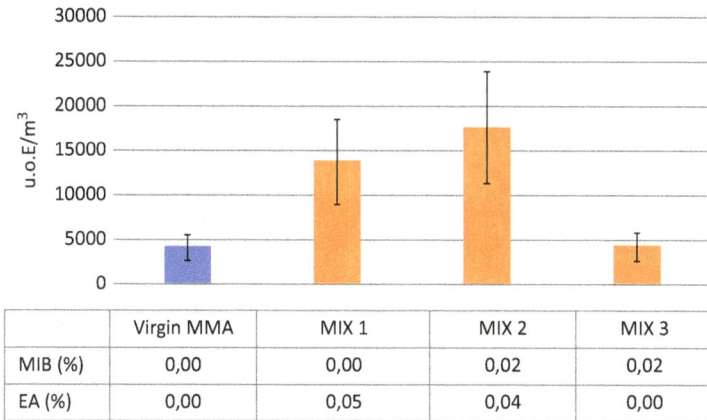

Figure 9.3: Odor concentration of virgin MMA and regenerated rMMA MIX1, MIX2, and MIX3 (dynamic olfactometry results; top) combined with MIB and EA concentration (GC-MS/FID data; bottom).

value. Correlating these observations with GC-MS/FID data (see insert in Figure 9.3) suggests that these high values are linked to the EA concentration. Therefore, the detection threshold of EA was determined in a separate dynamic olfactometry experiment, together with that of MIB and MMA, as shown in Table 9.4. It can be observed that the EA detection threshold is very low at 0.08 µg/m^3, while other substances have higher values (22.03 for MMA and 16.10 for MIB). This supports the interpretation that EA is responsible for the higher odor concentration of MIX1 and MIX2.

Table 9.4: Odor detection thresholds for MMA, MIB, and EA as determined by dynamic olfactometry.

Substance	Detection threshold (µg/m^3)
MMA	22.03
MIB	16.10
EA	0.08

9.6 Conclusions

The use of regenerated instead of virgin monomers for high-end applications likely necessitates the application of dedicated purification processes and characterization techniques. In this chapter, focus has been on the purification of rMMA as obtained by extruder-assisted depolymerization of various PMMA feedstocks, for which a two-step distillation process has been presented. The composition of the purified rMMAs has been analyzed by GC-MS/FID, and the odor has been assessed via dynamic olfactometry. In addition, safety and health aspects of using rMMAs as encountered for instance upon manufacturing large composite parts have been assessed.

The composition of various rMMA's as determined by GC-MS/FID ranges from 99.68 to 99.88% which is close to that of the virgin MMA benchmarked considered in this work (99.96%). All impurities are generally lower than 0.1%, among which EA appears to directly impact the rMMA odor concentration, due to its low detection threshold by the human nose, as determined through dynamic olfactometry. Calculations based on a worst case rMMA, with maximum concentration in impurities, show that these impurities do not generate additional safety and health risks compared to virgin MMA in applications in which significant exposure involving workers takes place.

References

[1]   Hann S, Connock T. Chemical Recycling: State of Play.
[2]   Ragaert K, Delva L, Van Geem K. Mechanical and chemical recycling of solid plastic waste. Waste Management 2017, 69, 24–58.
[3]   Al-Salem SM, Lettieri P, Baeyens J. Recycling and recovery routes of plastic solid waste (PSW): A review. Waste Management 2009, 29, 2625–2643.
[4]   Crippa M, De Wilde B, Koopmans R, Leyssens J, Muncke J, Ritschkoff A, et al. A Circular Economy for Plastics – Insights from Research and Innovation to Inform Policy and Funding Decisions, Brussels, 2019.
[5]   Angyal A, Miskolczi N, Bartha L. Petrochemical feedstock by thermal cracking of plastic waste. Journal of Analytical and Applied Pyrolysis 2007, 79, 409–414.
[6]   Dogu O, Pelucchi M, Van de Vijver R, Van Steenberge PHM, D'hooge DR, Cuoci A, et al. The chemistry of chemical recycling of solid plastic waste via pyrolysis and gasification: State-of-the-art, challenges, and future directions. Progress in Energy and Combustion Science 2021, 84, 100901.
[7]   Kaminsky W. Recycling of polymeric materials by pyrolysis. Makromolekulare Chemie. Macromolecular Symposia 1991, 48–49, 381–393.
[8]   Williams EA, Williams PT. Analysis of products derived from the fast pyrolysis of plastic waste. Journal of Analytical and Applied Pyrolysis 1997, 40–41, 347–363.
[9]   Westerhout RWJ, Kuipers JAM, van Swaaij WPM. Experimental determination of the yield of pyrolysis products of polyethene and polypropene. Influence of reaction conditions. Industrial & Engineering Chemistry Research 1998, 37, 841–847.
[10]   Bockhorn H, Hornung A, Hornung U. Stepwise pyrolysis for raw material recovery from plastic waste. Journal of Analytical and Applied Pyrolysis 1998, 46, 1–13.
[11]   Aguado R, Olazar M, Gaisán B, Prieto R, Bilbao J. Kinetics of polystyrene pyrolysis in a conical spouted bed reactor. Chemical Engineering Journal 2003, 92, 91–99.
[12]   Kaminsky W, Predel M, Sadiki A. Feedstock recycling of polymers by pyrolysis in a fluidised bed. Polymer Degradation and Stability 2004, 85, 1045–1050.
[13]   Grause G, Predel M, Kaminsky W. Monomer recovery from aluminium hydroxide high filled poly (methyl methacrylate) in a fluidized bed reactor. Journal of Analytical and Applied Pyrolysis 2006, 75, 236–239.

[14]   Moens EKC, De Smit K, Marien YW, Trigilio AD, Van Steenberge PHM, Van Geem KM, et al. Progress in reaction mechanisms and reactor technologies for thermochemical recycling of poly(methyl methacrylate). Polymers 2020, 12.

[15]   Kang BS, Kim SG, Kim JS. Thermal degradation of poly(methyl methacrylate) polymers: Kinetics and recovery of monomers using a fluidized bed reactor. Journal of Analytical and Applied Pyrolysis 2008, 81, 7–13.

[16]   Smolders K, Baeyens J. Thermal degradation of PMMA in fluidised beds. Waste Management 2004, 24, 849–857.

[17]   Kaminsky W, Menzel J, Sinn H. Recycling of plastics. Conservation & Recycling 1976, 1, 91–110.

[18]   Sasse F, Emig G. Chemical recycling of polymer materials. Chemical Engineering & Technology 1998, 21, 777–789.

[19]   Kaminsky W, Franck J. Monomer recovery by pyrolysis of poly(methyl methacrylate) (PMMA). Journal of Analytical and Applied Pyrolysis 1991, 19, 311–318.

[20]   Korobeinichev OP, Paletsky AA, Gonchikzhapov MB, Glaznev RK, Gerasimov IE, Naganovsky YK, et al. Kinetics of thermal decomposition of PMMA at different heating rates and in a wide temperature range. Thermochimica Acta 2019, 671, 17–25.

[21]   De Tommaso J, Dubois J-L. Risk analysis on PMMA recycling economics. Polymers 2021, 13.

[22]   Kaminsky W, Eger C. Pyrolysis of filled PMMA for monomer recovery. Journal of Analytical and Applied Pyrolysis 2001, 58–59, 781–787.

[23]   Achilias DS Chemical Recycling of Polymers. The case of poly (methyl methacrylate). International Conference on Energy and Environmental Systems 2006, 2006, 271–276.

[24]   Achilias DS. Chemical recycling of poly(methyl methacrylate) by pyrolysis. Potential use of the liquid fraction as a raw material for the reproduction of the polymer. European Polymer Journal 2007, 43, 2564–2575.

[25]   Liu Y, Qian J, Wang J. Pyrolysis of polystyrene waste in a fluidized-bed reactor to obtain styrene monomer and gasoline fraction. Fuel Processing Technology 2000, 63, 45–55.

[26]   Al-Salem SM, Lettieri P, Baeyens J. The valorization of plastic solid waste (PSW) by primary to quaternary routes: From re-use to energy and chemicals. Progress in Energy and Combustion Science 2010, 36, 103–129.

[27]   Fuels O. Feedstock Recycling and Pyrolysis of Waste Plastics, 2006, 2006.

[28]   Faravelli T, Bozzano G, Colombo M, Ranzi E, Dente M. Kinetic modeling of the thermal degradation of polyethylene and polystyrene mixtures. Journal of Analytical and Applied Pyrolysis 2003, 70, 761–777.

[29]   Marongiu A, Faravelli T, Ranzi E. Detailed kinetic modeling of the thermal degradation of vinyl polymers. Journal of Analytical and Applied Pyrolysis 2007, 78, 343–362.

[30]   Ranzi E, Dente M, Faravelli T, Bozzano G, Fabini S, Nava R, et al. Kinetic modeling of polyethylene and polypropylene thermal degradation. Journal of Analytical and Applied Pyrolysis 1997, 40–41, 305–319.

[31]   Walendziewski J, Steininger M. Thermal and catalytic conversion of waste polyolefines. Catalysis Today 2001, 65, 323–330.

[32]   Yan G, Jing X, Wen H, Xiang S. Thermal cracking of virgin and waste plastics of PP and LDPE in a semibatch reactor under atmospheric pressure. Energy & Fuels 2015, 29, 2289–2298.

[33]   Popescu V, Vasile C, Brebu M, Popescu GL, Moldovan M, Prejmerean C, et al. The characterization of recycled PMMA. Journal of Alloys and Compounds 2009, 483, 432–436.

[34]   Lopez G, Artetxe M, Amutio M, Elordi G, Aguado R, Olazar M, et al. Recycling poly-(methyl methacrylate) by pyrolysis in a conical spouted bed reactor. Chemical Engineering and Processing: Process Intensification 2010, 49, 1089–1094.

[35]   Szabo E, Olah M, Ronkay F, Miskolczi N, Blazso M. Characterization of the liquid product recovered through pyrolysis of PMMA-ABS waste. Journal of Analytical and Applied Pyrolysis 2011, 92, 19–24.

[36] Godiya CB, Gabrielli S, Materazzi S, Pianesi MS, Stefanini N, Marcantoni E. Depolymerization of waste poly(methyl methacrylate) scraps and purification of depolymerized products. Journal of Environmental Management 2019, 231, 1012–1020.

[37] Kikuchi Y, Hirao M, Ookubo T, Sasaki A. Design of recycling system for poly(methyl methacrylate) (PMMA). Part 1: Recycling scenario analysis. International Journal of Life Cycle Assessment 2014, 19, 120–129.

[38] Petlyuk FB. Distillation Theory and Its Application to Optimal Design of Separation Units, Cambridge University Press, Cambridge, 2004. [Online]. Available https://doi.org/10.1017/CBO9780511547102.

[39] Braido RS, Borges LEP, Pinto JC. Chemical recycling of crosslinked poly(methyl methacrylate) and characterization of polymers produced with the recycled monomer. Journal of Analytical and Applied Pyrolysis 2018, 132, 47–55.

Maria Savina Pianesi, Tommaso Compagnucci, Dagmar R. D'hooge, and Mark Bierens

10 Challenges for industrial repolymerization

10.1 Introduction

Polymerization processes in bulk, suspension, and emulsion are often characterized by a strong increase in the (dynamic) viscosity, due to either the increase of the polymer yield, the average chain length, or the solid content. This is specifically the case in radical polymerization, in which a gel effect can occur because of diffusional limitations on termination [1, 2]. Under such conditions, the relative importance of propagation increases so that the polymer synthesized at the start is not the same as the polymer synthesized later on. At very high monomer conversions and sufficiently low temperatures, propagation is even retained, which is known as the glass effect [3, 4]. Hence, a polymer synthesis product is by default heterogeneous in macromolecular composition or thus distributed [5–7], also bearing in mind a so-called cage effect that is related to diffusional limitations on the initiation step [8–10].

The most known distributed characteristic is the chain length with easily varies between 1 and 10^5 monomer units. This variation is also dynamic due to the aforementioned diffusional limitations or more generally a complex interplay of chemistry and transport phenomena. This means that even at lab scale, the polymerization kinetics are far from constant and therefore strongly time dependent. At any moment, there is competition of diffusion between non-macromolecules (e.g., solvent or monomer) and macromolecules to realize certain reactions, as conceptually illustrated in Figure 10.1 (left) [11].

In this figure, a so-called micro-scale is introduced, which is the scale at which concentrations can be defined. For such micro-scale, an apparent or observed reactivity can be utilized per elementary reaction, which for very fast; in the limit infinitely fast, diffusion becomes the intrinsic or chemical reactivity, and for very slow, diffusion makes the observed/apparent reaction rate diffusion controlled. Specifically, radical termination in chain-growth polymerization displays chain length and monomer conversion dependent rate coefficients [13], reflecting the lower mobility of thus decreased apparent termination reactivity for larger chains in a more viscous medium. For radical polymerization of methyl methacrylate (MMA), this double dependency is shown in Figure 10.1 (right) [12], displaying a very strong decrease in the apparent

Maria Savina Pianesi, Delta Srl, Via Tambroni Armaroli, 2, Montelupone MC 62010, Italy
Tommaso Compagnucci, School of Science and Technology, ChIP Unicam Research Center, University of Camerino, Via Madonna delle Carceri, Camerino, Italy
Dagmar R. D'hooge, Department of Materials, Textiles and Chemical Engineering, Technologiepark 125, Zwijnaarde 9052, Belgium
Mark Bierens, Delta Glass, Deltaweg 4, RX Tholen 4691, the Netherlands

https://doi.org/10.1515/9783111076997-010

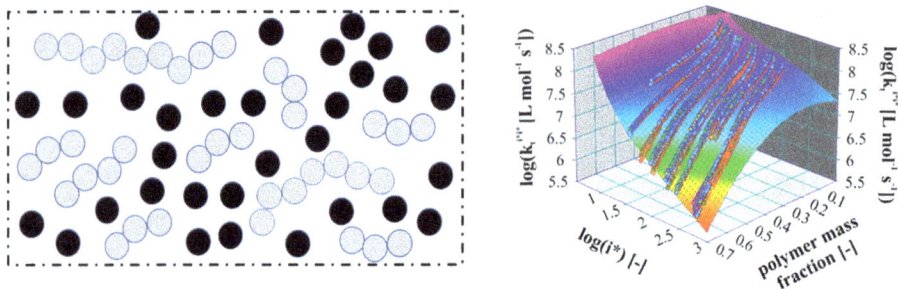

Figure 10.1: Left: The competition of non-macromolecules and macromolecules for diffusion and reaction in polymerization; right: the significant decrease in the apparent termination rate coefficient for radicals of equal length as a function of the chain length (i^*) and the polymer mass fraction. Adopted with permission from Derboven et al. [12] addressing radical polymerization of MMA at 353 K.

termination rate coefficient for termination between radicals of the same chain length with increasing chain length and polymer mass fraction.

A typical vinyl polymerization, such as radical polymerization of MMA, has this termination as its final reaction. On the molecular scale [14, 15], the so-called dead polymer chains are either formed by termination by recombination or disproportionation (see in detail in Chapter 5). The term dead is used to emphasize that these chains are unreactive once formed. In reality, they could be still involved in branching reactions, but for poly(methyl methacrylate) (PMMA) synthesis, the first approximation of these polymer-dependent reactions can be ignored. This is less the case for the related polymer family of polyacrylates in which both short-chain branches and long-chain branches are formed [16, 17].

The typical trajectory to come to these dead polymer chains is a chain-growth based cycle. One first has initiation via a conventional radical initiator delivering initiator radicals, the chain initiation of the latter with a certain efficiency, to subsequently witness propagation with many monomer units being incorporated, to conclude with a termination reaction. In parallel, chain transfer to monomer or to solvent or to a deliberately added chain transfer agent can take place, complexifying the overall kinetic scheme.

One should realize that industrial polymer synthesis generally deals with large reactor volumes (e.g., ton scale) so that besides a micro-scale, as defined in Figure 10.1, one also need to account for a so-called macro-scale. The latter scale is directly linked to the reactor geometry or the industrial synthesis environment. This explains why in the field of polymer reaction engineering, one utilizes the term multiscale modeling [18, 19]. In different regions of the reactor (e.g., a sufficiently large batch reactor with a given stirrer configuration or a mold), one can experience different macro-scale concentrations and temperatures. Within each such region, one has certain specific micro-scale kinetics, further complicating the heterogeneity level of an industrial polymer product. Practically, one can of course still aim at a desired overall or average behavior, and one needs to check if the reactor size and monomer conversion are below or above the threshold with respect to macro-scale variations.

The multiscale character is valid for both the polymerization as well as the industrial repolymerization, the latter relevant if one aims at polymer circularity as in the MMAtwo project addressing the transition from PMMA waste to MMA and then the resynthesis of PMMA. Specific recycled applications such as caravan windows/ sheets and kitchen sinks are discussed in the next chapter.

In what follows, the reader is made aware of challenges that need to be overcome during (re)polymerization. It is first explained that one should be aware of the relevance of the reactor temperature, and the relevance of impurities and the related purification. It is additionally explained that it is worthwhile to perform final part testing. The formulated insights are generally applicable but, to maintain a clear link with the other chapters, they are restricted to polymerization involving acrylate and methacrylate monomers. The main monomer with both monomer families is MMA and n-butyl acrylate (nBuA).

10.2 Challenge of reactor temperature control

Industrial polymer synthesis of vinyl monomers is often conducted at elevated temperature. More importantly, vinyl radical polymerization of (meth)acrylates is a strong exothermic process, as many propagation steps occur on a small time scale releasing significant reaction heat. Hence, temperature spikes are easily encountered, and appropriate cooling is necessary to at least mimic a desired set temperature (target 1) or a set temperature profile (e.g., peak value or pattern of peak values; target 2). In the present section, the first target is highlighted for free radical polymerization (FRP) of nBuA in a batch reactor and the second target for FRP of recycled MMA using a mold configuration.

Figure 10.2 displays a typical reactor-temperature profile for solution FRP of nBuA. The reactor temperature is sufficiently high so that β-scission reactions leading to the formation of macromonomer species become highly relevant [20–22]. In this way, two macromolecular populations are present, one being the conventional dead polymer population and one being the macromonomer population being shorter in chain length. Recent research [15] has shown that these β-scission reactions are occurring with mid-chain radicals (MCRs) formed through backbiting and not chain transfer to polymer. This means that the radical center of the MCR is still very close to the end so that one of the β-scissions delivers a macromonomer of chain length 3.

For solution FRP of nBuA with 2,2′-azobis(2-methylpropionitrile) (AIBN), starting at room temperature, the "$U = 0$" lines in Figure 10.3 highlight how the absence of reactor temperature control leads to an unsafe reactor operation. These lines have been obtained through kinetic Monte Carlo simulations, starting from the principles out-

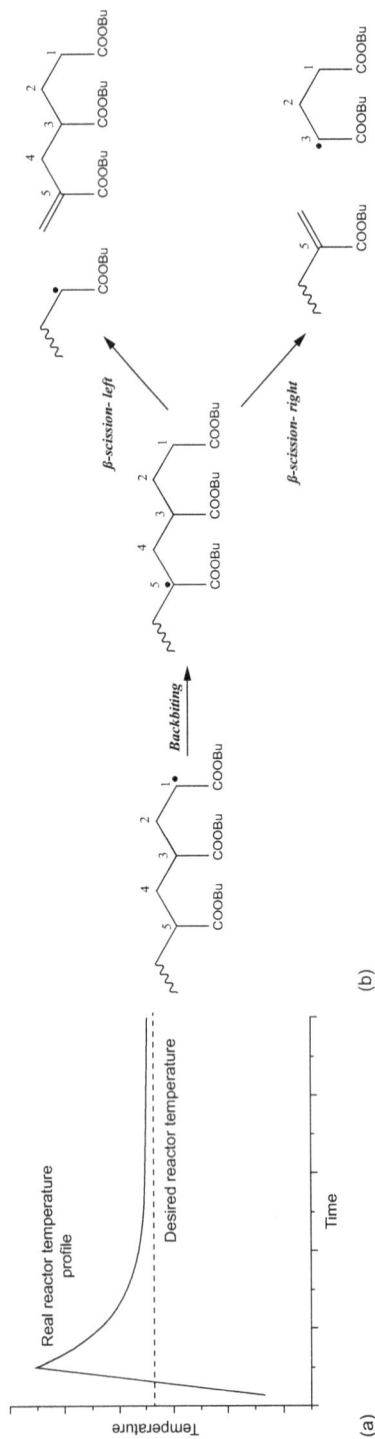

Figure 10.2: (a) Illustration of reactor temperature excursion/spike in industrially applied batch free radical polymerization (FRP); (b) backbiting and β-scission as important side reactions in FRP of acrylates (case of *n*-butyl acrylate), starting from conventional secondary end-chain radicals (ECRs) (adopted with permission from Edeleva et al. [15]).

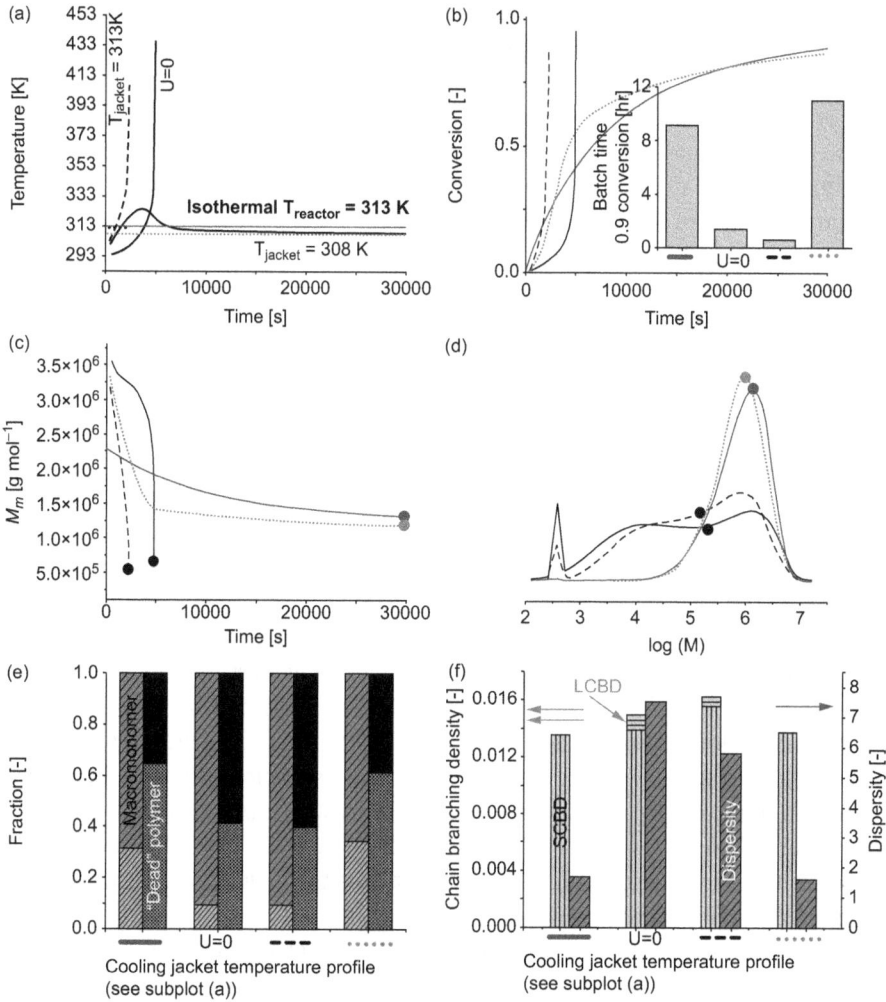

Figure 10.3: Comparing adiabatic and non-adiabatic reactor operations for solution free radical polymerization (FRP) of n-butyl acrylate (nBuA) with $[nBuA]_0 = 3.3$ mol/L; $[AIBN]_0 = 3 \times 10^{-3}$ mol/L and $T_{r,0} = 298$ K. Evolution of (a) reactor temperature (left axis) and jacket temperature (T_{jacket}; right) axis asa function of time, (b) monomer conversion as a function of time as well the final batch time for 0.9 (monomer) conversion, (c) mass average molar mass M_m as a function of time and (d) molar mass distribution (w(log M) format) at a monomer conversion of 0.90; (e) fractions of macromonomer and dead polymer at a monomer conversion of 0.90 with light green/red number fraction and dark green/red: mass fraction, (f) left axis: orange and blue: short- and long-chain branching density at a monomer conversion of 0.90, (f) right axis in dark-purple dispersity at 0.90 monomer conversion; light blue: $T_{jacket} = 313$ K, orange: $T_{jacket} = 308$ K, orange dashed line: $T_{jacket} = 308$ K, black: adiabatic operation with U equal to 0 instead of 100 W/m^2 K (the dimension of U is omitted in the figure), light green: isothermal (theoretical) at 313 K (adopted with permission from Edeleva et al. [15]).

lined in Trigilio et al. [23]. In the absence of cooling, the reactor temperature rapidly increases and reactor explosion is unavoidable. Very high reactor temperatures imply a very strong β-scission so that a very broad molar mass distribution results with a clear spike of the macromonomer product with a chain length 3 (black line in Figure 10.3). This strong relevance of β-scission is also evident with very low mass average molar mass values, high dispersities, high branching levels, and high macromonomer amounts, as illustrated in Figure 10.3.

It is further clear from Figure 10.3 that a sufficiently low jacket temperature is needed to cool down the polymerization process. At first sight, a low temperature of 313 K is not recommended as runaway still takes place. Only by selecting a temperature as low as 308 K, the (theoretical) isothermal case can be mimicked with a limited reactor temperature increase. Note that in this case the average molar masses are higher, the dispersities are lower and the macromonomer amounts lower.

Temperature monitoring can also be utilized for a kind of quality control for the (re)polymerization of MMA in molds, as explored in the MMAtwo project for the production of caravan windows or sheets. The casting process for PMMA sheet production uses glass as a mold. Especially for optical applications, glass is the only material with the smoothest surface compared to other materials. Polymerization of MMA takes place in the mold, ultimately resulting in a high average molar mass polymer.

The mold consists of two toughened glass panels of at least 10 mm thickness that are combined as a pair. The interior is filled with a partially polymerized MMA, a so-called prepolymer that is further polymerized at elevated temperatures. A gasket made of PVC is used as a spacer between the glass and seal to contain the liquid prepolymer in the mold. This prepolymer contains at least 90% MMA and at least 3–10% of polymer which results in an increased viscosity. The viscosity is critical, as very low viscosity leads to leakage of the prepolymer from the mold. In contrast, very high viscosity results in trapped air bubbles in the sheet.

The prepolymer is prepared in a stirred batch reactor with a hot water jacket. In this reactor, the MMA, initiator, UV additives, and color dyes are mixed at 70 °C. The temperature of the mixture increases due to exothermicity of the polymerization. At a certain temperature set point, the water jacket of the reactor is switched over to cooling water. The batch is finally cooled down to ambient temperature and a vacuum is applied on the reactor for 20 min to release dissolved oxygen.

After the molds are filled with prepolymer, 10 to 20 molds are collected on a rack, which is placed into an autoclave. The process time depends on the thickness of the sheet, with the most produced thickness being 3 mm at company Delta Glass. The autoclave is completely filled with hot water, as water is an effective direct source of heat. The heat transfer from hot water to the mold is effective, better, and faster than hot air. The heat of reaction is also released more effectively with water. The process time is only 2 h, starting at 68 °C and ending at 120 °C, reaching a monomer conversion of at least 98%. During several stages, the autoclave is pressurized at 4 bar. This is to suppress the boiling of MMA (101 °C under atmospheric conditions), as due of the gel effect the internal temperature in the mold can rise sharply above this boiling point. Due to the prevailing pressure in the autoclave, the boiling point of MMA is increased, preventing the formation of entrapped gas bubbles in the PMMA sheet. In Figure 10.4, a typical temperature profile of the interior of the mold is demonstrated.

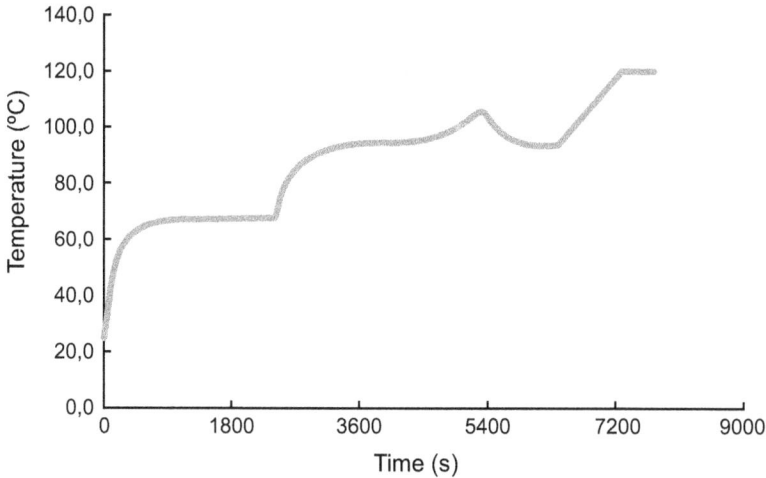

Figure 10.4: A typical temperature profile for polymerizing MMA in a mold in a pressurized autoclave, as relevant in the MMAtwo project.

Sheets with a thickness of 5 mm and more require a longer process time at lower temperatures. The reaction heat is more difficult to release to the water with a higher sheet thickness. From 8 mm thickness and more, retarding agents such as monoterpenes are used to slow down the reaction, especially to suppress the gel effect. The optical stability of a sheet is also more critical at higher thickness. If the temperatures are too high, the sheet will display distortion, which is an optical disturbance in the sheet showing a pattern of lines. Presumably, this pattern is caused by a disturbed refractive index. Distortion will be more visible after thermoforming of the PMMA sheet. Hence, the control of temperature and a controlled release of heat during the polymerization process are critical to guaranty a good quality optical PMMA sheet.

The MMAtwo project has also explored the (re)polymerization of MMA in molds for PMMA-based composites materials, particularly in the production of kitchen sinks by company Delta Plados. In this nonoptical application, the prepolymer is still used as organic viscous medium, with inclusion of both mineral fillers such as quartz, silica, wollastonite, and milled fiber glass in different granulometries, shape, and color, with additives like cross-linkers, pigment, and lubricant. The need of mineral filler is strictly due to the better mechanical properties reached by PMMA composites than pure polymer, required for the specific application. Moreover, fillers lead to a significant reduction of the exothermicity of the polymerization process, assisting the dissipation of reaction heat.

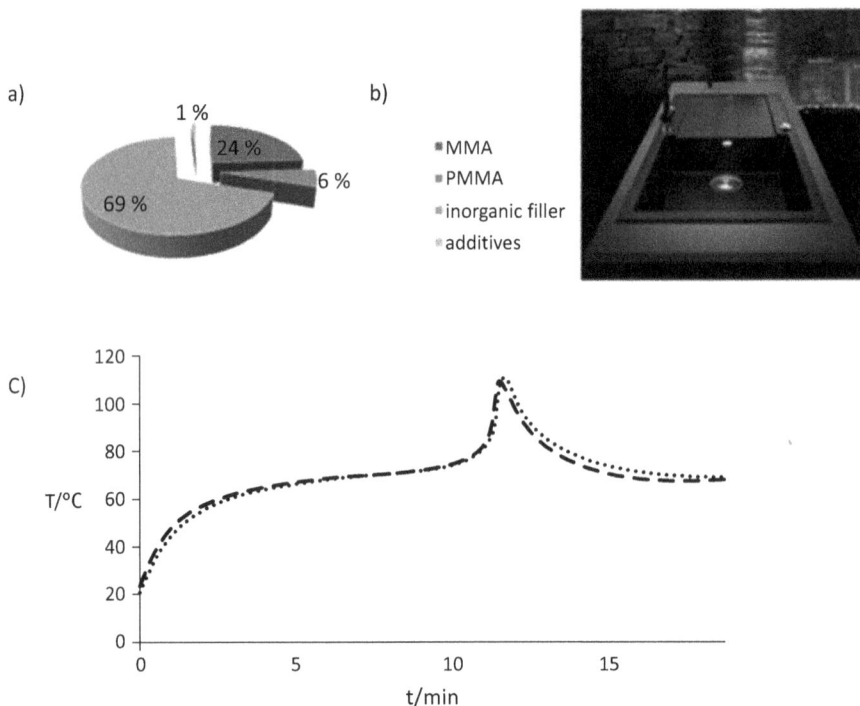

Figure 10.5: (a) The percentage mass composition of a dispersion, used in the production of kitchen sinks; (b) example of PMMA-based composite kitchen sink; (c) a typical temperature profile for PMMA composite polymerization in a mold, as relevant in the MMAtwo project; Dashed line: polymerization curve with reference virgin materials; dotted lines: polymerization curve for recovered MMAtwo materials.

More in detail, the industrial process starts again in a first step from the formation of the prepolymer. However, the Delta Plados technology is based on the dissolution of preformed PMMA beads inside liquid MMA, exploiting the uncommon solubility of the polymer in the monomer. The dissolution is performed in a stirred batch reactor with a hot water jacket. The addition of solid beads is performed under continuous stirring and at room temperature. Then, the reactor temperature is gently raised to 38 °C, keeping this temperature for a period of 1–3 h or in any case until the beads are completely dissolved. The amount and type of PMMA lead to prepolymers with different viscosity and rheological properties.

The second step is the gravity sedimentation of fillers in the prepolymer and homogenization, producing a more viscous dispersion. The mold consists of two pieces mass composition of a dispersion is 70% inorganic fillers and 30% organic materials, as shown in Figure 10.5(a). As a third step, a vacuum is applied on the dispersion, which is then cast inside combined as a pair. The interior is made of nickel and it is designed to be filled by the dispersion to match the kitchen sink shape after the molding. Hot water is necessary both as a source of heat-activating radical initiators and to dissipate the heat generated by the exothermic reaction. After the complete mold filling and pressurization, an important extra heating is applied (water temperature from 30 to 100 °C), leading to the formation of the kitchen sink composite product, with a medium thickness of 9 mm at the bottom and the walls (see dimensions in. Figure 10.5(b)). The cycle time of polymerization (see x-axis in. Figure 10.5(c)) is optimized to control the gravity precipitation of fillers inside the viscous medium during the polymerization, avoiding both unwanted separation of the inorganic and organic phase, typical for a very long process, and a low filler precipitation, if the reaction is excessively fast.

10.3 Challenge of removal of impurities and purification

An important aspect for repolymerization is the evaluation of the relevance of impurities. In the simplest situation, both practically and economically, one strives for direct repolymerization. One of the most pioneering contributions in the field is the one of Achilias [24], who performed pyrolysis experiments to subsequently repolymerize PMMA from the liquid fraction.

Figure 10.6 shows that with such liquid fraction, one can easily repolymerize under bulk conditions and this in a fast manner at the highest temperature of 95 °C [24]. At first sight, one would thus conclude that purification is not needed. However, one should always compare the polymerization behavior with the neat or virgin system, as explored in Figure 10.7(a). It follows from this figure that the variation of polymerization rate R_p, as defined as the monomer conversion derivative, with respect to the monomer conversion is higher utilizing neat MMA. A closer investigation reveals a different shape of the initial R_p values. The latter is made clearer in Figure 10.7(b), plotting the ratio of R_p and the fractional polymer yield $(1 - X)$, with X being the fractional monomer conversion. Upon using neat MMA, one directly jumps to a relatively high value for this ratio [24], whereas for the recycled case, a slow increase is obtained with the initial ratio of the neat case reached only at intermediate monomer conversions.

An explanation for these different variations of the aforementioned ratio lies directly in the presence of side products, as formed during the chemical recycling step [24]. Detailed analysis of the liquid fraction reveals that one has several small amounts of organic compounds such as methyl esters that can act as inhibitors therefore influencing the polymerization rate behavior, as also clear upon gel permeation chromatography analysis. As shown in Figure 10.7(c), at each polymerization temperature, the obtained number or mass average molar mass is lower, which is also reflected in a lower glass transition temperature, therefore complicating the direct

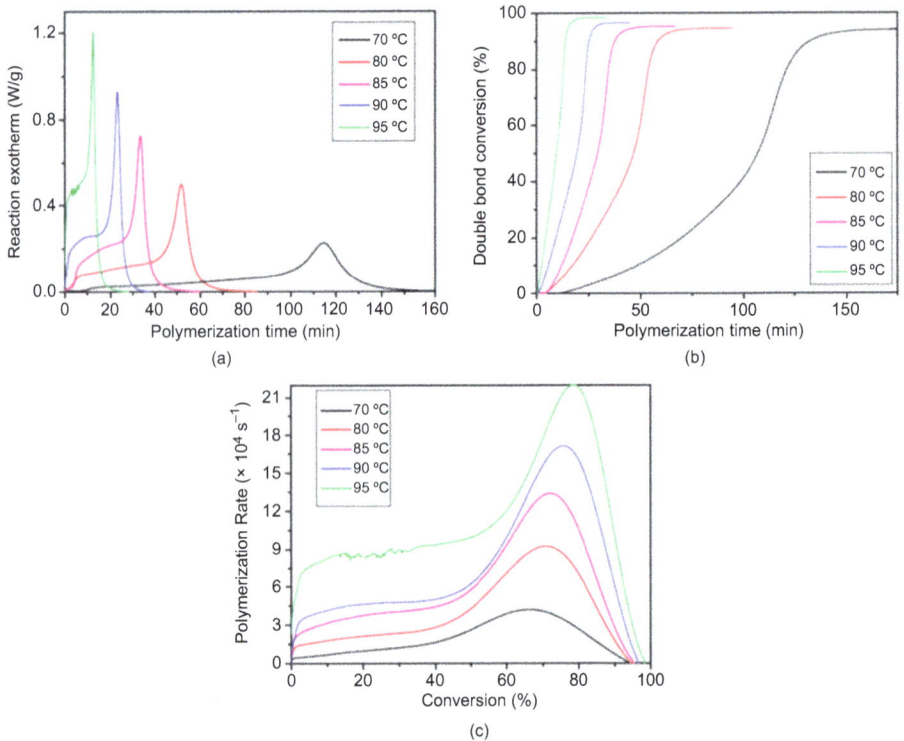

Figure 10.6: (a) Reaction exotherm of pyrolysis oil fraction for variation temperatures ([AIBN]$_0$ = 0.03 mol/L); (b) effect of polymerization temperature on the double-bond conversion (same conditions as in (a)); (c) previous subplot but with polymerization rate (adopted with permission from Achilias [24]).

reuse in application format [24]. A similar conclusion has been made by Godiya et al. [25], who studied both the depolymerization and repolymerization.

Notably this inhibition behavior has also been more fundamentally underpinned, as explained by Achilias [24], considering the work of Odian [26]. If one formally writes down all inhibitor molecules as Z, it has been put forward that under strong inhibiting conditions the polymerization rate R_p (defined as the monomer conversion derivative) becomes:

$$R_p = \frac{dX}{dt} = \frac{k_p(1-X)R_i}{k_z[Z]} \tag{10.1}$$

in which R_i is the initiation rate, k_p is the propagation rate coefficient, and k_z is the lumped rate coefficient for inhibition. Denoting y as the number of radicals stopped through an inhibition reaction, the inhibitor concentration [Z] decreases as a function of time according to [26]:

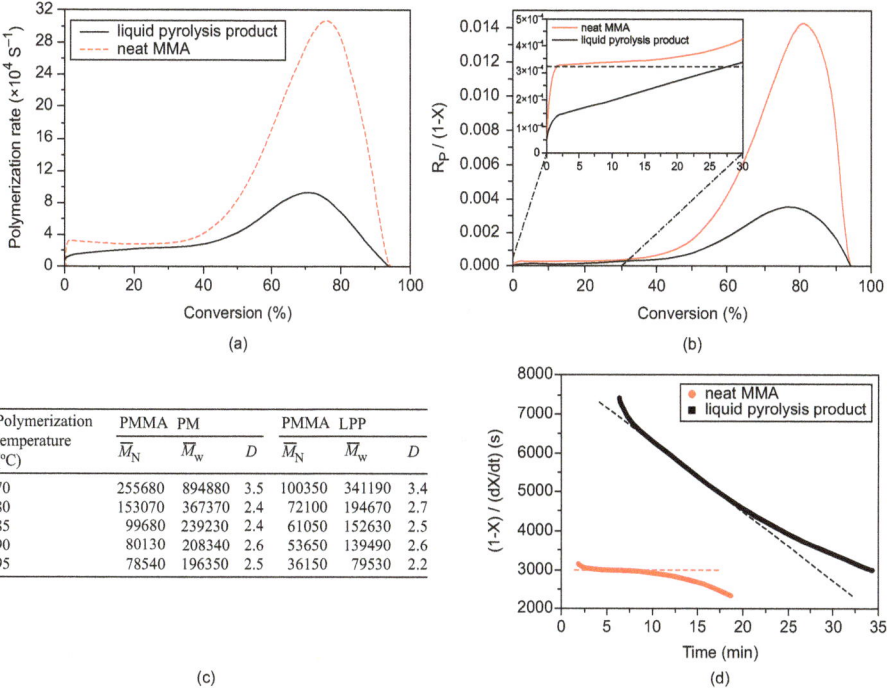

Figure 10.7: (a) Polymerization rate versus conversion of pyrolysis oil fraction compared to polymerization of neat MMA ([AIBN]$_0$ = 0.03 mol/L; polymerization temperature of 80 °C); (b) ratio of polymerization rate (R_p) and 1–X (polymer yield) for the polymerization of neat MMA at 80 °C; (c) number average molar mass and mass average molar mass, dispersity at different polymerization temperatures either from pure MMA (PPMA PM) or from the liquid pyrolysis product (PMMA LPP); and (d) ratio of and 1–X and R_p (dX/dt) during polymerization of neat MMA and the liquid pyrolysis product in order the evaluate the relevance of inhibition. Conversion range of 1–30% and temperature of 80 °C (adopted with permission from Achilias [24]).

$$[Z] = [Z]_0 - \frac{R_i t}{y} \tag{10.2}$$

Upon combining both equations it follows that:

$$\frac{(1-X)}{R_p} = \frac{\frac{k_z}{k_p}[Z]_0}{R_i} - \frac{\frac{k_z}{k_p}t}{y} \tag{10.3}$$

As illustrated in Figure 10.7(d), indeed, for the polymerization considering neat MMA k_z is more or less zero but for the polymerization utilizing the recycled unpurified MMA, this is not the case. There, a clear downward straight line can be visualized, as highlighted by the black extra dashed line. Hence, it is worthwhile to purify MMA after recycling of PMMA, as fully addressed in Chapter 9.

Godiya et al. [25] also focused on the residual monomer content. This was done via a dissolution/reprecipitation technique. Purified PMMA did not contain residual monomer, showing the effectiveness of the technique with purification levels of 100–500 ppm.

10.4 Challenge of final part testing

Ultimately, one strives for the reuse of the recycled monomer in a wide range of applications. Certain applications require certain macroscopic properties and it is thus worthwhile to test the final parts. Here, one can distinguish between thermal and mechanical properties. In the present section, a concise overview is given of such experimental techniques, again mainly focusing on polyacrylics. Emphasis is put in a first step on differential scanning calorimetry (DSC), and then in a second step on tensile and flexural strength testing as well as impact testing.

In DSC, a heating is applied to obtain a physical characteristic such as a glass transition temperature or crystallinity. More in detail, two pans are placed in a chamber that is heated. One pan contains the sample and one pan is a reference pan that can be empty and is employed to monitor the temperature and allow for regulation.

Table 10.1: Data from differential scanning calorimetry (DSC) analysis for blends composed of poly (methyl methacrylate) (PMMA) and ethylene vinyl acetate (EVA) polymer.

Sample (PMMA/EVA)	$T_{g,1}$ (°C)	T_c (°C)	$T_{m,1}$ (°C)	$T_{m,2}$ (°C)	$T_{g,2}$ (°C)	ΔH (J/g)
100(/0)	–	–	–	–	100.6	–
95/5	24.1	50.1	–	64.3	107.8	0.47
90/10	14.9	35.1	46.5	65.2	107.8	3.25
85/15	11.7	36.1	45.3	66.4	109.5	7.71
80/20	8.7	37.7	44.9	69.4	110.8	9.84
(0/)100	−26.9	36.1	47.9	70.8	–	38.3

T_g, glass transition temperature; T_c, crystallization temperature; T_m, melting temperature; ΔH, melting enthalpy. Adopted with permission from Poomalai et al. [27].

For mechanical testing, the preparation of test specimens according to standards is the recommended pathway, including stabilization in a controlled room, meaning a given temperature and humidity content. Examples are dog bone type 1A specimens (ISO 527-2; tensile testing) and unnotched rectangular specimens (ISO 197-1; flexure and Charpy impact test). Practically an important parameter is the deformation rate, and for the impact test, the initial pendulum energy.

Table 10.1 gives typical data as recorded by DSC. These data are taken from the work of Poomalai et al. [27] studying the relevance of blending ethylene vinyl acetate polymer (EVA) with PMMA. For the blends, one does not observe a single glass transition temperature highlighting the presence of two phases. By blending, benefiting from the basic chain configuration of EVA, an increase of the crystallinity is recorded with two melting temperatures, highlighting the presence of syndiotactic and non-syndiotactic segments in the EVA molecules.

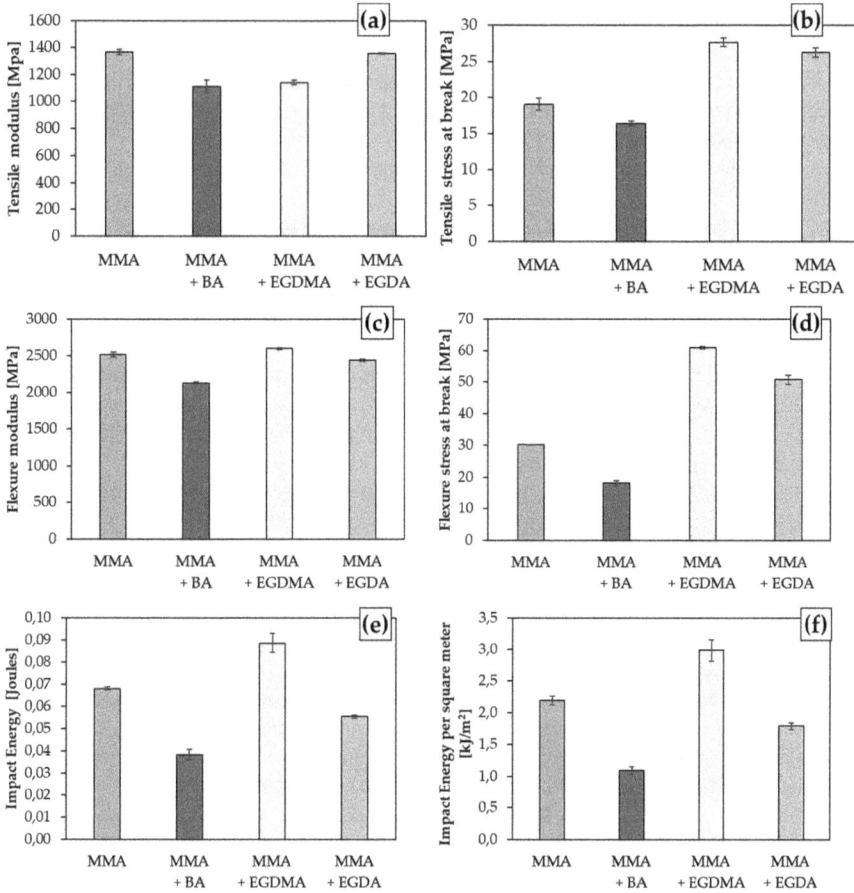

Figure 10.8: Results of mechanical screening testing with (a) tensile modulus in MPa, (b) tensile stress at break in MPa, (c) flexure modulus in MPa, (d) flexure stress at break in MPa, (e) impact Energy in joules, and (f) impact energy per square meter at the impact face. MMA, *n*-butyl acrylate (*n*BuA), ethylene glycol dimethacrylate (EGDMA), and ethylene glycol diacrylate (EGMA) (adopted with permission from Reyes et al. [28]).

Figure 10.8 gives typical results for mechanical testing for casts based on MMA and coformulations with MMA as main comonomer [28]. Besides the additional use of *n*BuA as comonomer, focus is on the use of the cross-linkers ethylene glycol dimethacrylate (EGDMA) and ethylene glycol diacrylate. It follows that a joint optimization of the tensile modulus, tensile stress at break, flexural modulus, flexural stress at break, and impact energy is far from trivial. Globally, the best performance is obtained with the EGDMA cross-linker, highlighting the relevance of coformulation design during (re)polymerization.

References

[1]   De Keer L, Van Steenberge PHM, Reyniers MF, Marin GB, Hungenberg KD, Seda L, et al. A complete understanding of the reaction kinetics for the industrial production process of expandable polystyrene. AIChE Journal 2017, 63, 2043–2059.

[2]   Tulig TJ, Tirrell M. Toward a molecular theory of the Trommsdorff effect. Macromolecules 1981, 14, 1501–1511.

[3]   Achilias DS. A review of modeling of diffusion controlled polymerization reactions. Macromolecular Theory and Simulations 2007, 16, 319–347.

[4]   Delgadillo-Velazquez O, Vivaldo-Lima E, Quintero-Ortega IA, Zhu SP. Effects of diffusion-controlled reactions on atom-transfer radical polymerization. AIChE Journal 2002, 48, 2597–2608.

[5]   De Keer L, Kilic KI, Van Steenberge PHM, Daelemans L, Kodura D, Frisch H, et al. Computational prediction of the molecular configuration of three-dimensional network polymers. Nature Materials 2021, 20, 1422–1430.

[6]   Gody G, Zetterlund PB, Perrier S, Harrisson S. The limits of precision monomer placement in chain growth polymerization. Nature Communications 2016, 7, 10514.

[7]   Zhou YN, Luo ZH. State-of-the-art and progress in method of moments for the model-based reversible-deactivation radical polymerization. Macromolecular Reaction Engineering 2016, 10, 516–534.

[8]   Achilias DS, Kiparissides C. Development of a general mathematical framework for modeling diffusion-controlled free-radical polymerization reactions. Macromolecules 1992, 25, 3739–3750.

[9]   Wieme J, Reyniers MF, Marin GB. Initiator efficiency modeling for vinyl chloride suspension polymerization. Chemical Engineering Journal 2009, 154, 203–210.

[10]   Kurdikar DL, Peppas NA. Method of determination of initiator efficiency – Application to UV polymerizations using 2,2-dimethoxy-2-phenylacetophenone. Macromolecules 1994, 27, 733–738.

[11]   D'hooge DR, Reyniers MF, Marin GB. The crucial role of diffusional limitations in controlled radical polymerization. Macromolecular Reaction Engineering 2013, 7, 362–379.

[12]   Derboven P, D'Hooge DR, Reyniers MF, Marin GB, Barner-Kowollik C. The long and the short of radical polymerization. Macromolecules 2015, 48, 492–501.

[13]   Barner-Kowollik C, Russell GT. Chain-length-dependent termination in radical polymerization: Subtle revolution in tackling a long-standing challenge. Progress in Polymer Science 2009, 34, 1211–1259.

[14]   Edeleva M, Van Steenberge PHM, Sabbe MK, D'Hooge DR. Connecting gas-phase computational chemistry to condensed phase kinetic modeling: The state-of-the-art. Polymers 2021, 13, 3027–3066.

[15]   Edeleva M, Marien YW, Van Steenberge PHM, D'Hooge DR. Jacket temperature regulation allowing well-defined non-adiabatic lab-scale solution free radical polymerization of acrylates. Reaction Chemistry & Engineering 2021, 6, 1053–1069.

[16]   Junkers T, Barner-Kowollik C. The role of mid-chain radicals in acrylate free radical polymerization: Branching and scission. Journal of Polymer Science Part a-Polymer Chemistry 2008, 46, 7585–7605.

[17]   Plessis C, Arzamendi G, Alberdi JM, van Herk AM, Leiza JR, Asua JM. Evidence of branching in poly (butyl acrylate) produced in pulsed-laser polymerization experiments. Macromolecular Rapid Communications 2003, 24, 173–177.

[18]   D'hooge DR, Van Steenberge PHM, Reyniers MF, Marin GB. The strength of multi-scale modeling to unveil the complexity of radical polymerization. Progress in Polymer Science 2016, 58, 59–89.

[19]   Xie L, Luo Z-H. Multiscale computational fluid dynamics–population balance model coupled system of atom transfer radical suspension polymerization in stirred tank reactors. Industrial & Engineering Chemistry Research 2017, 56, 4690–4702.

[20]   Marien YW, Van Steenberge PHM, Barner-Kowollik C, Reyniers MF, Marin GB, D'hooge DR. Kinetic monte carlo modeling extracts information on chain initiation and termination from complete PLP-SEC traces. Macromolecules 2017, 50, 1371–1385.

[21]   Vandenbergh J, Junkers T. Macromonomers from AGET activation of poly(n-butyl acrylate) precursors: Radical transfer pathways and midchain radical migration. Macromolecules 2012, 45, 6850–6856.

[22]   Reyes Y, Asua JM. Revisiting chain transfer to polymer and branching in controlled radical polymerization of butyl acrylate. Macromolecular Rapid Communications 2011, 32, 63–67.

[23]   Trigilio AD, Marien YW, Van Steenberge PHM, D'Hooge DR. Gillespie-driven kinetic Monte Carlo algorithms to model events for bulk or solution (bio)chemical systems containing elemental and distributed species. Industrial & Engineering Chemistry Research 2020, 59, 18357–18386.

[24]   Achilias DS. Chemical recycling of poly(methyl methacrylate) by pyrolysis. Potential use of the liquid fraction as a raw material for the reproduction of the polymer. European Polymer Journal 2007, 43, 2564–2575.

[25]   Godiya CB, Gabrielli S, Materazzi S, Pianesi MS, Stefanini N, Marcantoni E. Depolymerization of waste poly(methyl methacrylate) scraps and purification of depolymerized products. Journal of Environmental Management 2019, 231, 1012–1020.

[26]   Radical chain polymerization. In Principles of Polymerization. John Wiley & Sons, Inc., 2004, 198–349. Online ISBN:9780471478751. 10.1002/047147875X.

[27]   Poomalai P, Ramaraj BS. Thermal and mechanical properties of poly (methylmethacrylate) and ethylene vinyl acetate copolymer blends. Journal of Applied Polymer Science 2007, 106, 684–691.

[28]   Reyes P, Edeleva M, D'hooge DR, Cardon L, Cornillie P. Combining chromatographic, rheological, and mechanical analysis to study the manufacturing potential of acrylic blends into polyacrylic casts. Materials 2021, 14, 6939.

Mark Bierens, Jean-François Devaux, Clément Lemenu,
Tommaso Compagnucci, Juanjo Bermejo, Pascal Lakeman,
Maria Savina Pianesi, and Philippe De Groote

11 Applications of regenerated MMA

11.1 Introduction

Poly(methyl methacrylate), PMMA, is widely used for its interesting mix of macroscopic product properties, including high optical clarity and resistance to UV, chemicals, and abrasion. As a result, PMMA is used for a variety of applications, such as transportation (e.g., caravan windows, automotive rear light covers and windshields, and also blends with ABS for automotive interior cladding), construction (e.g., door and window profiles, facades, and skylights, often as cap layer on materials such as polycarbonate), sanitary and kitchen equipment (e.g., bathtubs made with PMMA capped ABS sheet and solid surface baths and sinks), visual communication (e.g., signage, museum casings and exhibition booths), and electronics (e.g., diffuser sheets in LCD screens and monitors) [1–4].

Currently, PMMA is mainly synthesized via free radical polymerization of virgin thus fossil-based methyl methacrylate (MMA). In Europe, the most used process for the production of virgin MMA is the acetone cyanohydrin route [5, 6], which consists of three steps, with methane, ammonia, acetone, methanol, and sulfuric acid (which can be recycled) as reagents, as illustrated in Figure 11.1. In the first step, hydrogen cyanide is produced from methane and ammonia via the Andrussow or Degussa process [7, 8]. In the second step, acetone cyanohydrin is produced from acetone and the hydrogen cyanide produced in the first step. In the third and final step, the acetone cyanohydrin produced in the second step is reacted with an excess of sulfuric acid, forming methacrylamide sulfate. The latter compound is subsequently converted to MMA and ammonium sulfate by bringing it into contact with an excess of aqueous methanol. The sulfuric acid can be recycled or neutralized using ammonia, leading to the coproduct ammonium sulfate, which can be sold to the fertilizer market [6].

As an alternative to polymerizing virgin MMA, one can repolymerize (purified) regenerated MMA (rMMA) into regenerated PMMA (rPMMA), as obtained via thermo-

Mark Bierens, Delta Glass, Deltaweg 4, Tholen 4691RX, the Netherlands
Jean-François Devaux, Arkema France, Rue Henri Moissan, Pierre Bénite F-69491, France
Clément Lemenu, Philippe De Groote, Certech, Rue Jules Bordet 45, Seneffe B-7180, Belgium
Maria Savina Pianesi, Delta Srl, Via Tambroni Armaroli, 2, Montelupone MC 62010, Italy
Tommaso Compagnucci, School of Science and Technology, ChIP Unicam Research Center, University of Camerino, Via Madonna delle Carceri, Camerino, Italy
Juanjo Bermejo, ProCoat, Avda de la Industria 4, Castellgalí 08297, Spain
Pascal Lakeman, Trinseo, Innovatieweg 14, Hoek, the Netherlands

https://doi.org/10.1515/9783111076997-011

chemical recycling of PMMA waste (Chapters 4–10), allowing to reduce the environmental impact of the PMMA end-product (Chapter 12). In the MMAtwo project, several potential applications of rPMMA have been successfully investigated, including cast sheets for optical applications, composites for nonoptical applications, and coatings for surface treatments. In this chapter, focus is on three exemplary applications, namely (i) cast sheet caravan windows as optical application, (ii) composites for wind energy, and (iii) kitchen sinks as two nonoptical applications.

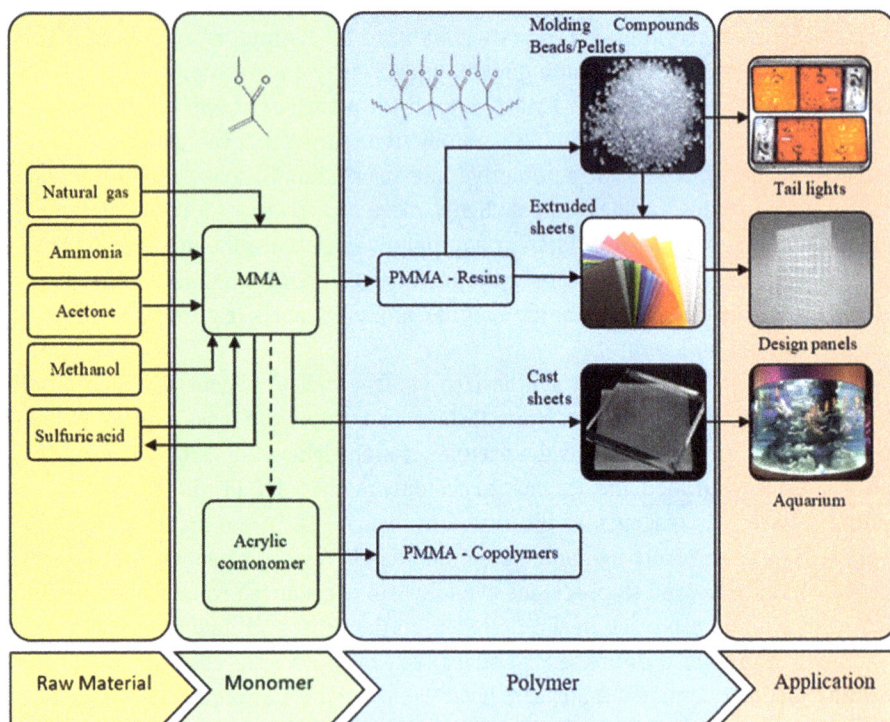

Figure 11.1: Applications of PMMA and typical production routes considering virgin thus fossil-based MMA. In this chapter, focus is on replacing virgin MMA with regenerated MMA (rMMA) (figure reproduced from Moens et al. [4] with permission from MDPI).

11.2 rPMMA cast sheets for optical applications

As already described in detail in Section 10.2 of Chapter 10, rPMMA cast sheets can be produced by polymerizing an rMMA solution containing prepolymer in a mold consisting of two toughened glass panels which are placed into an autoclave. In the MMAtwo project, the application potential has been illustrated considering caravan windows as final product. The rPMMA cast sheets have been produced at Delta Glass, considering several purified rMMAs as obtained by extruder-assisted depolymeriza-tion of different PMMA feedstocks. The rPMMA cast sheets have been subjected to ac-celerated weathering/aging tests with XenoBeta + according to the ISO 11341 norm. As illustrated in Figure 11.2(a), considering a total test duration of 10 500 h, which corre-sponds to 11 years under outside conditions, and considering several rMMAs as well as a virgin MMA reference, the color evolution is satisfactory for all samples. This is demon-strated by all curves being below the red line which indicates the maximum ΔE value. Also the yellow index, according to ASTM E313, and transmission are satisfactory, as demonstrated by all curves being below the maximum allowable yellow index (red line in Figure 11.2(b)) and above the minimum transmission value (red line in Figure 11.2(c)). It can be observed that the yellowness disappears over time (Figure 11.2(b)); the same trend is observed for ΔE (Figure 11.2(a)). The transmission is observed to remain constant over time (Figure 11.2(c)). Overall, it can be concluded that the trend for rPMMA is analo-gous to virgin PMMA, thus demonstrating the application potential of the considered cast rPMMA sheets for outdoor applications.

In addition to satisfactory optical properties over a long period of time upon expo-sure to outdoor conditions, automotive applications such as caravan windows require a sufficiently low odor of the cast sheets. This has been investigated in depth using the field of odors methodology (Champ des odeurs®) and dynamic olfactometry, of which the principle has been described in Chapter 9. Four samples have been considered: (i) a PMMA sheet obtained by polymerizing virgin MMA, (ii) a commercial rPMMA sheet, (iii) a rPMMA sheet obtained by polymerizing PIL1, that is, a purified rMMA stream ob-tained from extruder-assisted depolymerization of post-production clear transparent cast PMMA sheets, and (iv) a rPMMA sheet obtained by polymerizing MIX1, that is, a purified rMMA stream obtained from extruder-assisted depolymerization of a stream consisting of post-production mixed color injection PMMA grades from car lights, end-of-life (EOL) PMMA from LCD scraps, post-production white extrusion PMMA sheets, and post-production PMMA dust with polyethylene contamination.

The results of the odor assessment at room temperature are shown in Figure 11.3(a) (global intensity) and Figure 11.3(b) (field of odors). It follows that only the rPMMA cast sheet based on MIX1 releases an odor at room temperature, with an intensity of 1.7, which turns out to be exclusively an acrid note. All other materials have zero intensity odor at room temperature, including the commercial rPMMA material. It can be ob-served that all materials (including the cast sheet based on MIX1) qualify for automotive interior application, which requires a global intensity lower than 3, which is indeed achieved for all materials.

A further investigation at elevated temperatures is performed via dynamic olfac-tometry. Samples were conditioned for 2 h at 40 °C in order to generate the gaseous phase for evaluation. The results are shown in Figure 11.4. It is important to recall that a difference between olfactometric results is considered as significant in case at least a factor 2 is observed between 2 evaluations. It follows that the PMMA based on virgin MMA shows a zero odor concentration. Interestingly, the odor concentration of the rPMMA based on PIL1 (100 odor units) and MIX1 (350 odor units) are significantly lower than the odor concentration released by the commercial rPMMA sheet (close to 800 units), which expands the application potential of these rPMMA sheets.

11.3 Nonoptical applications

11.3.1 Composites for wind turbine applications

A key challenge for more sustainable energy production via wind is to increase the recyclability of wind turbine blades. Recycling of current wind turbine blades based on thermoset epoxy or polyester resins is very challenging, with typically a low-quality recyclate being obtained [9]. As a result, landfilling of wind turbine blades is often applied, although this practice has already been banned in several European countries. EOL blades can also be incinerated in cement kilns, but this is only a waste-to-energy treatment.

A way to overcome these challenges is to replace thermoset resins by thermoplastic resins. Arkema introduced in 2012 a MMA-based liquid resin for the manufacturing of thermoplastic composites [10]. A full product range has been developed under the trade name Elium® and more than 40 patent families have been issued. These liquid resins are used in many applications such as wind turbine blades, marine, construction (e.g., rebars for concrete), transportation (hydrogen tank or automotive parts), and consumer goods. These resins have the advantages of processability with low viscosity (50 to 1,500 mPa.s), adjustable reactivity (from 2 min to 3 h), and are polymerized in situ at desired temperature (from room temperature up to 120 °C). The current status is that the mechanical properties of these new systems are similar or even better compared to standard continuous fiber-reinforced thermoset resins [11]. These thermoplastic matrix composites could offer important advantages in cycle time reduction, assembly by welding, repair, and recyclability [12]. The main recycling options can be broadly divided into mechanical (by thermoforming compounding or hot compression), physico-chemical (by dissolution, which is a process unique to thermoplastic matrices, allowing recovery of both the polymer matrix and full-length fibers), and chemical (by thermal depolymerization) methodologies.

A demo-scale thermoplastic composite wind turbine blade of 13 m has been made using Elium® and has been compared to a thermoset epoxy blade of identical geometry [13]. Notably, the structural performance of both blades was similar, with the thermoplastic blade even leading to improved structural damping. The production of a commercial-scale 62 m Elium® wind blade was announced early 2022 [14].

MMA-based thermoplastic resins are thus promising for wind turbine applications. To further increase the sustainability of these wind turbines, the use of rMMA instead of virgin MMA has been investigated in the MMAtwo project. Seven purified

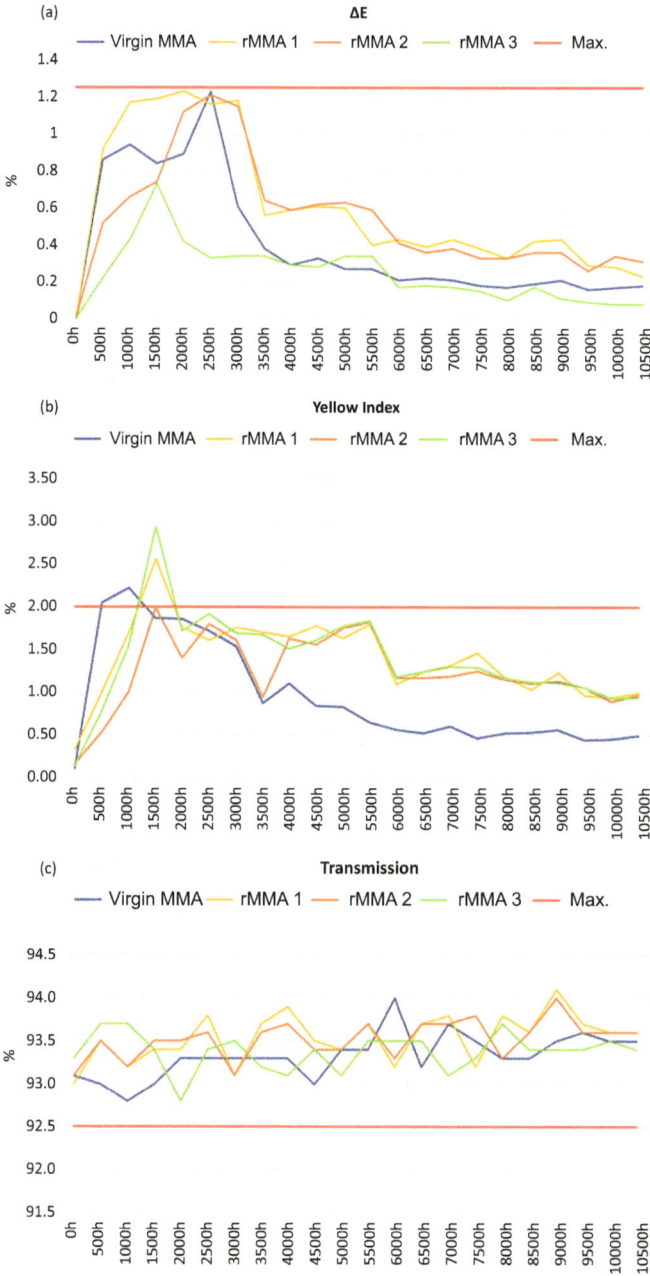

Figure 11.2: Evaluation of optical properties of cast rPMMA sheets during accelerated weathering/ aging tests (10,500 h = 11 years under outside conditions), considering (a) ΔE, (b) yellow index, and (c) transmission value. All sheets are satisfactory (ΔE and yellow index below the red line in (a) and (b); transmission value above the red line in (c)). Virgin and regenerated materials display similar optical properties.

(a)

(b)

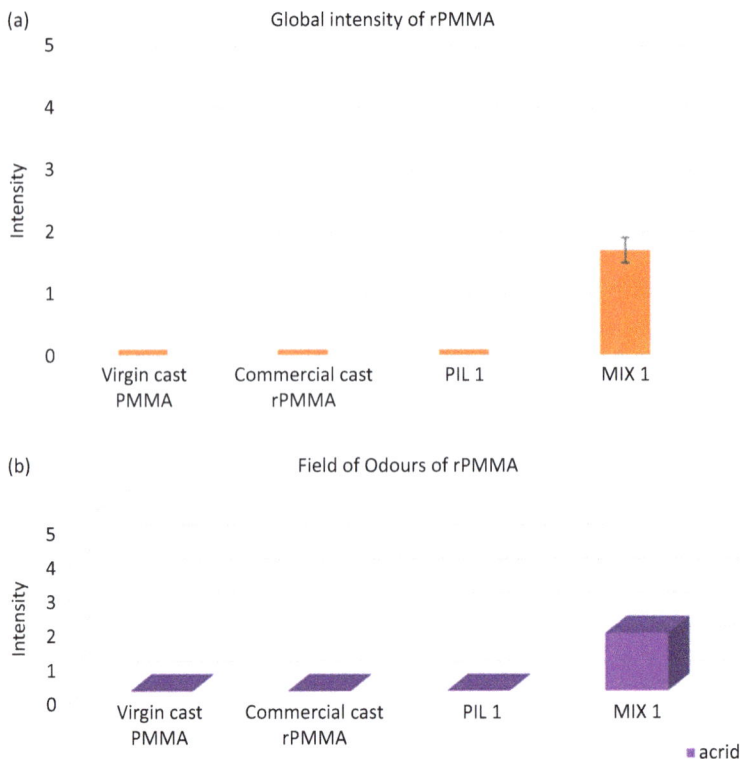

Figure 11.3: Odor assessment of cast optical rPMMA samples (typical panel member deviation = 0.5 intensity unit, panel consisting of three persons): (a) global intensity, (b) field of odors®.

rMMAs have been obtained by extruder-assisted depolymerization of various post-production (PP) or EOL PMMA: PP clear transparent cast (sample A), PP red car lights and EOL LCD screens (sample B), EOL LCD screens and dust of extrusion PMMA cutting (sample C), dust of extrusion cutting and PP white extrusion cast (sample D), PP mixture of cast sheet and extrusion grade (sample E), PP clear extrusion sheet (sample F), and PP smokey cast sheet and PP sanitary white cast sheet (sample G). These purified rMMAs have then been formulated to make liquid rElium® that has been used to prepare 40 × 40 cm glass fiber rElium® composites via vacuum-assisted resin infusion (VARI) (Figure 11.5(a)). For each considered rMMA, the thickness, fiber volume fraction, and flexural stress and modulus have been measured for three test specimens. The results shown in Figure 11.5(b–d) demonstrate that no statistical differences can be observed between the different rMMAs and the virgin MMA reference.

This study successfully demonstrates the potential of using rMMA in the production of liquid thermoplastic resin used in the manufacture of composites by VARI. It constitutes a major technological innovation for the wind energy sector since it paves the way to the use of recycled material for the manufacturing of wind turbine blades.

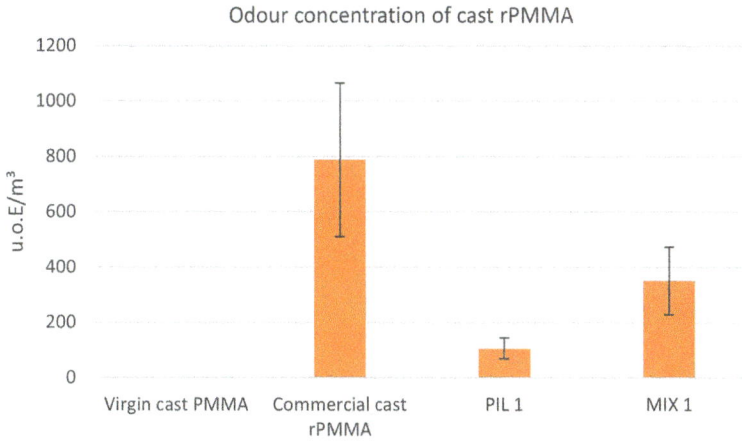

Figure 11.4: Dynamic olfactometry results of cast optical rPMMA samples. Samples have been conditioned for 2 h at 40 °C.

11.3.2 Composites for kitchen sink applications

A second application of rPMMA-based composites investigated in the MMAtwo project is an acrylic composite kitchen sink. As already explained in detail in Section 5.2 in Chapter 10, PMMA-based composite kitchen sinks, sometimes also referred to as "solid surfaces," are produced at Delta Plados via a three-step procedure. In a first step, PMMA beads are dissolved in liquid MMA. In a second step, inorganic fillers are added to the prepolymer containing MMA solution, leading to a dispersion of ca. 70/30 inorganic/organic content. In a third step, a vacuum is put on the dispersion, which is then cast inside a mold with a hot water jacket. After the mold is completely filled and pressurized, extra heating is applied leading to the formation of the composite kitchen sink.

Notably, in the MMAtwo project, the regenerated material content of composite kitchen sinks has been maximized by considering rMMA and rPMMA beads as obtained via extruder-assisted depolymerization, and commercially available regenerated milled glass fibers as component of the mineral filler mix. Physical, chemical, and mechanical properties of the acrylic composite kitchen sinks as listed in Table 11.1 have been evaluated and compared to standard products based on virgin materials.

It is reasonable to assume that the physical–mechanical properties like abrasion resistance, scratch resistance, Rockwell hardness, and Charpy resilience are more linkable to the inorganic components, while the polymerization peak temperature, glass transition temperature (T_g) and the resistance to chemicals, dry heat, and thermal shock are more related to the organic part. Consequently, the applicability of rMMA/rPMMA for composite kitchen sink applications has been evaluated by manufacturing composites with a prepolymer dispersion in which MMA and PMMA

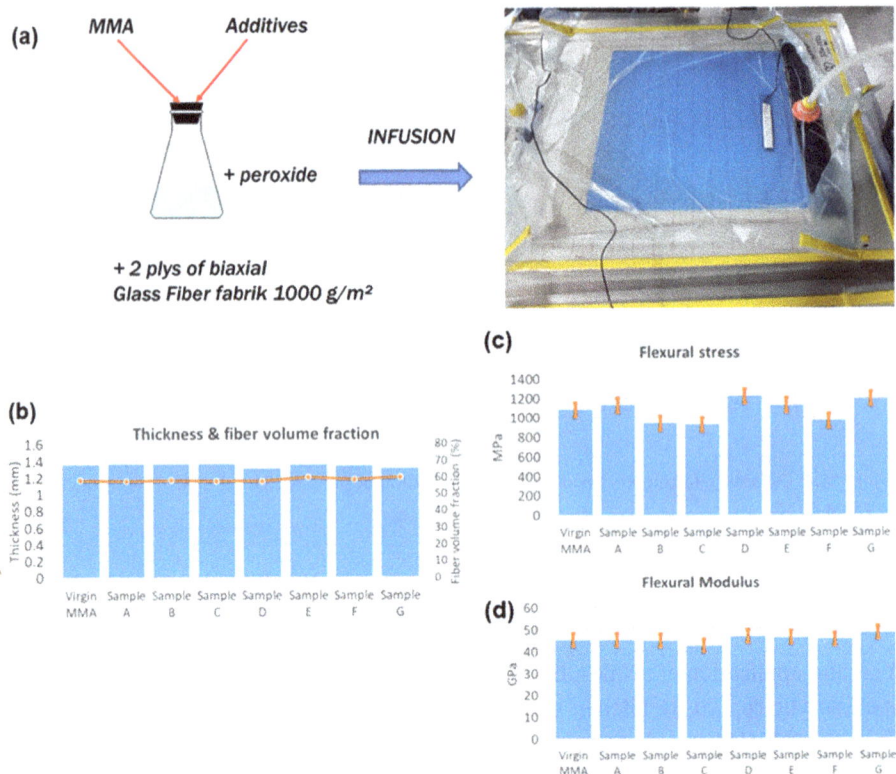

Figure 11.5: Nonoptical rPMMA-based composites: (a) preparation of composites via lab-scale infusion polymerization of rMMA; (b) thickness and fiber volume fraction; (c) flexural strength; and (d) flexural modulus. Results are averages of measurement of three samples. No statistical differences are obtained between virgin and regenerated samples.

were fully replaced by rMMA and rPMMA. Interestingly, the resistance to chemicals, dry heat, and thermal shock, studied in terms of discoloration (ΔE before and after the test), was highly satisfactory and comparable to virgin standards. Moreover, T_g (Figure 11.6) and the peak polymerization time (Figure 10.5 (c)) are in accordance with virgin standards.

In addition to physical, chemical, and mechanical properties, the odor impact has been investigated using the field of odors® methodology. Two formulations have been considered, namely a commercial reference formulation, as well as a formulation made from rMMA PIL1 and an extrusion post-production rPMMA recyclate identified as 38-P15, both processed as lab plates and kitchen sinks samples. The results are given in Figure 11.7(a) (global intensity) and Figure 11.7(b) (field of odors). It is observed that the lab plates exhibit a higher odor intensity than their kitchen sink counterparts, which is due to differences in the respective processes. Moreover, it is observed that the kitchen sink reference has virtually no odor (global intensity zero),

Table 11.1: Functional requirements for rPMMA-based composite kitchen sinks.

Functional requirements for kitchen sinks (UNI EN 13310:2018)	
Evaluation criteria	**Condition to be matched**
Dry heat resistance	No color should be observed
Chemicals resistance	No color should be observed
Thermal shock resistance	No break after the test
Scratch resistance (20 N)	Scratch size should be < 200 μm*
Abrasion resistance	Material loss should be < 100 mg*
Mechanical properties of composite (UNI EN ISO 2039-2: 2001;UNI EN ISO 179-1:2010; UNI EN ISO 178:2019)	
Evaluation criteria	Condition to be matched
Rockwell hardness	> 80 HRM*
Unnotched charpy impact strength	Resilience should be > 3 KJ/m^2*
Flexural test	Elastic modulus should be > 8,000 N/mm^2*
Study of the time and the temperature of the polymerization peak	
Evaluation criteria	Condition to be matched
Polymerization peak	Comparable to delta plados standards
Thermal analysis by DSC	
Evaluation criteria	Condition to be matched
Glass transition temperature (T_g)	>100 °C*
Aesthetic evaluation of the final kitchen sink	
Evaluation criteria	Condition to be matched
Aesthetic judgment	Absence of defects

while the formulation based on rMMA and rPMMA is higher with a value of 1.5. This value is still below the currently considered limit at 3.0, which is also used for automotive interior applications. The intensity of the acrid note as determined by the field of odors® follows more or less the global intensity. Hence, this note is dominating the smell.

Another pending point investigated in the MMAtwo project is the evaluation of the needed rMMA purity. For optical applications, a very high purity grade (e.g., 99.8%) is typically needed, while for composites like kitchen sinks, a lower (although still sufficiently high) purity grade monomer can be used without compromising the

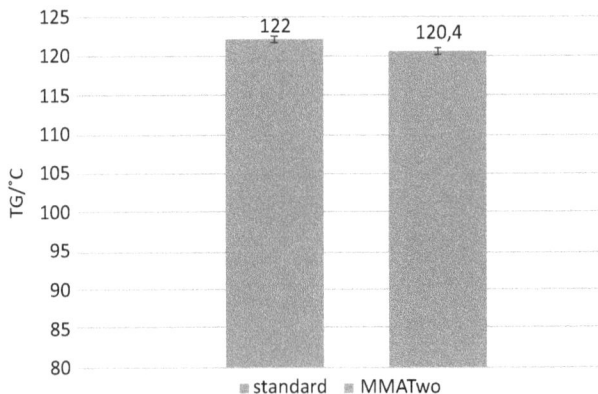

Figure 11.6: Glass transition temperature of a reference and rPMMA-based composite kitchen sink.

final product quality. As a less pure rMMA requires a milder purification of the depolymerized crude, resulting in cost savings, three rMMA purity grades have been investigated for the targeted application in this subsection.

In Figure 11.8 the polymerization peak of three samples in which the only difference is the rMMA purity grade (99.8% vs. 99.5% vs. 99%) is shown. The impurities in the monomer are principally methyl isobutyrate, methyl acrylate, and ethyl acrylate, complemented with traces of in for example methanol. From Figure 11.8, it is clear that a less pure monomer causes a polymerization peak at higher reaction times. However, the time difference is not so large (ca. 30 s), which one could try to adjust, for example, by using different additives. A further evaluation with lower rMMA purities could thus be considered, bearing in mind that lower rMMA qualities will likely lead to higher odor emissions, and thus at some point an optimum configuration needs to be selected.

11.4 Conclusions

rMMA can be used as raw material for both optical as well as nonoptical applications. Optical cast rPMMA sheets obtained in the MMAtwo project by polymerizing different rMMAs from extruder-assisted depolymerization showed fully comparable optical properties as their virgin reference during accelerated weathering/aging tests. Moreover, cast rPMMA sheets with a sufficiently low global odor concentration complying with automotive indoor applications have been obtained by polymerizing an rMMA steam from extruder-assisted depolymerization of (i) post-production clear transparent cast PMMA sheets (PIL1) and (ii) a mix of post-production mixed color injection PMMA grades from car lights, EOL PMMA from LCD scraps, post-production white ex-

(a)

Global intensity of rPMMA

(b)

Field of Odours® of rPMMA

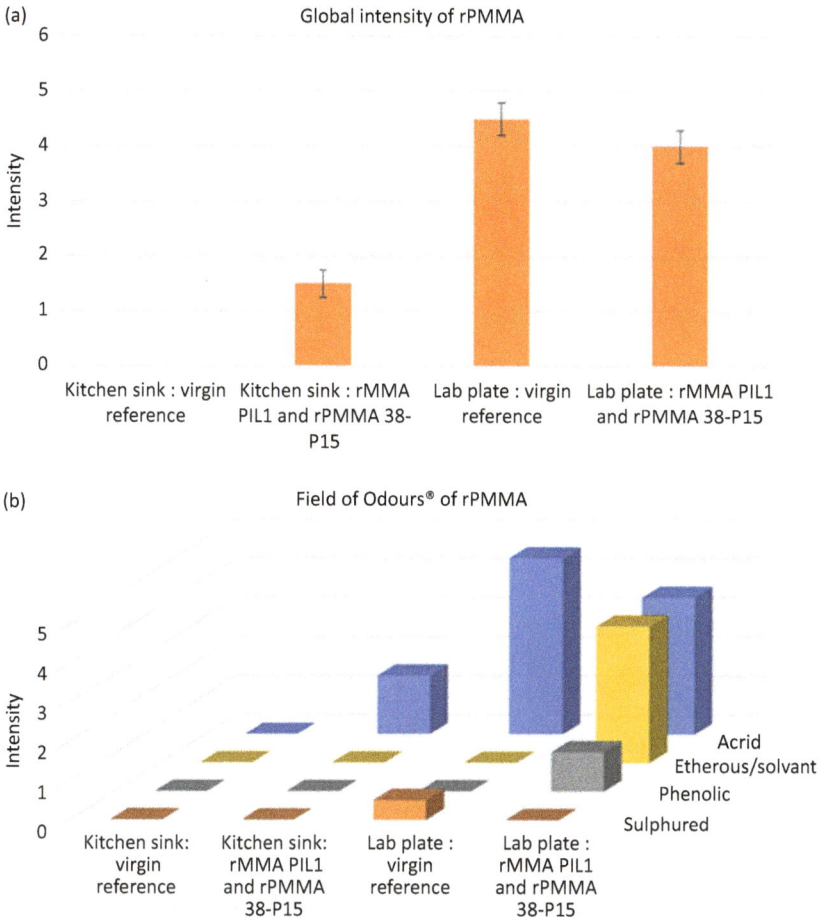

Figure 11.7: Odor assessment of acrylic kitchen sink composite samples (typical panel member deviation = 0.5 intensity unit, panel consisting of three persons): (a) global intensity; (b) field of odors®.

trusion PMMA sheets, and post-production PMMA dust with polyethylene contamination (MIX1). In dynamic olfactometry analyses with sample conditioning at 40 °C, a slightly higher odor concentration was measured with MIX1 compared to a virgin MMA reference, although still sufficiently low and significantly lower compared to a commercial rPMMA reference.

rPMMA composites have been investigated in detail for kitchen sink applications. It has been demonstrated that the MMA and PMMA in current commercial recipes can be fully replaced by rMMA and rPMMA, while maintaining all physical, chemical, and mechanical requirements, and resulting in a sufficiently low odor concentration for indoor applications. It has been further demonstrated that rMMA thermoplastic

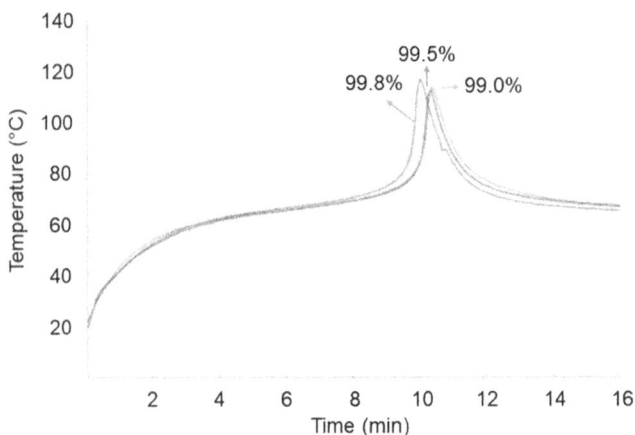

Figure 11.8: Effect of rMMA purity on the polymerization peak.

infusionable resins are promising to increase the sustainability of wind turbine blades.

From the interesting results reported in this chapter for the considered exemplary applications, it can be concluded that there is a very large application potential for rMMA. Extra applications can be here coatings for corrosion protection, as currently being further explored within the MMAtwo consortium.

References

[1] The Methacrylates Sector Group. PMMA Science: Types and Production. 2018. (Accessed 01/04/2022, at https://www.pmma-online.eu/pmma-science/types-and-production/.)

[2] Ali U, Karim KJBA, Buang NA. A review of the properties and applications of poly (methyl methacrylate)(PMMA). Polymer Reviews 2015, 55, 678–705.

[3] Albrecht K, Stickler M, Rhein T. Polymethacrylates. Ullmann's Encyclopedia of Industrial Chemistry 2013.

[4] Moens EKC, De Smit K, Marien YW, Trigilio AD, Van Steenberge PHM, Van Geem KM, et al. Progress in reaction mechanisms and reactor technologies for thermochemical recycling of poly(methyl methacrylate). Polymers 2020, 12.

[5] CEFIC. Methyl Methacrylate (MMA): Eco-profiles and Environmental Product Declarations of the European Plastics Manufacturers, Brussels, 2013, 26.

[6] Mahboub MJD, Dubois J-L, Cavani F, Rostamizadeh M, Patience GS. Catalysis for the synthesis of methacrylic acid and methyl methacrylate. Chemical Society Reviews 2018, 47, 7703–7738.

[7] Wang Z. Andrussow process. Comprehensive Organic Name Reactions and Reagents 2010, 1, 80–82.

[8] Gail E, Gos S, Kulzer R, Lorösch J, Rubo A, Sauer M, et al. Cyano compounds, inorganic. Ullmann's Encyclopedia of Industrial Chemistry 2000, 10, 38.

[9] Ramirez-Tejeda K, Turcotte DA, Pike S. Unsustainable wind turbine blade disposal practices in the United States: A case for policy intervention and technological innovation. New Solutions: A Journal of Environmental and Occupational Health Policy 2016, 26, 581–598.

[10] Gerard P, Glotin M, Hochstetter G. Composite material via in-situ polymerization of thermoplastic (meth) acrylic resins and its use. EP Patent 2 768 656 B1.

[11] Kazemi ME, Shanmugam L, Lu D, Wang X, Wang B, Yang J. Mechanical properties and failure modes of hybrid fiber reinforced polymer composites with a novel liquid thermoplastic resin, Elium®. Composites Part A, Applied Science and Manufacturing 2019, 125, 105523.

[12] Murray RE, Jenne S, Snowberg D, Berry D, Cousins D. Techno-economic analysis of a megawatt-scale thermoplastic resin wind turbine blade. Renewable Energy 2019, 131, 111–119.

[13] Murray RE, Beach R, Barnes D, Snowberg D, Berry D, Rooney S, et al. Structural validation of a thermoplastic composite wind turbine blade with comparison to a thermoset composite blade. Renewable Energy 2021, 164, 1100–1107.

[14] https://www.lmwindpower.com/en/stories-and-press/stories/news-from-lm-places/zebra-project-achieves-key-milestone-with-first-prototype-of-recyclable-blade. Accessed May 15, 2022.

Jonathan Ouziel, Elnaz Zeinali, and Andrea Corona

12 Life cycle approaches as decision tools for technology development

12.1 Introduction

This chapter presents an introduction to the life cycle analysis approach, covering both life cycle assessment (LCA) and life cycle costing (LCC). The first part of the chapter presents the LCA definitions and methodologies, and introduces the footprinting approach as well. The second part of the chapter presents the various types of LCC. The third part aims at applying both life cycle approaches (LCA and LCC) to recycling of polymers in general and also specifically to poly(methyl methacrylate) (PMMA) thermochemical recycling.

12.2 Life cycle assessment

12.2.1 General aspects

The life cycle of consumer products is illustrated in Figure 12.1, from which a circle between the various stages of the life of a product can be observed [1]. The assessment of this life cycle has been developed based on different environmental impacts and is based on the human health and ecological goal for the product life cycle.

The first definition of an LCA was introduced by the Society of Environmental Toxicology and Chemistry (SETAC), which limited the scope of LCA studies to the environmental impact of a product throughout its life cycle [2]. Afterward, the International Organization for Standardization (ISO) has extended the definition to also evaluate important impacts of a product on the natural environment, human health, and resources during its life cycle [3].

The methodology to conduct an LCA has been further given shape by running a series of workshops by SETAC and SETAC-Europe over the years 1990–1993, with a specific focus on developing an international method for LCA [4–6]. In 1993, a schematic triangle was provided by SETAC as a best practice recommendation to conduct an LCA (see Figure 12.2a) [7]. In this triangle, the main phases are: (i) goal and scope

Jonathan Ouziel, Andrea Corona, Quantis Rue de la Gare de Triage 5, Renens 1020, Switzerland
Elnaz Zeinali, Laboratory for Chemical Technology, Ghent University, Technologiepark 125, Zwijnaarde 9052, Belgium

https://doi.org/10.1515/9783111076997-012

definition, (ii) inventory analysis, (iii) impact assessment, and (iv) improvement assessment. This structure has been modified by ISO standards. In particular, the last
phase "improvement assessment" was replaced by "interpretation" in the international standard ISO 14040 [2], as illustrated in Figure 12.2b. A description of the four
steps in the ISO methodology can be found in ISO 14041–14043 [8–10], and additional
requirements and guidelines are provided in ISO 14044 [11].

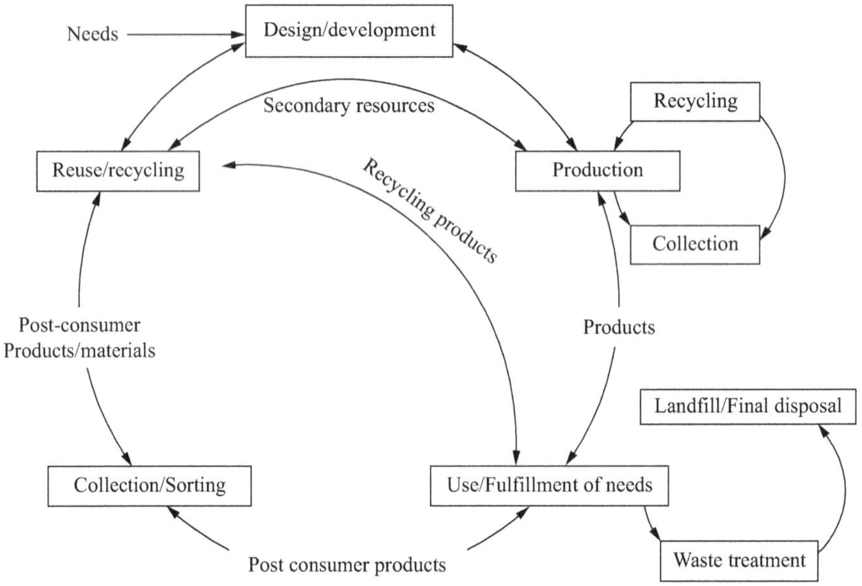

Figure 12.1: Schematic view of the life cycle of the product based on [1].

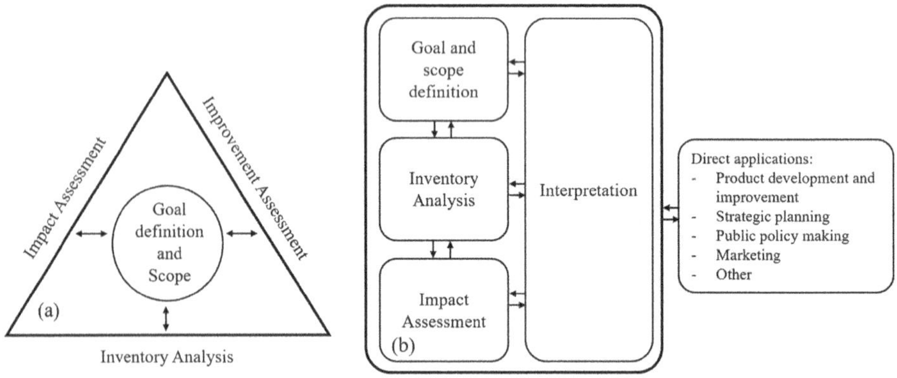

Figure 12.2: (a) The general methodology for LCA (SETAC) [7] and (b) LCA structure based on ISO 14040 [2].

12.2.2 Goal and scope definition

The goal and scope definition is the first phase of an LCA. Within this phase, the first substep is the goal definition which encompasses the purpose and the reason of conducting the LCA, that is, an analysis to determine the intended application(s) (e.g., waste management) of the LCA. This phase discusses how to perfectly define the goal of an LCA, which then leads to defining the scope and assessing the boundaries of the product system under analysis. These boundaries are criteria to determine which unit process shall be included in the LCA.

A product system is a set of processes that operate together to achieve a certain production goal. Based on ISO 14040, the goal of an LCA contains: (i) the aim of usage of the product involved in LCA; (ii) providing the reasons for performing the LCA; (iii) the target audience, that is, to whom the results of the analysis are intended to be communicated (consumers, consumer organizations, and companies might be the target audience); (iv) announcement of publication for the use of the LCA by the public.

The second substep within the goal and scope definition of an LCA is the scope definition, which is presented in the ISO standard and includes the following items: (i) a study of the functions of every unit process in the product system (e.g., chemical plant) is viewed from the perspective of the user; (ii) the definition of the functional unit (e.g., the mass of hot air needed to dry) within the quantitative aspects of the product(s) for the identified function(s) (e.g., in a case of plastic recycling, the functional unit may be defined as a quantifier of waste amount [specific mass, volume]); (iii) the set of criteria specifying which unit processes are part of a product system is defined as system boundary. The boundary of the product system distinguishes the system from the surroundings and the environment; (iv) selection of (a) impact categories (e.g., climate change/greenhouse effect, stratospheric ozone depletion) and (b) methodology for impact assessment and interpretation; (v) quality requirements for data (such as minimum age of data, variability of the data values, sources of the data, and specific technology); (vi) assumptions and limitations of the LCA; and (vii) type of critical review and format of the report required for the analysis.

Three reporting levels have been identified by the International Reference Life Cycle Data System (ILCD) Handbook: (i) internal use, (ii) external use by the third party, and (iii) comparative studies to be published to the general public.

12.2.3 From life cycle inventory analysis to life cycle impact assessment (LCIA)

From a scientific point of view, a life cycle inventory (LCI) is a chemical, material, and energy analysis based on the input of resources, materials, products, and the output of emissions, waste, and valuable products for the product system. Unit process as a smallest element is considered in LCI for which the environmental inputs and outputs are quantified. The laws of conservation of mass and energy can be used in every physical and chemical unit process (e.g., purification or polymerization). In general, the mass and energy of the inputs should equal the outputs due to the fundamental conservation laws applied to each unit process.

System boundaries should be identified and determined in the product system to quantify resources, energy use, and emissions to the environment. To perform an LCI analysis, the following four steps have been first carried out [11]: (i) evaluation of flow diagram (with boundary) of the process; (ii) develop a data collection plan; (iii) carry out the data collection; and (iv) validation of the collected data and report.

Then the inventory analysis is used as input for the life cycle impact assessment analysis (LCIA). The LCI analysis (flow diagram including resource uses, land uses, and chemical emissions) is linked to one or more impacts in the LCIA. The main aim of the LCIA is providing further information to improve the results of inventory analysis to ultimately obtain a better knowledge of the environmental impact on product system throughout the product life cycle.

As shown in Figure 12.3(a), LCIA consists of mandatory and optional elements [13]. The two mandatory steps (according to ISO) are the selection of impact categories and classification. The *selection* of impact categories and the method of selection chosen for each impact category appear in the goal and scope definition of the analysis. *Classification*, the primary flow of inventory, for example, resources, consumption, and emission into air, water, and soil, is related to the impact categories selected in the previous step, based on their contribution to different environmental impacts (Figure 12.3b). Other steps are characterization, normalization, and weighting.

The quantification of the various impact categories is the main purpose of the *characterization* step. By looking into the different global, regional, and local impacts on humans or the environment, various authors and SETAC working groups have described these environmental impacts from a life cycle perspective. Generally, in an LCIA, at both the midpoint level (in terms of potential impacts) and at the endpoint level (in terms of years of lost life), the environmental impacts are defined (Figure 12.3b) [13]. Labeled as typical midpoints are: climate change, stratospheric ozone depletion, acidification, eutrophication, photochemical ozone formation, ecotoxicity, human toxicity, particulate matter formation, ionizing radiation, land use, erosion, water use, abiotic resource use, biotic resource use, and noise. In addition, human health, natural environment, and natural resources are assessed as typical endpoint indicators, although some distinctions exist between them (Figure 12.4) [14].

Midpoint and endpoint methods with different stages in the cause–effect chain can be used to determine LCIA (Figure 12.4). The characterization at the midpoint level of the elementary flows in the LCI analysis in a collection of "midpoint impact indicator scores" is jointly referred to as "the characterized impact profile" of the product system at the midpoint level. Furthermore, the characterization at the endpoint level looks at environmental impact at the end of the cause–effect chain. In this respect, applying midpoint or endpoint method could cause various conclusions in

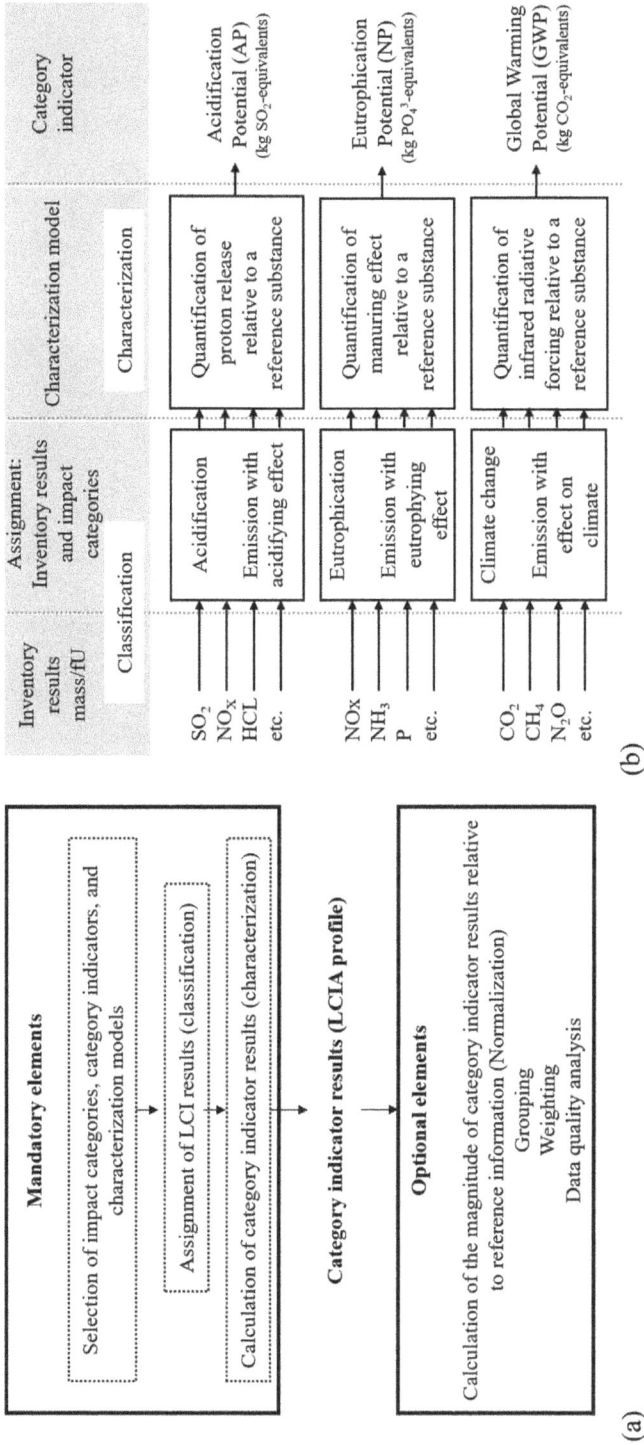

Figure 12.3: (a) Elements of LCIA ISO 14042 [9], and (b) classification and characterization in the phase life cycle impact assessment [12].

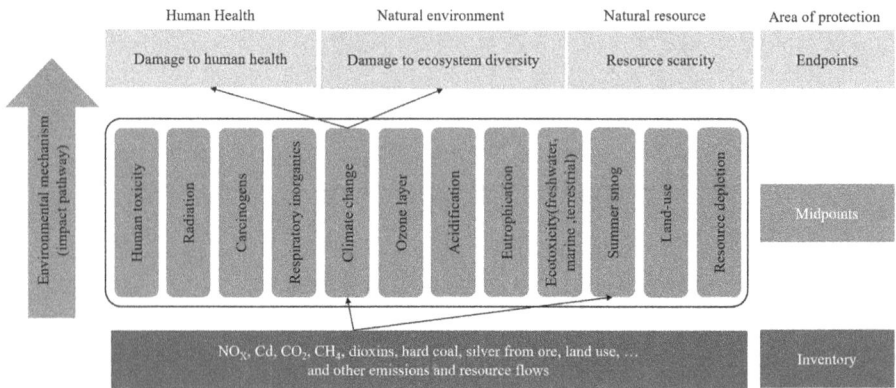

Figure 12.4: Linking elementary flows from the inventory results to indicator results at midpoint level and endpoint level [14].

the LCIA. To expand or connect the midpoint indicators to the endpoint ones, normalization can be performed.

Hence, the list of impact categories for midpoint indicators is similar to the endpoint ones. *Normalization* is an optional element that can be used as preparation for comparing the indicator results. The normalization would bring the scores for the different impact categories together on a common scale or ranking, which improves the consistency of the results and helps to communicate the results to the target audience.

Finally, data *weighting* can be considered to support a final comparison of indicator results across impact categories. To determine the most important impact categories and how important they are, the weighting can be applied after the normalization step.

12.2.4 Life cycle interpretation: Improvement assessment

The final phase of an LCA discussed the results of both the inventory (LCI) and the impact assessment (LCIA) as well as the evaluation of the uncertainties and the assumptions. In the life cycle interpretation, conclusions are drawn from the results of these previous phases (LCI and LCIA), and recommendations are made based on the objective of the analysis. The interpretation proceeds through three steps, according to ISO 14043 [10].

The aim of the first step of the life cycle interpretation, which is denoted as "The identification of significant issues," is to evaluate the results from earlier phases, to identify the most environmentally important topics, that is, those topics that potentially change the final results of the LCA. These so-called significant issues can be determined from (i) a sensitivity check in the evaluation step of the interpretation as well as (ii) information from the assumptions and uncertainty ranges from LCI and LCIA.

After identifying the significant issues, they are evaluated with the aim of assessing (i) their effect on the overall results of the LCA and (ii) the completeness and consistency with which they have been handled in the LCA. This is denoted as the second step: "The evaluation of the significant issues." Evaluating the validity of the conclusions of the LCA and the recommendations from the LCA is a challenge, and it must therefore be presented in a way that gives both the commissioner and the user of the LCA a clear understanding of the outcome.

The evaluation involves (i) completeness check: The significance of LCI and LCIA results should be determined. The sufficiency of results from LCI and LCIA should be examined through a completeness check to draw conclusions in accordance with the goal and scope definition. (ii) Sensitivity check: This check serves to determine the most important processes and elements that considerably influence the product system. (iii) Consistency check: Evaluating whether the methods, assumptions, and used data agree with the earlier-defined goal of the LCIA/LCI and remain within the scope of the LCIA/LCI.

In the third step, conclusions are drawn and limitations in the LCA are identified. In addition, we implement logical recommendations based on the results and objective of the LCA.

12.3 Footprinting

An approach linked to LCA is footprinting. The LCA approach takes various impact categories (e.g., environmental and health categories, energy consumption, and emission) into account, whereas the footprinting approach focuses on one environmental impact category at a time: for example, greenhouse gas (GHG) emissions such as CO_2, CH_4, N_2O, and SF_6 inducing climate change impacts, also commonly called "carbon footprint." The "footprint" approach thus uses LCA tools to focus on specific environmental impacts.

Footprints are thus always connected to a resource. Specifically, the following footprints are often analyzed: carbon footprint; water footprint (ISO 14046) [15]; energy footprint (renewables or nonrenewables); emission footprint (emissions to air, water, and soil); and work environment footprint. Analysis of these footprints is required to consider their effect on the natural environment, human health, and other (nonfootprint) resources related to carbon, water, and energy. In the following, we consider the first three fingerprints.

The measurement of the total amount of GHG emissions during the product life cycle is called the *carbon footprint*. A product system is a set of processes that operate together to achieve a certain production goal in LCA. By "total," we refer to both direct emissions (e.g., production of plastics) and indirect emissions (e.g., incineration of plastic waste). The total amount of GHG emissions is measured and expressed in the unit of mass (kilogram or ton) of carbon dioxide equivalent (CO_2-eq), and thus emissions are not converted to real quantities (e.g., ton per square km). Two major methodologies, process analysis and input–output analysis, are used for assessing the carbon footprint [16]. They are calculated over the product life cycle by summation of the amount of GHG emissions in each step of the product (raw materials, manufacturing, transportation, usage, and recycle and disposal stage).

The *water footprint* method quantifies one or more metric(s) of water usage for a product, process, or organization, with the expressed aim to assess its impact on the environment. The types of water used in a product system are divided into blue, green, and gray footprinting categories [17].

The *energy footprint* is a quantification of different energy supplies such as heating water or hot water, process energy, and electricity used in the product life cycle. Also related footprints exist, the so-called energy-related footprints, in which the use of energy from different carriers is multiplied by their land/area requirements. The energy associated with GHG emissions from various incineration processes is included in the energy-related footprint, while the production/emission of those GHG emissions themselves are accounted for in the carbon footprint.

12.4 Life cycle costing

12.4.1 General aspects

Compared to LCA and footprinting, LCC is another life cycle approach. Instead of being an examination of the environmental impact, an LCC is generally defined as an analysis of all costs related to a certain product or process over the product life cycle. Hence, LCC is a tool that accounts for the operations of a product or service. It consists of the cumulative cost or economic analysis of a product or process over its life cycle. Evidently, reliable cost data must be available to perform an effective life cycle cost analysis. Thus, reliable cost databases are essential requirements for an LCC. The result of the LCC is useful for optimizing the design, scheduling maintenance of the process, and planning of purchasing products [18].

The major methodological differences between LCA and LCC are summarized by Norris in Table 12.1 [19]. Whereas LCA is mostly an environmental analysis, LCC calculates the cost of construction, operation, and maintenance throughout product life cycle (see Table 12.1, entry 1). The processes that have an impact on the direct costs or benefit should be included in the scope of the LCC. On the other hand, in the LCA scope, all process effects on the product might be examined (see Table 12.1, entry 2).

Another important difference between LCA and LCC is related to flow (Table 12.1, entry 3) because an LCC only focuses on cost flow. In contrast, an LCA includes the energy and material flows for each process. Although the various times for the input, output, consumption, and emission are needed in LCA (e.g., most LCA use 100 years as the time to assess GHG emissions), timing is critical and is also present in LCC. Additionally, the LCC can be defined as an analysis of total cost for the product over fixed time horizon and the length of time is required for investment analysis (Table 12.1, entry 5). Furthermore, while LCA is conducted using comprehensive international standard methodology [2, 8–11] to analyze the effect of environmental impacts, no such uniform approach exists or has been used for LCC.

The three types currently employed are conventional, environmental, and societal LCC (Table 12.2). Conventional LCC should assess the traditional costing or internal costs by individual companies. This analysis is based on real, internal costs, sometimes without end-of-life or use costs accounted for. Environmental LCC represents the additional assessment of external costs and LCA which is included in conventional LCC. Societal LCC further includes externality costs using accounting prices [20]. The societal LCC is an analysis from the broadest sense of society, that is, nationally and

Table 12.1: The differences in purpose and approach for LCA and LCC [15].

Tool/method	LCA	LCC
Purpose	Compare relative environmental performance of alternative product systems for meeting the same end-use function, from a broad, societal perspective	Determine cost-effectiveness of alternative investments and business decisions, from the perspective of an economic decision maker such as a manufacturing firm or a consumer
Activities which are considered part of the "life cycle"	All processes causally connected to the physical life cycle of the product; including the entire pre-usage supply chain; use and the processes supplying use; end-of-life and the processes supplying and-of-life steps	Activities causing direct costs or benefits to the decision maker during the economic life of the investment, as a result of the investment
Flows considered	Pollutants, resources, and interprocess flows of materials and energy	Cost and benefit monetary flows directly impacting decision maker
Units for tracking flows	Primarily mass and energy; occasionally volume, other physical units	Monetary units (e.g., dollars and euro)
Time treatment and scope	The timing of processes and their release or consumption flows is traditionally ignored; impact assessment may address a fixed time window of impacts (e.g., 100 year time horizon costs or benefits occurring outside that scope are ignored for assessing climate change potential impacts) but future impacts are generally not discounted	Timing is critical. Present valuing (discounting) of costs and benefits. Specific time horizon scope is adopted, and any costs or benefits occurring outside that scope are ignored

internationally, including governments. Note that taxes and subsidies do not influence this LCC, so they are not included in a societal LCC.

In the simplest model, all technologies are modeled as being in steady state, that is, their behavior does not change with time, which is called a steady-state model. In contrast, dynamic models predict the time evolution of variables. The quasi-state model shares features of both the steady-state and dynamic models. Most variables remain constant in time, and one or more important variables change with time. This last category of models is typically used in conventional and societal LCC (see Table 12.2, entry 6).

Table 12.2: The differences in purpose and approach for LCA and LCC [15].

Aspect	Conventional LCC	Environmental LCC	Societal LCC
Product system (model)	Life cycle, without end-of-life phase	Complete life cycle	Complete life cycle
System boundaries	Only internal costs[a]	Internal costs, plus external[b] costs expected to be internalized	Internal plus external cost
Perspectives: actors	Mainly 1 actor, either the manufacturer or the user or consumer	Manufacturers, supply chain, and end user or consumer	Society overall, including governments
LCA "ISO14040/44"	No	Yes	Not recommended
Reference unit	Item or product	Functional unit	System
Cost model	Quasi dynamic	Steady state	Quasi-dynamic

[a]Internal costs include the production, use, or end-of-life expenses that are related to a business cost for the direct stakeholder.
[b]The environmental and social impacts such as the consumers, or the government bodies that are producing, using, or handling the products are considered in the external costs.

12.4.2 Procedures

Several LCC procedures have been introduced [21, 22]. Six common steps are essential for each procedure: problem definition, cost element definition, system modeling, data collection, cost profile development, and evaluation.

The *problem definition* of an LCC contains both (i) the problem statement and (ii) the envisaged scope of the LCC. The cost of the element, as well as all assumptions in the LCC, should be defined in this step.

All *cost elements* that significantly affect the LCC of the system should be listed, defined, and evaluated in this step. According to the international standard of LCC IEC60300-3-3, it is important to define the cost element in the cost breakdown structure (CBS) with three independent axes: (i) "life cycle phases," (ii) "product/work breakdown structure," and (iii) "cost categories" (Figure 12.5). This three-dimensional matrix is composed of the following three aspects: (i) the product/work breakdown (decomposition of the analysis into phases and work packages); (ii) the life cycle phase (the research and development costs, manufacturing and construction costs, and system operation costs should be considered); and (iii) the cost categories (the cost for important resources and activities, e.g., labor, materials, energy, and transportation).

The *system model* is required to quantify the cost for elements in the LCC. Although the available existing model (e.g., RBD, FTA, and Markov [23, 24]) should be applied to estimate the cost of elements, the new model for the cost elements needs to be established. To build the system model, it is necessary to find convenient relations between input parameters and the cost elements. In addition, to develop the model, the availability, maintainability, logistics, risk, and human error in the system should be evaluated. The availability and the maintainability of equipment are the most significant factors that influence the categorized cost elements.

The input data and access to reliable data sources related to LCC are required for the *data collection*. To quantify the cost element in the CBS, reliable data should be

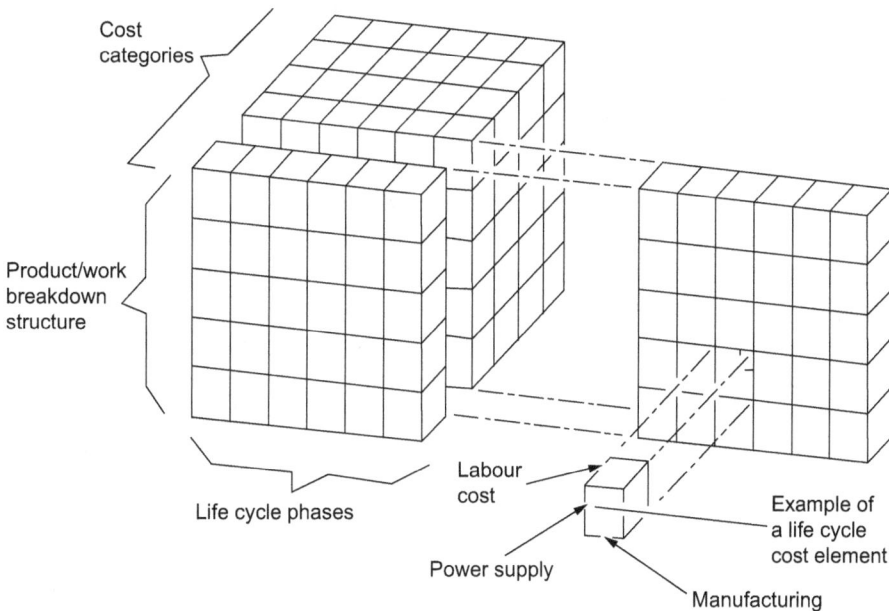

Figure 12.5: Cost element concept IEC60300-3-3 [16].

applied for each cost element. A wide variety of reliable data have been acquired and made available to the public [25–27]. In particular, whenever suitable data is not available, then the cost of elements can be estimated based on expert judgments.

The *cost profile development* over the entire life cycle is one of the main steps of LCC. Comparing with a reference point for the cost profile is essential for financial judgments, because an economic analysis is required before modifying the system design or making an investment.

In the *evaluation step*, we require to examine whether the system meets the objective of problem definition in the first step of LCC. The sensitivity and uncertainties of the input data should be assessed in the evaluation step. The influence of changes in input parameters on the result of analysis should be examined through sensitivity analysis. Deterministic and stochastic approaches are the two main approaches used in sensitivity analysis. Possible ranges of the estimates and their effect on decisions are the focus of the uncertainty analysis. Parameter uncertainties, modeling uncertainties (such as model assumptions), and completeness uncertainties (to ensure an adequate input data in order to draw a conclusion) are estimated by the uncertainty analysis.

12.5 Case study

12.5.1 General aspects and overall strategy

While engineering polymeric materials provide many benefits such as lightweight compared to using wood, metal, and glass in most applications, the disposal of the associated waste is the main disadvantage, especially upon using large amounts of plastic in single-use applications. Different recycling procedures for various plastics are used as a solution for the waste problem. Hence, an evaluation of all environmental impacts generated by the recycling process should be considered. The selection of the procedure to conduct an LCA with the aim to identify environmental impacts is important. Recycling the most demanded polymers in Europe and improving the recycling procedure (e.g., by sorting) has caused a reduction in emission of GHG by approximately 73% [28].

Various LCA studies of plastic recycling processes have been reported [29–32]. A first example is the work of Shen et al. [33], who conducted LCA for the conversion of polyethylene terephthalate (PET) bottles to recycled fibers. The results indicate a reduction of nearly 40–85% in energy consumption and a decrease of global warming potential by 25–75% in comparison to preparing the virgin PET from fresh monomers. Hence, the environmental impact is reduced by recycling PET with different methods.

A second example relates to GHG emissions (e.g., CO_2, methane, and nitrous oxides) originating from chemical recycling technologies for polyolefin-based plastic that are considered as one environmental impact category in LCA [34]. The results clearly show the reduction in GHGs by relying on chemical recycling processes. The reduction in emission of GHGs can be studied by using LCA as a tool for the recycling process of PET as well [31]. Zhang et al. [35] reported the combination of LCA and LCC for the recycled waste PET. Specifically for PET, environmental impacts such as global warming, fossil, and water consumption were the main factors to consider.

A third example addresses the polymer in the MMAtwo project. The transparency of PMMA and its resistance to harsh weather conditions has led to an increase in its use as automobile lighting equipment, design panels, and so on [36]. Furthermore, PMMA is used in liquid crystal display (LCD) panels due to transparency and light-weight [37]. As a result of the increased popularity of LCD products, the recycling processes for PMMA become more important. The qualification of various environmental impacts related to the PMMA monomer recycling process is the aim of this section. Recycling PMMA via the pyrolysis method could further improve the efficiency compared to mechanical recycling (beyond conventional injection and extrusion grades), by applying LCA, as explained by Kikuchi et al. [38]. Firstly, GHG emission has been quantified for different recycling processes. The pyrolysis process has been selected due to considerable reduction in GHG emissions and other environmental impact.

To evaluate the environmental impact and investigate the economic importance of the PMMA recycling process, the LCA should be combined with LCC. Quantifying the environmental and economic analysis of production of recycled methyl methacrylate (rMMA) from PMMA waste via a coupled LCC and LCA is considered in this section. The system to be studied represents the value chain developed within the MMAtwo project. The MMAtwo system includes four steps: (i) collection of PMMA waste, (ii) pretreatment of PMMA waste, (iii) depolymerization, and (iv) purification.

For the LCA, a specific focus is set on the following key indicators, given their importance for the chemical/material production sector: (i) GHG emissions; (ii) nonrenewable energy use; and (iii) water use. Those indicators are taken from the Environmental Footprint method (European Commission, 2017b). This method assesses 16 different potential impact categories. These categories are the result of a project of the European Commission that analyzed several LCIA methodologies to reach consensus. This impact assessment method is the official method to be used in the Product Environmental Footprint (PEF) context of the Single Market for Green Products initiative (European Commission, 2013). In compliance to the PEF guidelines (European Commission, 2017a), two additional life cycle-based indicators are utilized to provide a more comprehensive overview of environmental impacts. The two indicators are endpoint indicators calculated using the IMPACT 2002 + method (i) human health and (ii) ecosystem quality. The rationale to include two endpoint indicators is to obtain a more comprehensive overview of the environmental impacts of the studied system. The selected indicators aim to cover the environmental impacts related to the emission of toxic substances to air and water.

The LCA model has been developed using SimaPro software [39], which is one of the most recognized software in the field of LCA. EcoInvent database v3 [40] in the cutoff by classification allocation model has been used as a main data source. EcoInvent is recognized as one of the most complete background LCI databases available from a quantitative (number of included processes) and a qualitative (quality of the validation processes, data completeness, etc.) perspective. Historically focused on European production activities, it has reached a global coverage of thousands of commodities and industrial processes.

The PMMA scraps have been collected in three different grades with a target volume of 30,000 ton/year of PMMA scraps. With this target volume, the average collection radius is estimated at 820 km from the central location. It is assumed that the collection is done with a 16–32 t Euro5 truck, and the central location is Utrecht, the Netherlands, where the pilot pretreatment process is located. Each grade has a different purity and needs a different pretreatment process before the depolymerization. Furthermore, the acquisition price differs as well.

In the pretreatment phase, the PMMA waste is separated from the non-PMMA parts to prepare a feedstock material suitable for the depolymerization stage. Depending on the grade considered, different pretreatment processes are needed, and each pretreatment process will produce different intermediate products and wastes. Those values represent pilot-scale conditions and were measured in the pretreatment pilot plant setup.

pretreatment and depolymerization are done in a different facility compared to the purification stage. Potential advantage of this decoupling is that the two parts of the process do not necessarily need to be operated on the same site. Liquid crude methyl methacrylate (MMA) can be collected and transported for further processing elsewhere. A hypothetical transportation of the liquid crude MMA of 500 km is assumed. This allows for multiple depolymerization facilities that can provide the feedstock for a central purification plant, which may be also external party subcontracted for the given task. The LCI has been calculated based on the technoeconomic assessment (TEA).

The TEA also focused on the capital and operating expenses for the depolymerization and purification part as shown in Figure 12.6. The cost calculation related to the collection and pretreatment is based on the final data of the capital expenditures from Heathland.

12.5.2 Main results and fingerprint for PMMA case

This section presents the final results of the LCA/LCC of the MMAtwo recycling technologies. These relate to the environmental impacts of the technology across five environmental indicators: climate change, resource use, water use, human health, and ecosystem damage. Interpretation of results is mostly focused on carbon footprint

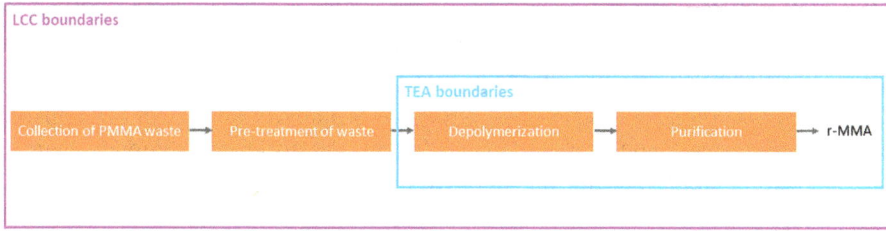

Figure 12.6: Capital and operating expensed for the depolymerization and purification part.

since the MMAtwo project aims at producing regenerated MMA (rMMA) with at least a 60% GHG reduction compared to virgin MMA. Furthermore, the carbon footprint is one of the most recognized, widely used indicators in LCA and a good proxy for other indicators.

The carbon footprint results of the process are shown in Figure 12.7. It shows that the carbon footprint of 1 ton of rMMA produced at a purity of 99,8% is 0,89 ton CO_2-eq; depolymerization and pretreatment are the largest contributors accounting respectively to 33% and 25% of the impacts. Collection and Purification are nearly equal contributors, each one accounting for approximately 21% of the impacts.

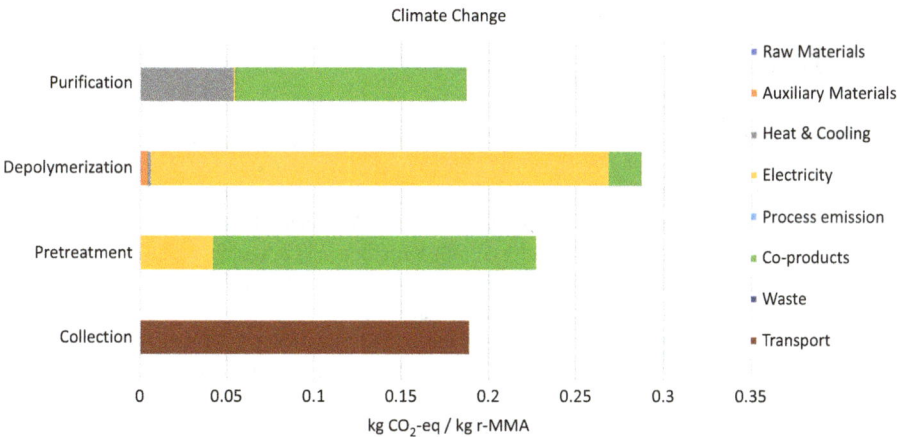

Figure 12.7: Carbon footprint of 1 kg of r-MMA (99.8% purity) produced from PMMA of grade 80/20. The figure shows the utility and waste and by product treatment contribution to each life cycle stage. The total carbon footprint of 0,89 kg CO2-eq / kg r-MMA is obtained by summing up the contribution of the 4 categories.

Additional analysis revealed that the highest contributor in the pretreatment stage is the waste and by-product treatment accounting for approximately 81% of the pretreatment impact. The remaining impacts are related to the electricity consumption for grinding and separating the PMMA scraps. The highest contribution in the depolymerization stage is the electricity consumption (91% of the depolymerization impacts). Most of the electricity consumed in this stage is due to the twin-screw configuration. The highest contribution in the purification stage is the waste and by-product treatment accounting for approximately 70% of the impact. The rest of the impacts are split between heat consumption and cooling for the distillation columns. Overall, there are two main hotspots for the MMAtwo process: (i) waste and by-product treatment accounting for 38% of the total carbon footprint impacts; and (ii) electricity accounting for 26% of the total carbon footprint impacts.

Regarding resource use, fossil results show that electricity use in the depolymerization stage and fuel consumption in the collection stage generates most of the impacts. Regarding water use, results show that the electricity in the depolymerization stage generates most of the impacts. Some credits can be seen at pretreatment and purification stages from the treatment of co-products. Regarding Human Health, results show that impacts are mostly arising in the depolymerization and collection stage due to the electricity and fuel use. The pretreatment and purification stages are net negative as credits generated from the co-product are bigger than the impact from utilities. Regarding Ecosystem quality, results show that impacts are mostly arising in the collection stage due to truck diesel/fuel consumption and in the depolymerization stage due to electricity use.

References

[1]   Rebitzer G, Ekvall T, Frischknecht R, Hunkeler D, Norris G, Rydberg T, et al. Life cycle assessment: Part 1: Framework, goal and scope definition, inventory analysis, and applications. Environment International 2004, 30, 701–720.
[2]   Guidelines for Life-Cycle Assessment: A. Code of practice. 31 March – 3 April 1993, Society of Environmental Toxicology and Chemistry (SETAC), Workshop held at Sesimbra, Portugal.
[3]   ISO 14040. Environmental Management – Life Cycle Assessment -Principles and Framework, International Organization for Standardization, Geneva, 2006.
[4]   Conceptual A. Framework for life-cycle assessment impact assessment. 1–6 February 1992, Society of Environmental Toxicology and Chemistry (SETAC), Workshop Held at Sandestin, Florida, USA.
[5]   Life-Cycle Assessment Data Quality. 4–9 October 1992, Society of Environmental Toxicology and Chemistry (SETAC), Workshop Held at Wintergreen, Virginia, USA.
[6]   A Technical Framework for [Product] Life-Cycle Assessments. 18–23 August 1991, Society of Environmental Toxicology and Chemistry (SETAC), Workshop Held at Smuggler's Notch, Vermont, USA.
[7]   Klöpffer W. Life cycle assessment. Environmental Science and Pollution Research 1997, 4, 223–228.
[8]   ISO 14041. Environmental Management – Life Cycle Assessment – Goal and Scope Definition and Inventory Analysis, International Organization for Standardization, Geneva, 1998.
[9]   ISO 14042. Environmental Management – Life Cycle Assessment – Life Cycle Impact Assessment, International Organization for Standardization, Geneva, 2000.
[10]   ISO 14043. Environmental Management – Life Cycle Assessment – Life Cycle Interpretation, International Organization for Standardization, Geneva, 2000.

[11]   ISO 14044. Environmental Management – Life Cycle Assessment -Requirements and Guidelines, International Standards Organization, Geneva, 2006.

[12]   Klöpffer W, Grahl B. Life Cycle Assessment (LCA): A Guide to Best Practice, Wiley, 2014.

[13]   Bare JC. Life cycle impact assessment research developments and needs. Clean Technologies and Environmental Policy 2010, 12, 341–351.

[14]   Handbook – General Guide for Life Cycle Assessment – Detailed Guidance. First edition March 2010 ed. Luxembourg, European Commission, Joint Research Centre, Institute for Environment and Sustainability, 2010.

[15]   ISO 14046. Environmental Management – Water Footprint – Principles, Requirements and Guidelines, International Standards Organization, Geneva, 2014.

[16]   Pertsova CC. Ecological Economics Research Trends, Nova Publishers, 2007.

[17]   Gu Y, Xu J, Wang H, Li F. Industrial Water Footprint Assessment: Methodologies in Need of Improvement, ACS Publications, 2014.

[18]   IEC60300-3-3. Application Guide – Life Cycle Costing, International Electrotechnical Commission, Geneva, Switzerland, 2004.

[19]   Norris GA. Integrating life cycle cost analysis and LCA. The International Journal of Life Cycle Assessment 2001, 6, 118–120.

[20]   Martinez-Sanchez V, Kromann MA, Astrup TF. Life cycle costing of waste management systems: Overview, calculation principles and case studies. Waste Management 2015, 36, 343–355.

[21]   Greene LE, Shaw BL. The steps for successful life cycle cost analysis. In IEEE Conference on Aerospace and Electronics. IEEE, 1990.

[22]   IEC60300-3-1. Application Guide – Analysis Techniques for Dependability –Guide on Methodology, International Electrotechnical Commission, Geneva, Switzerland, 2003.

[23]   Kawauchi Y, Rausand M. Life Cycle Cost (LCC) Analysis in Oil and Chemical Process Industries, Toyo Engineering Corp, Chiba, 1999.

[24]   Bouissou M, Bourgade E. Unavailability evaluation and allocation at the design stage for electric power plants: Methods and tools. In Annual Reliability and Maintainability Symposium. IEEE, 1997.

[25]   Van Sciver GR. Guidelines for process equipment reliability data by CCPS. In Reliability Data Collection and Use in Risk and Availability Assessment. Springer, 1989, 104–114.

[26]   Allan R, So T, Gazidellis D. An integrated data base and reliability assessment of electrical distribution systems. In Reliability Data Collection and Use in Risk and Availability Assessment. Springer, 1989, 840–847.

[27]   Tomic B, Lederman L. Data selection for probabilistics safety assessment. In Reliability Data Collection and Use in Risk and Availability Assessment. Springer, 1989, 80–89.

[28]   Schwarz A, Ligthart T, Bizarro DG, De Wild P, Vreugdenhil B, van Harmelen T. Plastic recycling in a circular economy; determining environmental performance through an LCA matrix model approach. Waste Management 2021, 121, 331–342.

[29]   Arena U, Mastellone ML, Perugini F. Life cycle assessment of a plastic packaging recycling system. The International Journal of Life Cycle Assessment 2003, 8, 92–98.

[30]   Arena U, Mastellone ML, Perugini F. The environmental performance of alternative solid waste management options: A life cycle assessment study. Chemical Engineering Journal 2003, 96, 207–222.

[31]   Romero-Hernández O, Hernández SR, Muñoz D, Detta-Silveira E, Palacios-Brun A, Laguna A. Environmental implications and market analysis of soft drink packaging systems in Mexico. A waste management approach. The International Journal of Life Cycle Assessment 2009, 14, 107–113.

[32]   Perugini F, Mastellone ML, Arena U. A life cycle assessment of mechanical and feedstock recycling options for management of plastic packaging wastes. Environmental Progress 2005, 24, 137–154.

[33]   Shen L, Worrell E, Patel MK. Open-loop recycling: A LCA case study of PET bottle-to-fibre recycling. Resources, Conservation and Recycling 2010, 55, 34–52.

[34] Chemical Recycling: Greenhouse gas emission reduction potential of an emerging waste management route. 2020. (Accessed 2020, at https://quantis-intl.com/wp-content/uploads/2020/11/cefic_paper_v8.pdf.)

[35] Zhang R, Ma X, Shen X, Zhai Y, Zhang T, Ji C, et al. PET bottles recycling in China: An LCA coupled with LCC case study of blanket production made of waste PET bottles. Journal of Environmental Management 2020, 260, 110062.

[36] Moens EK, De Smit K, Marien YW, Trigilio AD, Van Steenberge PH, Van Geem KM, et al. Progress in reaction mechanisms and reactor technologies for thermochemical recycling of poly (methyl methacrylate). Polymers 2020, 12, 1667.

[37] Kim G. A PMMA composite as an optical diffuser in a liquid crystal display backlighting unit (BLU). European Polymer Journal 2005, 41, 1729–1737.

[38] Kikuchi Y, Hirao M, Sugiyama H, Papadokonstantakis S, Hungerbühler K, Ookubo T, et al. Design of recycling system for poly(methyl methacrylate)(PMMA). Part 2: Process hazards and material flow analysis. The International Journal of Life Cycle Assessment 2014, 19, 307–319.

[39] SIMAPRO software. Accessed October 2013, Préconsultants. available at www.pre-sustainability.com/TheNetherlands.

[40] Weidema BPBC, Hischier R, Mutel C, Nemecek T, Reinhard J, Vadenbo CO, Wernet G. Overview and methodology. Data quality guideline for the Ecoinvent database version 3. In Ecoinvent Report 1 (v3). Swiss Centre for Life Cycle Inventories: The Ecoinvent Centre, St. Gallen, Switzerland, 2013.

Jean-Luc Dubois and Jonathan Ouziel

13 MMAtwo's Footprinter

The MMAtwo project came to an end by October 2022, and one of the major achievements has been the "Footprinter" developed collectively by the partners. That's a tool accessible on internet via the MMAtwo project website (mmatwo.eu) at mmatwo-footprinter.eu. The Footprinter can be used to compare various virgin and regenerated MMA technologies; and gives the impact of reduction of greenhouse gases (GHG) emissions, and water and energy consumptions. Cases have been built for different types of feedstocks (a "Cast" PMMA, which is in fact a mixed cast + PVC gasket type of scrap, "extrusion," "injection," and a "mixed" colored cast and extrusion wastes); and also for the three main virgin MMA technologies, and several depolymerization technologies (molten metal, dry distillation, rotating drum, fluid bed, and stirred tank) and, of course, for the twin-screw extruder process; and, most importantly, for different purities of regenerated MMA being produced. PMMA is known by its brand names of Altuglas, Plexiglas, or Perspex, for example. All the data included in the database represent cases seen on the market, in Europe, Asia, and Brazil. In a series of posts on LinkedIn, we illustrated with the examples described below, what the user can derive as information from the Footprinter.

13.1 Study parameters

13.1.1 Cradle to gate

This environmental Footprinter displays the four life cycle stages of the MMAtwo recycling technology (cradle to gate approach):
- Collection (of the PMMA waste in different locations in Europe);
- Pretreatment (of those waste to sort them in different waste grades), including separation of PVC and polyethylene film;
- Depolymerization step (to return the PMMA to a crude MMA monomer);
- Purification step (to purify the material, making the crude MMA "pure," at which point the final product is considered being r-MMA = regenerated MMA).

Jean-Luc Dubois, Altuglas International/Trinseo, Tour CB21,16 Place de l'Iris, 92400 Courbevoie, France
Jonathan Ouziel, Quantis Rue de la Gare de Triage 5, Renens 1020, Switzerland

https://doi.org/10.1515/9783111076997-013

13.1.2 MMA technologies

The MMAtwo technology will be compared with two different types of technologies: (i) virgin routes of MMA production and (ii) current recycling routes to produce rMMA from PMMA scraps. The tool measures and compares the environmental benefits of the rMMA using MMAtwo technology compared to the following virgin and recycling routes:
– Virgin production C2 (Europe and Asia)
– Virgin production C3 (Europe and Asia)
– Virgin production C4 (Europe and Asia)
– Recycling technology: molten lead (Europe)
– Recycling technology: fluid bed (Asia)
– Recycling technology: rotating drum (Asia)
– Recycling technology: dry distillation (Asia)
– Recycling technology: stirred tank (Europe)

The technologies displayed vary according to the selected market, the waste quality grade (quality of scraps) and the final *r*-MMA purity achieved. Only the technologies that fulfil the three variables are displayed in the benchmark dropdown menu list.

13.1.3 Parameters

Throughout the lifecycle stages, these variable parameters are included in the calculation, meaning environmental impact results will be different, based on the choice of the user who will use the Footprinter. The user will be able to choose:
– The "market area" of the business: Europe or Asia
– "Quantity of PMMA waste to recycle" (amount to be disposed of) or "quantity of rMMA to purchase" (giving the yield of rMMA to be purchased)
– The "Quality of scraps" to be treated:
  – 80/20 mix grade (representing EU baseline mix, 80% cast and 20% extrusion)
  – Cast grade (the yield considered in this Footprinter includes the removal of potential PE film and PVC gaskets, and it explains why a larger amount of scrap is needed)
  – Extrusion grade
  – Injection grade

The "rMMA purity" to be achieved:
– ≥99.8% (achievable with virgin routes, MMAtwo technology and twin fluid bed technology);
– ≥99.5% (achievable with virgin routes, MMAtwo technology, twin fluid bed technology and the stirred tank process);

- ≥99.0% (achievable with virgin routes, MMAtwo technology, twin fluid bed technology, Stirred tank process and molten lead technology);
- ≥98.0% (achievable with virgin routes, MMAtwo technology, twin fluid bed technology, Stirred tank process, molten lead technology, and dry distillation technology);
- ≥95.0% (achievable with all technologies listed above).

13.1.4 Results

The calculated results of the environmental footprint of the batch are estimates for
both MMAtwo and other virgin/recycling technologies. Calculations have a margin of
error, depending on the environmental indicator considered (e.g., 10% for carbon
footprint). These environmental indicators are the computed values of cumulated
emissions generated by the technologies being discounted of cumulated emissions
due to energy valorization of side products. When the resulting number is small, the
uncertainty on the net value is large due to the subtraction of two large numbers, and
comparison should be done with caution. The environmental indicators considered in
the tool are:
- Climate change (or carbon footprint, expressed in t CO_2-eq or kg CO_2-eq)
- Water consumption (expressed in m^3)
- Resource use, fossil (expressed in MJ)

The results displayed in this tool are based on LCA performed by Quantis of the
MMAtwo technology and the other virgin/recycling technologies available. For more
information, see: https://www.mmatwo.eu/.

13.2 Example 1: MMAtwo technology (twin-screw extruder) versus the C3 (acetone cyanhydrin) process

There are several established processes to produce virgin methyl methacrylate. The
common process in Europe and North America is the "Acetone Cyanhydrine" process,
also called C3 process (three carbons in acetone). In this process, acetone is reacted
with hydrogen cyanide to produce acetone cyanhydrine. Then, it is further reacted to
produce methyl methacrylate and ammonium sulfate. In some processes, ammonium
sulfate is upgraded to fertilizer grade; while in other processes, it is directed to a sulfuric acid regeneration unit where ammonia will be lost. One of the major problems of
this process is the need for ammonia, which impacts severely the carbon footprint. So,

a variation of the process was developed in Japan, where hydrogen cyanide is recycled internally. For the various processes, the reader can find details in the reference [1].

The Figure 13.1 illustrates a case where a mixed waste of 80% cast PMMA and 20% extrusion sheets is depolymerized to a purity of 99.8% (achieved in the MMAtwo project). One ton of scrap will give 0.75 ton of regenerated MMA. We can compare the rMMA with virgin MMA using the C3 process, either per ton (or per kg) of scrap or per ton (or per kg) of MMA (regenerated or virgin), shown in Figures 13.1 and 13.2, respectively.

On the top right corner, the respective carbon footprint, and water consumption and energy consumption are given. This illustrates the large impact of regenerated MMA using the MMAtwo depolymerization process.

The carbon and water footprints, and energy consumption were computed from process models, in order to generate an average for the variations of the C3 processes. All of them consume large amount of energy and have a high carbon footprint. The MMA production processes require diverse natural resources, from fossil carbon (natural gas and petroleum) to minerals (for catalysts, reactors, energy production, etc.). So, the water consumption is also quite large as the calculation is done on a cradle to gate basis, meaning, including the production of raw materials and up to the production of virgin MMA in the producing plant.

For the recycling processes, we considered the collection (or PMMA scraps), sorting, and pretreatment steps, but also the depolymerization and the purification up to the selected level. To have a fair comparison, the purity level is selected to be equivalent to virgin-MMA (99.8% vs 99.9%). PMMA scraps are collected with other polymers, such as the polyethylene film, which is protecting the PMMA sheets. In the model here, the coproducts are considered to be valorized and to substitute other products. In the recycling process, side products are also generated such as solid residue and gases at the depolymerization, but also light and heavy boilers at the purification as distillation was considered here. Whenever possible, these side products were valorized to produce energy, and then substitute fossil fuels.

However, in these calculations, we never took into account the avoided emissions from the incineration of PMMA, which would have been a possible end of life for the PMMA scraps. In the future, incineration and land-filling should be prohibited, so that it is not a fair benchmark for plastic recycling operations.

In this first example, the carbon footprint reduction reaches 87% while the water consumption reduction is even higher. Indeed, virgin MMA C3 process consumes a lot of energy and water to be produced (and a lot of natural resources), while for the recycling process, to simplify: the carbon footprint is nearly equally distributed among the collection, pretreatment, depolymerization, and purification steps, and is much lower than for the virgin MMA. It also means that even if we collect PMMA scraps from a larger distance, it will have a minor impact on the global carbon footprint. The water consumption is low in the recycling process, as water is only needed in the cooling steps as heat carrier fluid.

Calculator

Calculate the **environmental impacts** of MMAtwo technology and value chain to **recycle PMMA** and the environmental benefits compared to other technologies.

● Learn more about the study

Quantity (between 1 and 1'000'000 kg)

1 kg ↕ **0,75** kg ↕

PMMA scraps to recycle   r-MMA to purchase

r-MMA purity   Market/Region

99,8% ›   **Europe**   Asia

Quality of scraps

80/20 mix grade (EU baseline mix: 80% Cast & 20% Extrusion) ›

Climate change

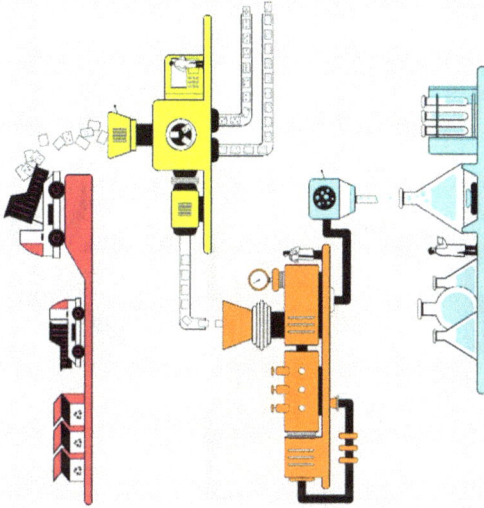

■ Collection — 21.5%   ■ Depolymerization — 32.6%
■ Pre-treatment — 25.3%   ● Purification — 20.6%

Environmental impacts comparison

	r-MMA	Virgin C3
purity of r-MMA	99.8%	99.9%
climate change	0.66 kg CO_2-eq	4.78 kg CO_2-eq
water use	0.05 m³ depriv.	22.7 m³ depriv.
resource use, fossil	4.43 MJ	88 MJ

Environmental benefits of **MMAtwo** vs. **Virgin C3**

–87%	–100%	–95%
climate change	water use	resource use, fossil

*These environmental indicators are the computed values of cumulated emissions generated by the technologies being discounted of cumulated emissions due to energy valorisation of side products. When the resulting number is small, the uncertainty on the net value is large due to the subtraction of 2 large numbers, and comparison should be done with caution.

Figures 13.1 and 13.2: Comparison of virgin MMA (C3 process) with regenerated MMA (99.8% purity) from the MMAtwo process (twin-screw extruder) – available also at https://lnkd.in/gZb2fzsZ.

Calculator

Calculate the **environmental impacts** of MMAtwo technology and value chain to **recycle PMMA** and the environmental benefits compared to other technologies.

🔵 Learn more about the study

Quantity (between 1 and 1'000'000 kg)

| 1,33 | kg | ⬍ | **1** | kg | ⬍ |

PMMA scraps to recycle   r-MMA to purchase

r-MMA purity   Market/Region

99,8% | **Europe** | Asia

Quality of scraps

80/20 mix grade (EU baseline mix: 80% Cast & 20% Extrusion)

Climate change

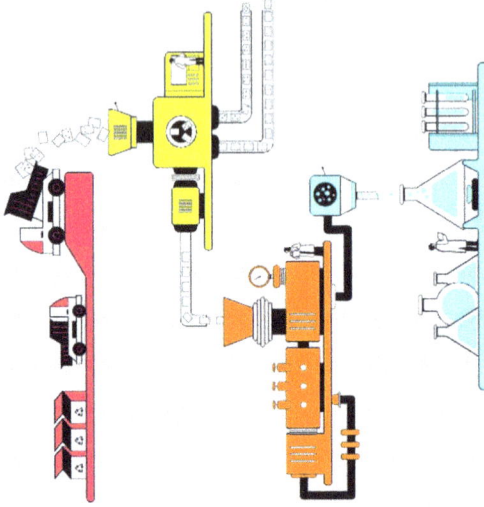

■ Collection — **21.5%**   ■ Depolymerization — **32.6%**
■ Pre-treatment — **25.3%**   ■ Purification — **20.6%**

Environmental impacts comparison

	MMA	Virgin C3
purity of r-MMA	99.8%	99.9%
climate change	0.88 kg CO_2-eq	6.37 kg CO_2-eq
water use	0.07 m³ depriv.	30.2 m³ depriv.
resource use, fossil	5.9 MJ	117 MJ

Environmental benefits of **MMAtwo** vs. **Virgin C3**

| **−87%** climate change | **−100%** water use | **−95%** resource use, fossil |

*These environmental indicators are the computed values of cumulated emissions generated by the technologies being discounted of cumulated emissions due to energy valorisation of side products. When the resulting number is small, the uncertainty on the net value is large due to the subtraction of 2 large numbers, and comparison should be done with caution.

Figures 13.1 and 13.2 (continued)

13.3 Example 2: MMAtwo technology (twin-screw extruder) versus the C4 (isobutene and tertiobutanol oxidation) process

In Japan and other Asian countries, a common process to produce MMA is using iso-butene (isobutylene) or tertiobutanol oxidation. Here also, there are several variations of processes. When tertio-butanol is used as feedstock, it is dehydrated on a first catalyst layer to isobutene and water, which then reach an oxidation catalyst to produce methacrolein. When isobutene is the feedstock, it is mixed with steam and oxidized directly on the same type of catalyst. Methacrolein can be oxidized to methacrylic acid, which is further esterified to MMA, or it can be oxidized in the presence of methanol at low temperature to directly produce MMA (oxyesterification process).

For a fair benchmark, the C4 process, which is common in Asia, is compared with the MMAtwo PMMA depolymerization in an Asian environment, i.e., with the carbon footprint of the electricity production average in Asia. Although several variations exist of the C4 process, an average is taken into account. Some improvements, both for the virgin and the recycled processes, are possible when using a more sustainable electricity source Figure 13.3.

The feedstock considered this time is "Cast PMMA," but one has to consider that in fact the waste is contaminated with PVC. The material of choice is PMMA cast sheet off-cuts, which still contains the PVC gasket. In cast PMMA sheet production, a syrup is poured between two glass sheets surrounded by a PVC gasket. When the sheets have polymerized, they are cut to dimensions, and there is always a fraction of "good" cast PMMA that remains contaminated with the PVC gasket. The separation of the PVC is needed to avoid corrosion of the depolymerization reactor or downstream purification. Some technologies are being developed [2, 3] to avoid using intensive labor that manually cut the PVC, but the PVC recovered can be valorized and sold to PVC recyclers. The sheets also come with a polyethylene protective film, which has to be removed and recycled, generating also some value (and improving the environmental footprint). So in this example, we consume 1.95 kg of scraps to generate 1 kg of rMMA, at 99.8% purity. At this level of purity, the regenerated MMA competes with the virgin MMA. The C4 route has a lower carbon footprint than the C3 route seen in Europe, but still a higher carbon footprint than the depolymerization process.

It also means that it is more difficult to implement a PMMA recycling plant in Japan because the "State-of-the-art" process consumes less energy than the European C3 process. Indeed, we should consider that any marketed rMMA will substitute the locally state-of-the-art virgin MMA.

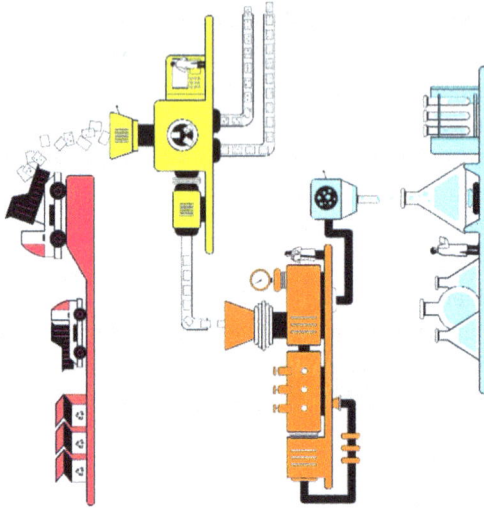

Figure 13.3: Comparison of the MMAtwo technology in an Asian environment with the C4 virgin process.

13.4 Example 3: MMAtwo technology (twin-screw extruder) versus the dry distillation process in Asia

One of the oldest depolymerization process and common in Asia (India, China, etc.) is Dry distillation. The technology is very simple and is used in batch processes, as explained in other chapters of this book. However, it is not possible to achieve a high purity, and most often, the purity seen on the market is 96.5%. When the operator selects the PMMA scraps and operates in best conditions, it could be possible to achieve about 98% purity level. So, for the fair benchmark here, we compare the MMAtwo technology above 95%. The purity level has a significant impact, as to achieve higher purity, one must accept to have a lower yield. This means less rMMA is produced and more energy is consumed, and so there is a higher carbon footprint.

The dry distillation technology is common in Asia and South America to regenerate MMA from PMMA scraps. In this process, a pile of PMMA scraps is loaded in a tank in batch operations. Fire is lighted under the tank and whatever can distil (evaporate) is collected in a condenser and further purified. Depending on the location, different energy sources would be used (fuel, gas, coal or wood). So, there is a wide dispersion of the environmental footprint, and in the analysis, a most representative case is taken into account Figure 13.4.

For the example, a mix of 80% cast sheet and 20% extrusion sheets scraps is used as feedstock. These are off-cuts from larger sheets, and so it is usually not contaminated with PVC, but still has Polyethylene protective film, which is separated and valorized. For 1 kg of rMMA, we need 1.26 kg of scraps. The comparison is done in Asia, and we require MMA of a purity above 95%.

The lowest purity we achieved with MMAtwo was 98%, and this is selected for a fair benchmark with the 96.5% purity for the dry distillation representative case. The electricity supply is taken from an Asian grid mix.

What we can see from the environmental impact parameters is that there is not much difference between the two cases (although the MMAtwo technology is able to deliver a much higher purity for the rMMA). So, the rMMA produced with dry distillation can make PMMA products that satisfy the local markets, with also a significant impact reduction versus the virgin MMA. The difference between the two technologies is within calculation and assumption errors and should not be considered as discriminating. However, when the dry distillation operators would want to produce a higher quality product, they would significantly degrade their environmental footprint. In Europe, the market is expecting better performances and purity.

Calculator

Calculate the **environmental impacts** of MMAtwo technology and value chain to **recycle PMMA** and the environmental benefits compared to other technologies.

🔶 Learn more about the study

Quantity (between 1 and 1'000'000 kg)

1,26   kg ⇅   **1**   kg ⇅

PMMA scraps to recycle   r-MMA to purchase

r-MMA purity   Market/Region

95,0% ›   Europe   **Asia**

Quality of scraps

80/20 mix grade (EU baseline mix: 80% Cast & 20% Extrusion) ›

Climate change

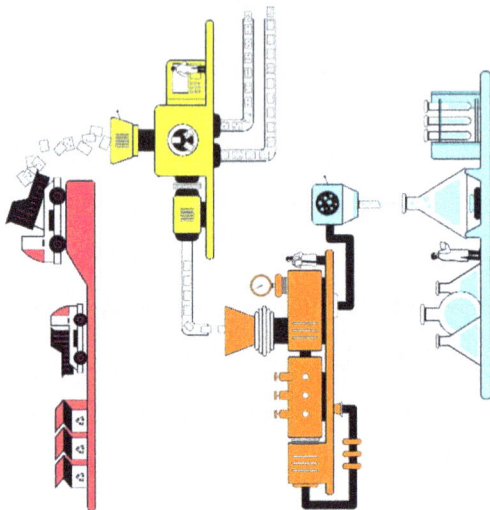

■ Collection — 15.1%   ■ Depolymerization — 57.5%
■ Pre-treatment — 18.9%   ● Purification — 8.4%

Environmental impacts comparison

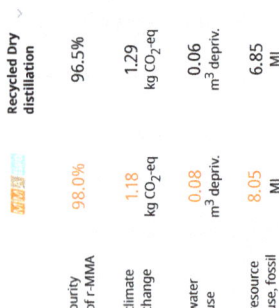

	MMAtwo	Recycled Dry distillation
purity of r-MMA	98.0%	96.5%
climate change	1.18 kg CO_2-eq	1.29 kg CO_2-eq
water use	0.08 m^3 depriv.	0.06 m^3 depriv.
resource use, fossil	8.05 MJ	6.85 MJ

Environmental benefits of **MMAtwo** vs. **Recycled Dry distillation**

–9% climate change	+33% water use	+17% resource use, fossil

*These environmental indicators are the computed values of cumulated emissions generated by the technologies being discounted of cumulated emissions due to energy valorisation of side products. When the resulting number is small, the uncertainty on the net value is large due to the subtraction of 2 large numbers, and comparison should be done with caution.

Figure 13.4: MMAtwo technology versus dry distillation process in an Asian environment.

13.5 Example 4: MMAtwo technology (twin-screw extruder) versus the molten lead process in Europe

Historically, in Europe, the leading process for PMMA depolymerization has been the molten lead process. In this process, PMMA scraps are poured in the bath of molten lead. The process uses a low melting point metal, which serves as a heat transfer medium. Lead could be replaced by other metals or eutectics, but so far it has been the best performer. Data have been collected from several plants using this technology, which is still operated in several countries in a continuous mode. The purity of the rMMA is usually around 98.5% but can reach 99.0% with selected scraps. That's the assumption taken here, and to reach that purity level, "Cast" PMMA scrap (i.e. Cast contaminated with PVC, but sorted in the pretreatment stage) was assumed to be used. Like in the previous example, the difference between the MMAtwo technology and the molten lead process is within assumption and calculation errors. Especially in this case, there is valorization of the side products and PVC, which have a contribution to reduce the carbon footprint of the rMMAs Figure 13.5.

We need 1.33 kg of scraps (mix cast PMMA and extrusion sheets) to produce 1 kg or rMMA with the MMAtwo process at 99.0% purity. However, we need 1.89 kg of the PVC-contaminated PMMA per kg of rMMA because of the high PVC content in this type of waste. We selected that level of purity since this is the best we have seen with the molten metal process, while MMAtwo can do better, in order to have a fair comparison.

The molten metal process was widely used in Europe, and is still in use there and also in Africa, Asia, and South America. All the large PMMA producers once had a molten metal depolymerization plant, which have since been shut down, mostly for economic or strategic reasons. PMMA scraps are continuously fed to the molten metal reactor, where they are heated and depolymerized. MMA vapors produced are collected and condensed, to be further purified.

The molten metal process seems to have a lower carbon footprint than the MMAtwo technology, but it cannot achieve the highest purity, and cannot process all kind of wastes. Indeed, if the PMMA scraps are highly contaminated and generate a lot of solid residues, these residues would be contaminated by the metal used (lead) and would have to be disposed and treated properly. However, again in this case, one has to remember that the final differences between the processes are within the calculations uncertainties.

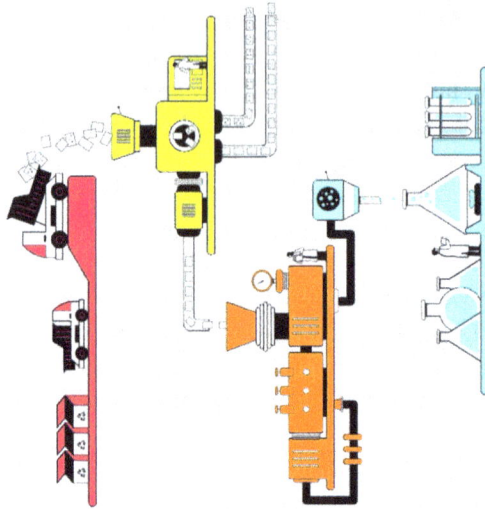

Figure 13.5: MMAtwo twin-screw extruder technology for PMMA depolymerization versus the molten metal technology.

13.6 Example 5: MMAtwo technology (twin-screw extruder) versus the twin-fluid bed process in Asia

PMMA depolymerization to rMMA in a twin-fluid-beds reactor has been heavily documented in the academic literature. It was operated at industrial scale in Japan. So, this MMA regeneration technology should best be compared with the C4 process, which uses isobutene or tertiobutanol as feedstock and which has already been described in Example 2 Figure 13.6.

In this example, the PMMA scraps are injection-grade PMMA particles, which mean a copolymer of MMA and an acrylate (methyl, ethyl or butyl acrylate). We need 1.25 kg of scraps per 1 kg of rMMA for the MMAtwo technology, and for a rMMA purity above 95%. For this type of scraps, the twin-fluid-bed process (one reactor and one regenerator to reheat the sand, which serves as heat transfer medium) gave 97.7% rMMA purity (and with methyl acrylate as comonomer). For a fair benchmark, we also selected 98% for the twin-screw extruder process (MMAtwo technology), although we could reach 99.8%.

The results show a high carbon footprint for the fluid bed technology and high energy consumption. This means that the fluid bed process would show a lower improvement versus the C4 process (Virgin MMA), and it is probably one of the reasons why the technology was ultimately decommissioned.

13.7 Example 6: MMAtwo technology (twin-screw extruder) versus the rotating drum process in Asia

The technology that we called "Rotating Drum" is common in plastics pyrolysis. In this technology, a batch is loaded in a big cylinder, which is put in rotation and heated externally. It is a variation of the dry distillation, with some agitation. It is more common in Asia, but it has several drawbacks. The rMMA purity achieved is more often around 95%, and in the benchmark below, we compare it with the lowest purity (98%) achieved in MMAtwo. A lower purity also means a higher yield, since we do not purify extensively the rMMA. There is also then lower energy consumption. The PMMA scrap we selected here is the mix of cast and extrusion sheets, which is most often the post-production mix with multiple colors and no PVC included. For 1 kg of rMMA, with the MMAtwo technology, we need to collect 1.25 kg of scraps Figure 13.7.

Calculator

Calculate the **environmental impacts** of MMAtwo technology and value chain to **recycle PMMA** and the environmental benefits compared to other technologies.

● Learn more about the study

Quantity (between 1 and 1'000'000 kg)

1,25   kg ⌄ **1**   r-MMA to purchase   kg ⌄

PMMA scraps to recycle

r-MMA purity   Market/Region

95,0% ⌄   Europe ⌄   **Asia**

Quality of scraps

Injection grade

Climate change

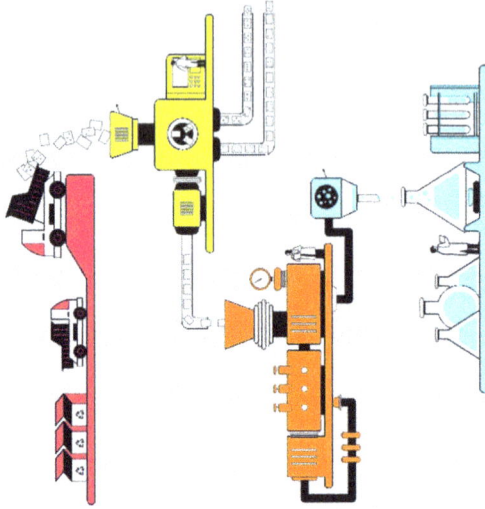

■ Collection — **15.7%**   ■ Depolymerization — **61.6%**
■ Pre-treatment — **10.8%**   ■ Purification — **12.0%**

Environmental impacts comparison

	MMA	Recycled Twin fluid bed
purity of r-MMA	98.0%	97.7%
climate change	1.39 kg CO_2-eq	2.69 kg CO_2-eq
water use	0.11 m^3 depriv.	0.46 m^3 depriv.
resource use, fossil	9.36 MJ	24.7 MJ

Environmental benefits of **MMAtwo** vs. **Recycled Twin fluid bed**

climate change	water use	resource use, fossil
−49%	**−77%**	**−63%**

*These environmental indicators are the computed values of cumulated emissions generated by the technologies being discounted of cumulated emissions due to energy valorisation of side products. When the resulting number is small, the uncertainty on the net value is large due to the subtraction of 2 large numbers, and comparison should be done with caution.

Figure 13.6: PMMA depolymerization to rMMA in fluid bed reactor compared with the MMAtwo technology with injection-grade wastes.

Calculator

Calculate the **environmental impacts** of MMAtwo technology and value chain to **recycle PMMA** and the environmental benefits compared to other technologies.

🔶 Learn more about the study

Quantity (between 1 and 1'000'000 kg)

1,25   kg ↕   **1**   kg ↕

PMMA scraps to recycle   r-MMA to purchase

r-MMA purity   Market/Region

95.0% ›   Europe   **Asia**

Quality of scraps

80/20 mix grade (EU baseline mix: 80% Cast & 20% Extrusion) ›

Climate change

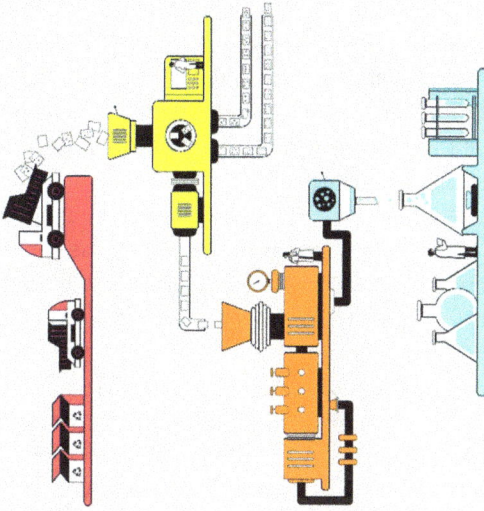

🔴 Collection — **15.1%**   🟠 Depolymerization — **57.5%**
🟡 Pre-treatment — **18.9%**   🔵 Purification — **8.4%**

Environmental impacts comparison

	MMA	Recycled Rotating drum
purity of r-MMA	98.0%	95.0%
climate change	1.18 kg CO_2-eq	1.14 kg CO_2-eq
water use	0.08 m³ depriv.	0.1 m³ depriv.
resource use, fossil	8.05 MJ	10.9 MJ

Environmental benefits of **MMAtwo** vs. **Recycled Rotating drum**

+3% climate change	**−21%** water use	**−26%** resource use, fossil

*These environmental indicators are the computed values of cumulated emissions generated by the technologies being discounted of cumulated emissions due to energy valorisation of side products. When the resulting number is small, the uncertainty on the net value is large due to the subtraction of 2 large numbers, and comparison should be done with caution.

Figure 13.7: Comparison of the MMAtwo technology with the rotating drum process.

The environmental footprint benchmark shows that both technologies are very close to each other, but with an advantage to the MMAtwo process that can deliver a much higher purity.

13.8 Example 7: MMAtwo technology (twin-screw extruder) versus the stirred tank process in Europe

For this example, we are back in Europe, which means with the average electricity grid of Europe. The benchmark technology is a stirred tank process. The data for that process are based on a mix of paper studies and operating technologies. The process benefits also of some improvements included in the paper study (the current reference plant does not achieve the same purity), so that a purity of 99.7% could be achieved with better yields. A technology of this type is illustrated in reference [4].

The feedstock is "cast PMMA," which as in previous cases, is a PMMA and PVC off-cuts, in which the PVC gasket contaminates the PMMA. Those gaskets must be separated before depolymerization. PVC is valorized separately. We therefore need 1.9 kg of scraps for 1 kg of rMMA produced with the MMAtwo technology Figure 13.8.

For a fair benchmark, we compare this "stirred tank" process with MMAtwo at 99.5% purity, although we could achieve a higher 99.8% purity.

The environmental footprint benchmark shows that the two technologies are very close to each other.

13.9 Conclusions

Do not hesitate to play with MMAtwo's Footprinter – Environmental Benefits Calculator (mmatwo-footprinter.eu), which has been constructed with a rather large database of plants all over the world. It is possible to compare different technologies, different scraps, and geographical environments.

One of the main conclusions is that most of the recycling processes require less energy and water than the virgin processes, whatever the purity level they can achieve. PMMA depolymerization has already a good track record of sustainable processes. But we can do better and continue to improve. If renewable electricity is used as an energy source the climate change parameter will further improve. New technologies should also pay more attention to the water consumption and focus their interest on low quality scraps, which are still landfilled or incinerated today.

Calculator

Calculate the **environmental impacts** of MMAtwo technology and value chain to **recycle PMMA** and the environmental benefits compared to other technologies.

🔸 Learn more about the study

Quantity (between 1 and 1'000'000 kg)

1,9   kg   ⇕   **1**   r-MMA to purchase   kg ⇕

PMMA scraps to recycle

r-MMA purity   Market/Region

99,5% › | **Europe** | Asia

Quality of scraps

Cast grade   ›

Climate change

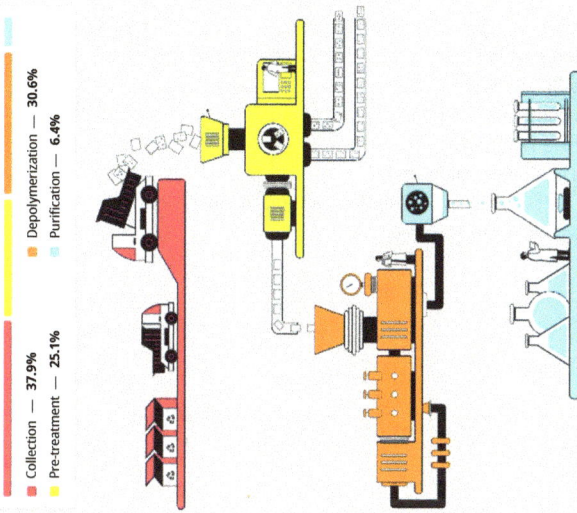

🔴 Collection — **37.9%**   🟠 Depolymerization — **30.6%**
🟡 Pre-treatment — **25.1%**   🟦 Purification — **6.4%**

Environmental impacts comparison

	MMAtwo	Recycled Stirred tank
purity of r-MMA	99.5%	99.7%
climate change	0.87 kg CO_2-eq	0.99 kg CO_2-eq
water use	n.a. m³ depriv.	n.a. m³ depriv.
resource use, fossil	6.81 MJ	4 MJ

Environmental benefits of **MMAtwo** vs. **Recycled Stirred tank**

−13% climate change   **n.a.** water use   **+70%** resource use, fossil

*These environmental indicators are the computed values of cumulated emissions generated by the technologies being discounted of cumulated emissions due to energy valorisation of side products. When the resulting number is small, the uncertainty on the net value is large due to the subtraction of 2 large numbers, and comparison should be done with caution.

Figure 13.8: Comparison of the MMAtwo technology with the stirred tank process.

13.10 Acknowledgments

The MMAtwo project has received funding from the European Union's Horizon 2020 research and innovation program under grant agreement no. 820687.

The MMAtwo project results presented reflect only the authors' views. The Commission is not responsible for any use that may be made of the information it contains.

References

[1] Darabi Mahboub MJ, Dubois JL, Cavani F, Rostamizadeh M, Patience GS. Catalysis for the synthesis of methacrylic acid and methyl methacrylate. Chemical Society Reviews 2018, 47, 7703–7738. https://doi.org/10.1039/C8CS00117K.

[2] Dubois JL, Bentaj A, Couchot N, PCT Patent application WO2022123176 – Method For Separation Of A Plastic Article, Priority date 11/12/2020.

[3] Triboelectricity (various equipment suppliers).

[4] De Tommaso J, Dubois JL. Risk analysis on PMMA recycling economics. Polymers 2021, 13, 2724. https://doi.org/10.3390/polym13162724.

Jacopo de Tommaso, Gregory S. Patience, and Jean-Luc Dubois

14 Recycling of artificial marble

Artificial marble, also known as solid surface, is a composite material of an inorganic pigment, usually about 60 mass % aluminum trihydroxide (or ATH) and the rest is a polymeric matrix of methyl methacrylate (MMA), comonomers, and additives. The technology developed by R&E in Korea is the sole technology implemented in the world to recycle this material. The other polymer recycling technologies investigated are incapable of processing this material because of the high fraction of solid residue that accumulate in the reactor (e.g., dry distillation and rotating drum), or the need to clean the heat transfer medium continuously (e.g., molten lead or fluid bed).

The process developed by R&E is a stirred tank reactor, especially designed for solid surface products. The company operates three sites in Korea in a cluster of solid surface producers (Figure 14.1). They produce MMA and alumina. In the process, they dehydrate aluminum trihydrate below 250 °C, and then depolymerize PMMA. In this way, they aim to minimize the hydrolysis of the MMA. The coked alumina is then calcined at high temperature to produce a low-surface area alumina.

This process treats composites so that the value generated by the alumina fraction is good enough to take a share of the cost and of the CO_2 emissions, and the overall process is sustainable. However, the impurities generated during the R&E's process limit the quality of the rMMA, together with those inherently present in the formula of the polymeric matrix.

Acronyms

MMA	Methyl methacrylate
Crude MMA	Crude methyl methacrylate
rMMA	Regenerated methyl methacrylate (purified)
PMMA	Poly(methyl methacrylate)
$Al(OH)_3$	Aluminum hydroxide
ATH	Aluminum hydroxide
Al_2O_3	Alumina
MAA	Methacrylic acid
MA	Methyl acrylate
EA	Ethyl acrylate
MIB	Methyl isobutyrate
E/M propionate	Ethyl/methyl propionate

Jacopo de Tommaso, Gregory S. Patience, Ecole Polytechnique de Montréal, Polytechnique Montréal, 2500 ch. de polytechnique, Montréal, Québec, Canada
Jean-Luc Dubois, Trinseo France SAS – Altuglas International SAS, Tour CB21, 16 place de l'Iris, 92400 Courbevoie, France

https://doi.org/10.1515/9783111076997-014

14.1 Introduction

In this chapter, we analyze the waste artificial marble recycling process, a mixture of PMMA, aluminum trihydroxide (ATH) or Al(OH)$_3$, and additives. The products are also known as "Solid Surface" or "CORIAN" (DuPont's brand name). In Korea, there is a recycling cluster of this waste, with many patents filed by several companies and research institutes. We will focus on the technology developed by R&E, a Korean company which claims to recycle 35,000 t/year of waste artificial marble (35% PMMA, 1% additives, and 64% Al(OH)$_3$ (by mass)) in three different plants [1]. They collect waste artificial marble scraps and dust from nearby companies such as LG Hausys, Hyundai, or Dupont. Based on their website, patents, and a publication, we recreated a conceptual process design to identify features and limitations of this process: a semicontinuous stirred tank reactor, fed with an auger screw. The depolymerization operates in two stages, with a first dehydration around 250 °C and a depolymerization at higher temperature.

Dehydrated alumina is semicontinuously removed from the reactor with residual carbon and directed to a calcination furnace, where the combustion of the carbon residue provides energy for calcination. The crude MMA enters a complex purification process, combining decantation, chemical treatments, filtrations, and distillations. Purified MMA is produced at a yield of about 54% (by mass) (from initial PMMA content). The artificial marble is made from Al(OH)$_3$ and MMA syrup. This latter derives from

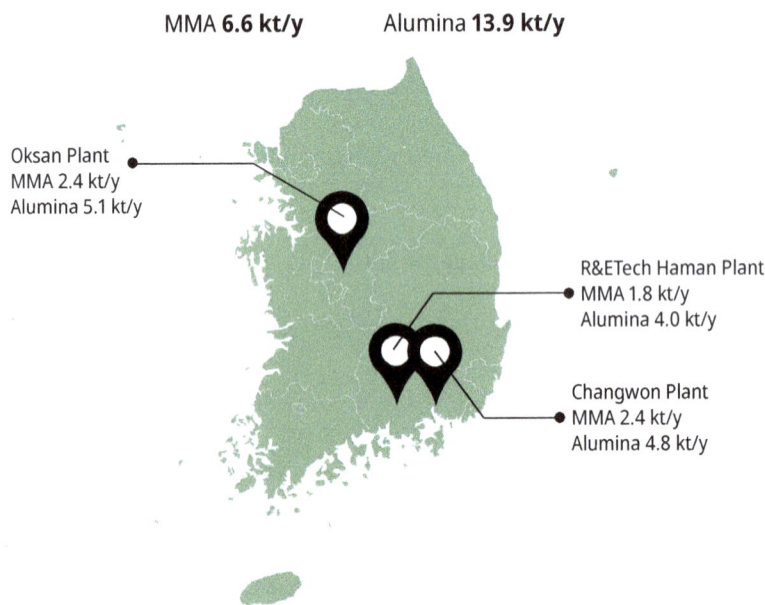

MMA 6.6 kt/y **Alumina 13.9 kt/y**

Oksan Plant
MMA 2.4 kt/y
Alumina 5.1 kt/y

R&ETech Haman Plant
MMA 1.8 kt/y
Alumina 4.0 kt/y

Changwon Plant
MMA 2.4 kt/y
Alumina 4.8 kt/y

Figure 14.1: R&E solid surface chemical recycling plant in Korea [1].

MMA and PMMA beads or crushed extruded grade PMMA containing some acrylates, because low molecular PMMA dissolve more easily in MMA. Therefore, during depolymerization, acrylates are to be expected. Acrylates will delay the depolymerization during the heating stage at 250 °C.

14.1.1 Company presentation

R&E is a Korean recycling company, now part of Veolia (acquisition in 2019), a French multinational group that operates in water, waste management, and energy. R&E was originally established in 1999 in Korea; and since 2009, has patented several technologies for the recycling of waste artificial marble. The flowsheet [2] (corresponding to US patent [3]), and the technology [4] were probably acquired by R&E in 2010 from LOTTE Advanced Materials Co. Ltd [5]. Similarly, the pretreatment of the waste artificial marble is described in [6], the filler recovery section in [7], and the MMA purification in [8], which are all patents owned by R&E but built on, or referencing, former Lotte's patents.

14.2 Process steps

We assumed the process to consist in four sections [3] (Figure 14.2):
1. Pretreatment of the waste artificial marble
2. Pyrolysis and condensation
3. Crude purification
4. Filler calcination

14.2.1 Pretreatment (section 100)

Section 100 (Figure 14.3) processes artificial marble waste, dry and wet dust. The artificial marble wet dust is dried in a furnace (114), and then stored in a dust storage tank along with the dry dust marble (111). The marble scraps are pulverized in a train of pulverizers (116) (120 mm to 12 mm to 2–3 mm), sorted into dust and granules by a separator, and then stored. The dust and the granules are sent to the respective reactors (211) to be depolymerized (section 200, Figure 14.3).

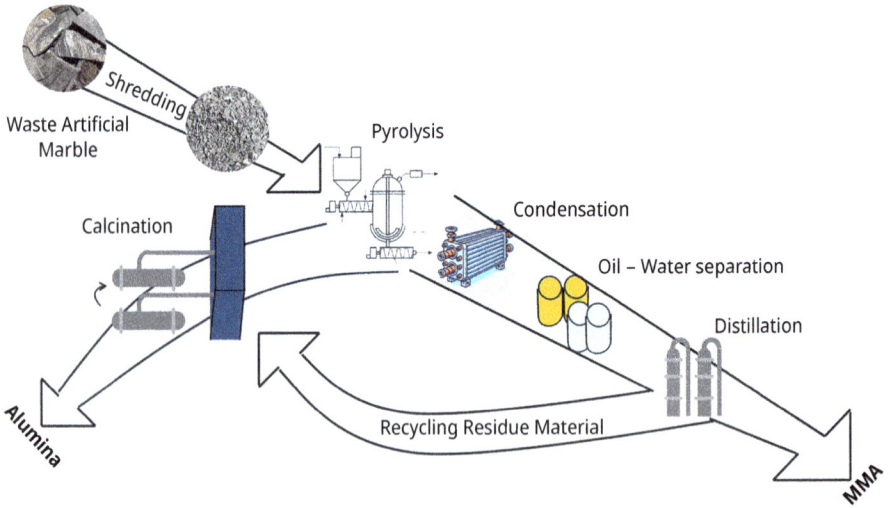

Figure 14.2: Simplified diagram of R&E process in four sections. First, the artificial marble is crushed and shred shredded, then it is pyrolyzed, producing a gas, and a solid product. The pyrolysis gases condense and then decant into an oil rich and a water rich phase. The oil-rich phase containing MMA, the target product, is distilled and stored. The solid residues from the reactor contain the inorganic filler, plus some polymer in the form of coke. Calcination transforms these residues to low surface area alpha-alumina.

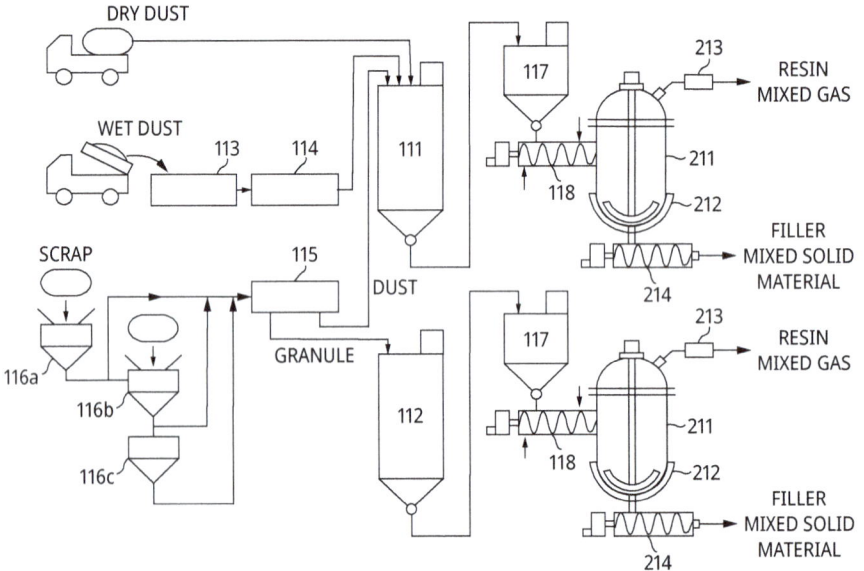

Figure 14.3: Pretreatment (100 sections) and depolymerization (200 sections) [3].

14.2.2 Depolymerization (section 200)

The depolymerization step is quantitatively described in [5], and the reactor concept (Figure 14.4) resembles the Plastic Energy mixed-plastics recycling configuration, i.e., a stirred-tank reactor. It thermally decomposes the artificial marble wastes while stirring and heating, and then condenses (311, Figure 14.7) the pyrolysis gases and recovers the alumina (214).

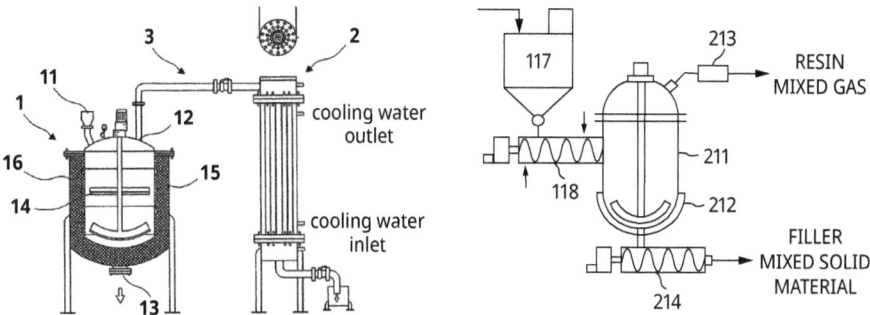

Figure 14.4: Depolymerization reactor and condensation (from [5] left, [3] right). The reactor (1) pyrolyzes particles coming from a buffer tank (117) via a preheated furnace (118) and nozzle (11). During the reaction, the particles are stirred (14) and externally heated (15). The gaseous products (12) are connected to a condenser (2) through a simple pipe (3) with a de-duster (213). The solid products are instead released by gravity (13) and conveyed (214) to the calcination section.

Because $Al(OH)_3$ in the artificial marble has the known dehydration reaction,

$$2Al(OH)_3 \Leftrightarrow Al_2O_3 + 3H_2O$$

Operating the depolymerization into two steps reduces the tendency of PMMA/MMA to hydrolyze.

First, the reactor heats the scraps (or dust) to about 250 °C, routing whatever vaporizes during the first heating process to a wastewater treatment step. $Al(OH)_3$ starts to decompose to aluminum oxyhydroxyde/alumina and water at around 220 °C (on-set dehydration temperature). About 70% of the water is removed during the heating step at 250 °C. Then, when the temperature is progressively increased from 250 °C to 400–450 °C, the PMMA part of the composite starts to depolymerize to MMA. Whatever vaporizes in this step – MMA, light gases, water, light impurities (acrylates, propionates, and isobutyrate), methacrylic acid and other organic acids, and some heavy-ends – is condensed. At least two reactors are operated semicontinuously on the reaction side [5]; and continuously on the purification side [3], eventually switching the reactors after each batch (Figure 14.5b).

(a)

(b)

(c)

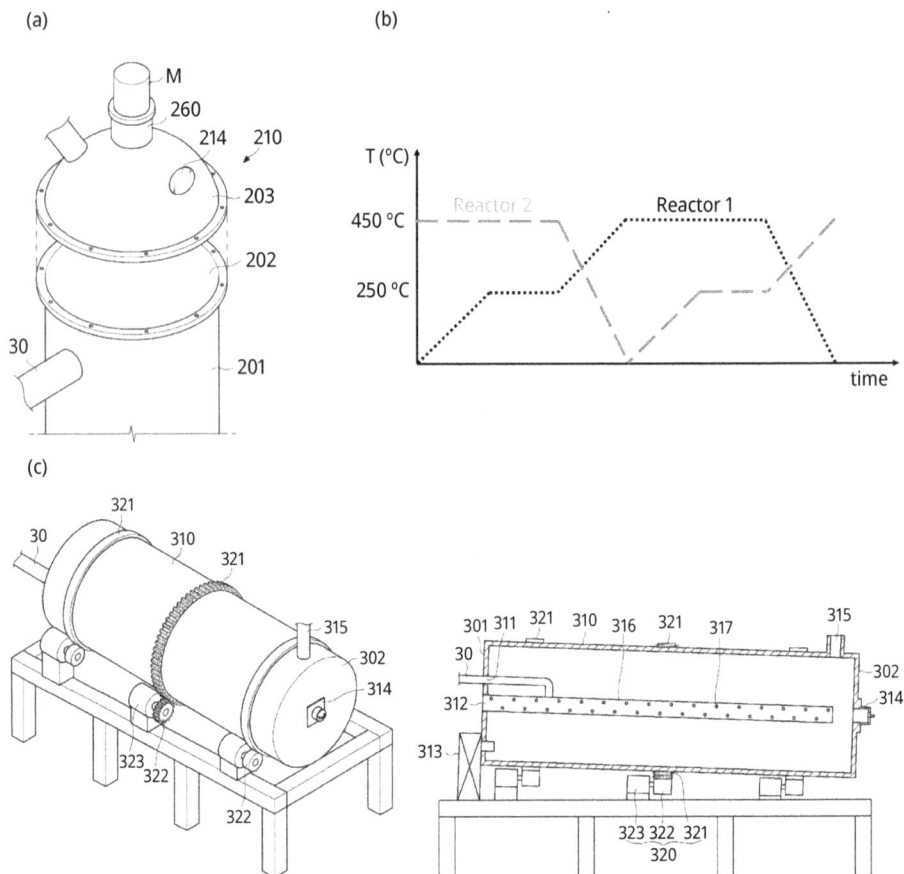

Figure 14.5: Detail of the pyrolysis zone (a) [9]; example of reaction cycles, with two reactions operating asynchronously in parallel. While one reactor dehydrates, the other pyrolyzes and vice versa. (b); and alumina calcination rotating drum kiln, where oxygen inflow (312) and an external burner (313) control the calcination temperature (c) [2].

The pyrolysis gases are condensed and purified by a "three-stage distillation tower" [3]. Accordingly, R&E claims that 1,000 kg of artificial marble (PMMA + Al(OH)$_3$) produce 440 kg of inorganic alumina (maybe as α-alumina) and a crude MMA, which is giving 370 kg of purified r-MMA.

The PMMA/Al(OH)$_3$ ratio (67:33) is consistent with a 64:35:1 ratio, where 1 is the additives (by mass). Since it is impossible to get 370 kg of MMA from only 326.7 kg of PMMA, we assume that the 370 kg corresponds to the MMA–water azeotrope (boiling point: 83 °C at 1 atm, 14% water, 86% MMA (by mass)). To produce 370 kg of azeotrope, the stream contains 318 kg of MMA and 52 kg of water. However, most probably, the stream contains more water than the simple azeotrope.

14.2.3 Purification (section 300)

In the purification section (Figure 14.7) [3], the pyrolysis vapors from the depolymerization reactor are first filtered (equipment 213, Figure 14.7) to remove some elutriated alumina dust (and maybe some carbon) and condensed (311). Then, the crude liquid passes through a first tri-phase separator (312), a cooler (313) to keep the crude at 10–15 °C, a second tri-phase separator (314) to degas the crude from the uncondensed lights, an additional cleaner (315) for the solid particles, and an oil–water separator (316) to remove part of the water. Prior to the first distillation, the crude is first treated chemically (318), with active carbon/clay to adsorb and remove some impurities originally present in the polymer (e.g., pigment), and then filtered (319) to separate and recover the inorganic material. The crude is then first distilled (321) to remove some heavy residue, part of the water and methacrylic acid (MAA), and other organic acids. The distillate is the MMA/water azeotrope, methanol, and the light impurities (MA, EA, MIB, E/M propionate, and organic acids). After partial condensation (324), the distillate is flashed/decanted (325), to degas the methanol (with some MMA, at azeotropic composition [10]), which is sent to a deodorant furnace (329). The cooled MMA–water stream decants in 325, and water is removed with some MMA.

After the first purification, the r-MMA is eventually treated chemically a second time (331) to remove some impurities, and then sent to a second purification section. Again, here the MMA and MMA/water azeotrope distils on top of (341) (for the remaining water traces in crude MMA), then is cooled (in 342), flashed/decanted in (343), degassed from some impurities close to MMA/water azeotrope (with probably acrylates), cooled in (344), filtered and treated (346) on water adsorbent (maybe $CaCl_2$), down to 0.35% water (by mass), and then finally stored. According to R&E, the MMA content in the rMMA is above 96.5% (by mass), with less than 0.35% of water (by mass). R&E recycles 35,000 t/year of artificial marble, made of 35% PMMA, 1% additives, and 64% $Al(OH)_3$ (by mass), to produce 6,600 t/year purified r-MMA and 13,860 t/year of alumina; therefore, their global r-MMA yield is only 53.8% (by mass).

As a summary, aluminum trihydroxyde partially dehydrates at 250 °C. Some water is left with the alumina, together with carbon residue. The carbon left should be sufficient to provide the energy for the alumina calcination step. A small amount of MMA is lost during that initial dehydration. The core of the MMA is produced during the heating at high temperature, together with the remaining water and organic impurities. In the purification section, MMA is recovered and purified. The remaining water, decanted, removed, and treated on site in a wastewater treatment section, together with utilities waters, drains, and wastewater streams from the other sections of the process.

In Figure 14.6, we report a block flow diagram of the process with the overall mass flux; the rationale of the balance and our assumptions will be explained in the following sections.

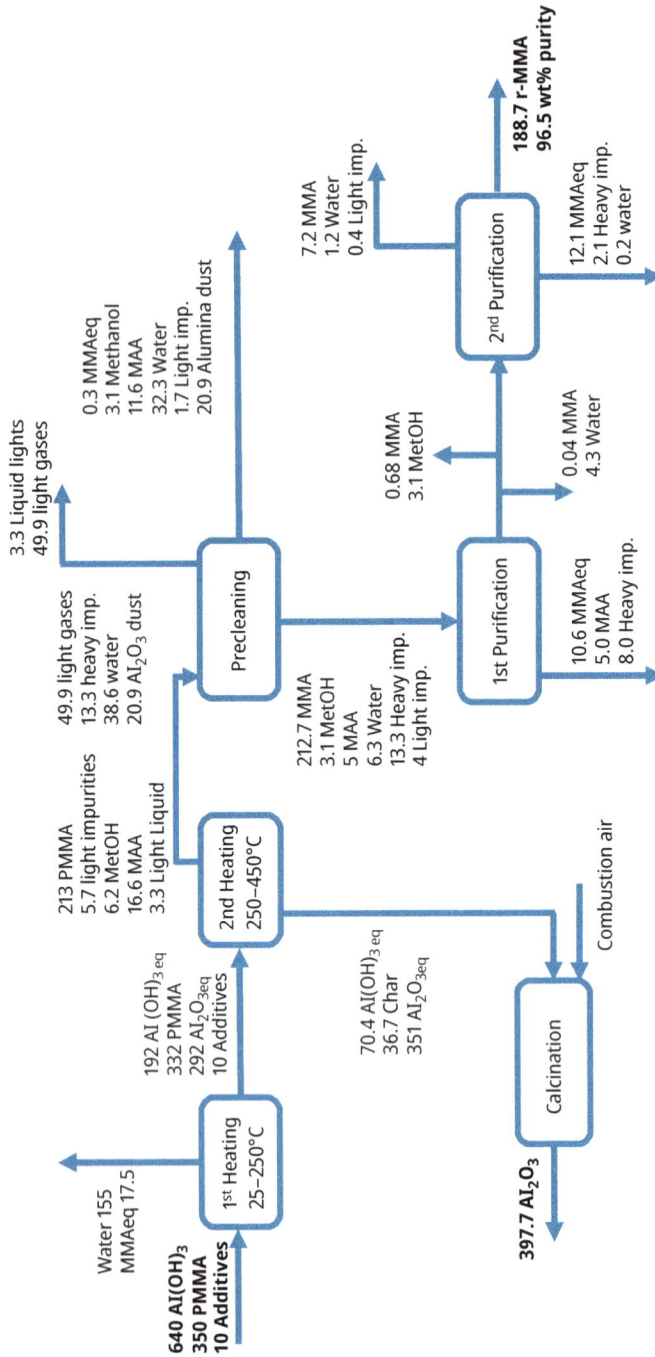

Figure 14.6: Block diagram of mass flows.

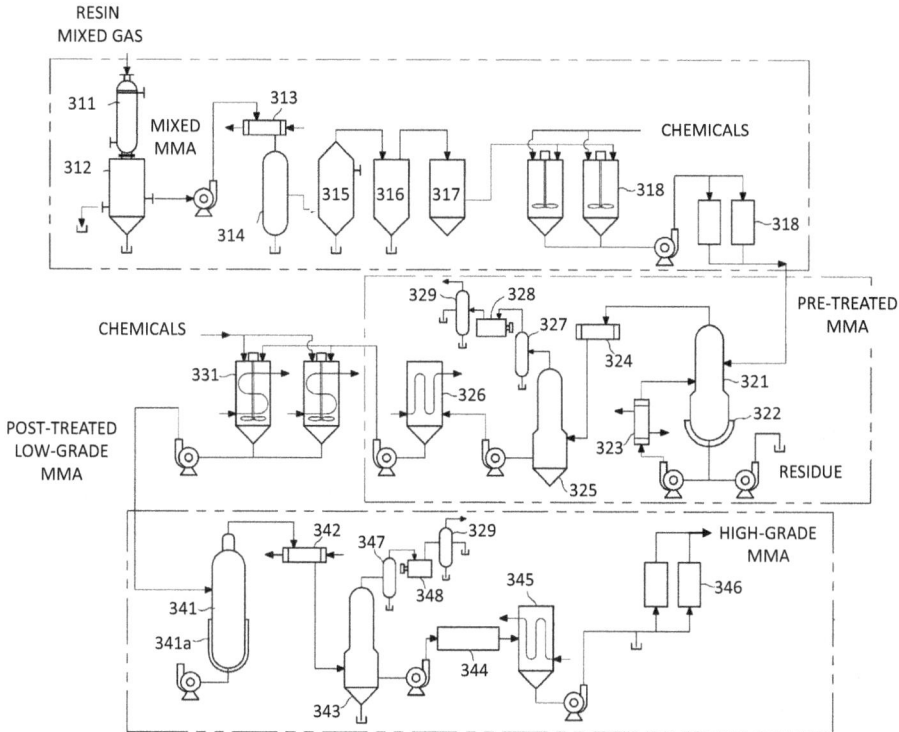

Figure 14.7: Purification section (section 300) [3].

14.3 Mass balance

We closed the mass balance of the process comprising the purification described in [3], as the data reported in [5] and [4] do not allow us to get a convincing data set. The global r-MMA yield is 53.8 by mass, and the r-MMA purity is 96.5%, with 0.35 by mass of water. The alumina yield of around 95%, with some losses due to the elutriation of the finer particles lost with the pyrolysis gases.

For the mass balance, the feed is therefore made of 350 kg/h PMMA, 10 kg/h additives, and 640 kg/h of Al(OH)$_3$, which is virtually able to release 222 kg of water.

We divide the process into four main sections:
- 100: Pretreatment
- 200: Pyrolysis
- 300: Condensation and purification
- 400: Alumina calcination

14.3.1 Pyrolysis

In the pyrolysis section, a first dehydration of the artificial marble at 250 °C generates gaseous products, which are routed to wastewater treatment. In a second heating stage from 250 to 400 °C [3] or 450 °C [5], the gases that are supposed to be mostly MMA are collected (Figure 14.5a and b).

The effect of pyrolysis temperature on the alumina quality suggests that the optimum in terms of chromaticity (color), surface area, and pore volume is a pyrolysis at 500 °C, which leaves 4–6% of carbon residue on the fillers (by mass) after pyrolysis [11]. A low carbon residue would mean a lower loss at the pyrolysis, less water is left with the alumina; and if the calcination temperature is properly controlled, one could theoretically get a high surface alumina. However, doing that, one might over-crack the PMMA, and concurrently needing a reactor that is able to reach and withstand this high pyrolysis temperature (500 °C).

The alumina produced has a low surface area (30–80 m^2/g), suggesting that the calcination is difficult to control, due to a higher carbon content, corresponding to a pyrolysis done at a lower temperature, more in the order of 450 °C, leaving a higher quantity (>7–8%) of carbon on the filler. 8% of residual PMMA on the alumina (by mass) would correspond to 10.2% of the initial PMMA (by mass), and this impacts the rMMA yield. This is in line with [3], where 8–15% PMMA (by mass) is left on fillers for the calcination. At lower pyrolysis temperatures (400–450 °C), the PMMA residue are not completely charcoaled, and therefore the resulting deposit is brownish.

In the current mass and energy balances, we opted for a pyrolysis temperature of 450 °C, corresponding to 8% carbon residue (by mass) and an alumina that could still be a high surface alumina, if properly calcined. In these operating conditions, the pyrolysis does not completely dehydrate the alumina. After the pyrolysis, the crude alumina is made of 397.3 kg of Al_2O_3, 24.4 kg of water (as alumina oxyhydroxide or $Al(OH)_3$ equivalent) and 36.7 kg of PMMA equivalent as carbon residue.

14.3.2 Alumina calcination

Depending on the type of hydroxide in the artificial marble, one could produce a different kind of alumina after calcination. When heated, gibbsite can either produce χ-alumina or boehmite ($AlO(OH)$), therefore only partially dehydrating. When calcined, boehmite will produce to γ, δ, or θ alumina, which are more interesting (especially, γ as it has high surface area and pore volume, and demand as catalyst support for various applications). Gibbsite does not completely dehydrate during the composite pyrolysis. At 250 °C, the fillers lose at least 10% of their weight in water (data from thermogravimetric analysis), but as the temperature is kept for some time at 250 °C, we assume a 70% loss (by mass) of the water at this stage (first heating), or 155 kg.

From the starting 640 kg of Al(OH)$_3$, after the preheating at 250 °C, we end up with 192 kg of Al(OH)$_3$ eq and 293 kg of alumina eq. (which is more alike a partially dehydrated/aluminum hydroxide).

After the first dehydration (at 250 °C), in the artificial marble, we still have:
– 95% of the PMMA, or 332.5 kg;
– 30% of the water to be released, or 66.5 kg;
– And all the potential alumina.

After the second heat treatment (250–450 °C), the alumina residue contains:
– water (24.4 kg of water as hydrate of alumina);
– carbon residue (36.7 kg of char, assuming 8% of carbon in the crude alumina (by mass)), and the complement of not fully dehydrated alumina 351 kg.

The crude MMA contains:
– 213 kg of MMA;
– 20.9 kg of Al$_2$O$_3$;
– 5.7 kg of light impurities (from MMA and additives);
– 6.2 kg of methanol (from MMA hydrolysis);
– 16.6 kg of methacrylic acid (from MMA hydrolysis);
– 3.3 kg of light liquids;
– 49.9 kg of light gases;
– 13.3 kg of heavy impurities; and
– 36.8 kg of water.

The carbonaceous residue on the crude alumina must provide heat for the calcination. To justify the choice of 8% of carbonaceous residue on the alumina (by mass), we analyzed the effect of the content of carbonaceous material left on alumina on the final calcination temperature. The adiabatic temperature rise is then calculated.

Assuming a lower heating value (LHV) of the carbonaceous material, equal to the one of solid PMMA (24.516 MJ/kg), the combustion of the carbon (or MMA) would have to provide heat for heating the air/diluted air stream, to heat the alumina, dehydrate alumina, and rearrange the alumina crystalline phase.

The heat of combustion of the char is consumed by the remaining alumina dehydration 169 MJ, the crystallization to alpha alumina 53 MJ, and 8.2 kg of stoichiometric air (O$_2$ + N$_2$) per kg of char (to burn the carbon residue assumed to be PMMA). 8% of char (by mass) (corresponding to 36.7 kg of carbon residue as remaining PMMA) generates 339 kg/h of flue gas (81 CO$_2$ + 26 H$_2$O + 232 N$_2$).

Therefore, assuming a constant specific heat of 1.2 kJ/kg/°C for the alumina and a constant specific heat for the flue gas of 0.8 kJ/kg/°C, the adiabatic temperature rise at calcination (ΔT) increases with the carbon residue. At 8 wt% residue, the adiabatic temperature rise is 890 °C, which is enough to carry the calcination.

For the mass balance, Table 14.1, we assumed 8% carbon residue on the crude alumina fillers (by mass), which seems reasonable considering the low surface alumina obtained by R&E.

In addition to these, for the pyrolysis we assumed:

- water released at first dehydration: 155 kg;
- MMA losses at first dehydration (by mass): 5% (17.5 kg);
- MMA/PMMA left after the first dehydration: 332.5;
- MMA losses at pyrolysis (by mass) = 30.6 kg PMMA as carbon char on the fillers, 15% as light gases (49.9 kg), 1% as liquid lights (3.3 kg), 4% as heavies (13.3 kg), 5% as MAA (16.6 kg), 0.5% as impurities (1.7 kg as EA, MA, EP, MP, MIB, odors, and organic acids); and
- 60% of the additives become carbon char, and 40% impurities (6 kg and 4 kg, respectively).

14.3.3 Purification

In the pre-cleaning part of the purification section (before the distillation), we assumed:

- Removing 30% (by mass) of the impurities from pyrolysis with the first chemical pretreatment (318/319) with clay/adsorbents.
- The solubility of MAA in water at 20 °C is 97 g/L, therefore, all the MAA should be soluble in the water in the first decanter (316). However, we assume that only 70% of the MAA (by mass) is separated by decantation, while the rest remains in the crude at this stage (i.e., in stream 10).
- The solubility of water in MMA is 1% (by mass) [12], but we assumed that the water entrained with the crude is three times its solubility limit, as droplets for example, or since methanol increases the water solubility.
- The solubility of MMA in water is 10.03 g/L at 15 °C [13] (operating temperature of the decanter 316), and we assumed that 100% of the soluble MMA is lost in the decantation.

In the first distillation section, we assumed:

- The column (321) distils on top the MMA/water azeotrope, along with the methanol, the light impurities, and 40% of the heavy impurities (by mass). The residues are the complement of the total water, 60% of the heavy impurities (by mass), the rest of the MAA (and organic acids) that was still in the crude, and 5% of the MMA of the crude (by mass) that is lost.

- The top of the column is cooled (324) (at around 50 °C), and then flashed in (325), what R&E calls "decanter tank." Here, we separate the methanol along with the MMA at the azeotropic composition [14], and water is decanted at the bottom from the MMA. We assumed that the water is entrained as droplets with the purified crude at its solubility limit in MMA (1% water in MMA (by mass)). The flash takes the heat from the liquid stream to evaporate the methanol azeotrope.
- We assume to remove some more light impurities in the second chemical treatment (331) (20% (by mass), a little less than the first chemical treatment). However, the nature of this treatment is unknown.

Finally, in the second distillation section, we assumed:
- The column (341) distils from top the MMA and MMA/water azeotrope, and the light impurities remain. The bottom residue is 6% of the MMA (by mass), and 40% of the remaining heavies (by mass).
- The top of the column is cooled (342) and flashed (343) to eliminate the MMA/water azeotrope and some odorous impurities (for instance acrylates), with some MMA loss to close the balance.
- The water still present (0.84 kg at this stage) is partially eliminated with a $CaCl_2$ (or magnesium sulfate) treatment (346), to bring it down to r-MMA specs (0.35% water in r-MMA (by mass)).

The purified r-MMA stream is then composed of (by mass):
- MMA: 96.5%;
- light impurities: 1.45%;
- heavy impurities: 1.7%;
- water: 0.35%.

Treating 1,000 kg/h of artificial marble, the process produces (see table 14.1):
- 299 kg of wastewater, with 34.2 kg (or 11.5% (by mass)) of organics in it;
- 91 kg of by-products to be burned, with (or without) energy recovery;
- 12.6 kg of light impurities/odors/MMAeq to be thermally oxidized on site;
- 188.7 kg of purified r-MMA (96.5% purity (by mass)) to be sold;
- 397.3 kg of Al_2O_3 to be sold.

Table 14.1: Mass balance.

Stream		1	2	3	4	5	6	7	8	9a	9
Name	Molecular weight	Artificial marble	Dehydration discharge	Preheated marble	Cracked gas	Fillers/crude alumina	Alumina	Crude MMA	Uncondensed gases	First precleaning purge	First decanter purge
Temperature	*(°C)*	*250*	*250*	*450*	*450*	*25*	*25*	*15*		*15*	*15*
PMMA		350									
MMA	100		17.5	332.5	213			213			0.3
Al(OH)$_3$	78	640		192		70.4					
Char						36.7					
α-Alumina	101.9			292.8	20.9	351.3	397.3			20.9	
Additives		10.0		10.0							
Light impurities (EA, MA, MIB, ketones, and pigments)					5.7			5.7		1.7	
Methanol	32.04				6.2			6.2			3.1
MAA	86.06				16.6			16.6			11.6
Light liquids					3.3				3.3		
Light gases (CO$_2$, CO, ethylene, C3, and CH$_4$)					49.9				49.9		
Heavy impurities					13.3			13.3			
Water	18		155.1		38.6	(24.4)		38.6			
Total		1,000		827.3	367.6	458.4	397.3	293.5	53.2	22.6	47.4

Stream		10	11	12	13a	13	22	14	15	16	17	18	19	20	21
Name	Molecular weight	Precleaned crude	Top first column	Residue first column	Flash to first odor removal	Second decanter purge	Second chemically treated purge	Chemically treated	Top second column	Residue second column	Decant. second column	Odor removal second column	Dirt rMMA	Water removal purge	rMMA
Temperature (*°C*)		83	83									83			
PMMA															
MMA	100	212.7	202.2	10.6	0.68	0.04		201.4	191.3	12.1		7.2	182.0		182.0
Al(OH)$_3$	78														
Char															
α-Alumina	101.9														
Additives															
Light impurities (EA, MA, MIB, ketones, and pigments)		4.0	4.0				0.8	3.2	3.2			0.4	2.8		2.8
Methanol	32.1	3.1	3.1		3.1										
MAA	86.1	5.0		5.0											
Light liquids															

(continued)

Table 14.1 (continued)

Stream	10	11	12	13a	13	22	14	15	16	17	18	19	20	21
Name	Precleaned crude	Top first column	Residue first column	Flash to first odor removal	Second decanter purge	Second chemically treated purge	Chemically treated	Top second column	Residue second column	Decant. second column	Odor removal second column	Dirt rMMA	Water removal purge	rMMA
Molecular weight	18													
Light gases (CO$_2$, CO, ethylene, C3, and CH$_4$)														
Heavy impurities	13.3	5.3	7.98				5.32		2.13			3.19		3.19
Water	6.3	6.3	0.0	3.8	4.3	0.8	2.0	2.0	0.00	0.0	1.2	0.84	0.18	0.66
Total	244	219	23.6	3.8	4.3	0.8	212	196	14.2	0.0	8.8	188.8	0.2	188.7

14.4 Energy balance

From our mass balance, we estimate the main energy consumption of the process. Assuming:

- $C_{pl,v}$, calorific value of liquid and gas MMA = 1.6 kJ/kg/K
- $C_{pl,v}$, calorific value of molten PMMA = 1.6 kJ/kg/K
- Heat of fusion (H_f) of PMMA = 50.2 kJ/kg
- Heat of depolymerization H_{dep} of PMMA = 540 kJ/kg
- Temperature of depolymerization = 450 °C
- Heat of condensation/(−vaporization) ($H_{cond,MMA}$) in E-201: 360 kJ/kg
- C_{pp}, calorific value (kJ/kg/K) of PMMA = 2.3 kJ/kg/K
- C_{pAl}, calorific value of Al(OH)$_3$/Alumina system = 1.2 kJ/kg/K [15, 16]
- $H_{dehydration}$ enthalpy of Al(OH)$_3$ dehydration to gamma/kappa/chi alumina = 2,400 kJ/kg Al(OH)$_3$ [15, 16]
- Reflux ratio in both distillation columns, $R = 3$
- Artificial marble "melting" temperature = 250 °C
- Dehydration temperature of Al(OH)$_3$ = 250 °C

14.4.1 Pyrolysis

To calculate the total energy needed to depolymerize and dehydrate the artificial marble, we split it into the energy needed for the two components of the solid: PMMA and Al(OH)$_3$, and then sum the two contributions. It will be also useful later when we need to split the energy and CO$_2$ emissions on the two products. For the PMMA, we need to heat it from 25 °C until the melting temperature (250 °C), melt it, heat the molten polymer up to the depolymerization temperature (450 °C), and then depolymerize it.

At the same time, we will heat up the Al(OH)$_3$ part of the artificial waste marble, from 25 °C to 250 °C (the dehydration temperature), and then heat up the Al$_2$O$_3$ formed to 450 °C. Therefore, the total energy required for the pyrolysis is:

For the PMMA part: 593 MJ, and for the alumina/Al(OH)$_3$: 1656 MJ. The total duty for the pyrolysis is 2,249 MJ. Splitting this energy in MMA and crude alumina, by mass allocation, we have:

- 878 MJ to crude MMA, and
- 1371 MJ to crude alumina.

14.4.2 Purification

The reboiler of the first column (321) consumes 264 MJ, while the second column consumes 229 MJ. So, the total energy required for the purification is 493 MJ, which we allocate fully to MMA.

The total overall energy for the process is 2,742 MJ.

14.4.3 Energy split by mass allocation

The MMA contribution to the total energy of the process is 878 + 493 = 1,371 MJ, corresponding to 7.26 MJ/kg purified r-MMA. This corresponds to the MMA contribution at cracking (by mass allocation) and the purification energy. The alumina contribution to the total energy of the process is also 1371 MJ, corresponding to 2.99 MJ/ kg crude alumina, or 3.45 MJ/kg of Al_2O_3. This energy represents the alumina contribution (by mass allocation) at the cracking.

Theoretically, part of this heat could be provided from the by-products combustion. In fact, stream 8, the uncondensed gases from pyrolysis, stream 12, and stream 16, the heavy residue of the two columns, could provide, respectively (assuming an LHV = 27.252 MJ/kg for all the streams): 1447 MJ, 639 MJ and 387 MJ. For a total of 2,476 MJ provided by by-product combustion, around 90% of the total duty is required by the process. The reason why the by-products could provide almost all the energy needed is also the very low MMA yield (54% (by mass)).

14.4.4 Water treatment

Although not mentioned directly in the process description, the high amount of wastewater produced (299 kg with 11.5% of organics in it (by mass)) calls for a water treatment unit on site. As a first approximation, the COD (chemical oxygen demand) equal to the pollutant concentration. In our case, the organics are mainly MMA equivalents, Methanol, and MAA, which means, for instance, to oxidize 34.4 kg/h of MMA, we would need 66.4 kg/h of O_2, and so the COD would be around twice the pollutant concentration.

Nevertheless, considering the broad nature of the pollutants (organics somehow close to MMA, MAA, and Methanol), we believe that an appropriate choice could be a catalytic wet air oxidation unit (CWAO). Air oxidizes the wastewater in a catalytic bed in an exothermic reaction. Then, the hot treated water from the reactor preheats the incoming polluted water. At the same time, the treated water cools down before degassing in a liquid–water separator. Here, we recover the treated water, separating the off-gases on top.

We split the water produced by mass allocation in r-MMA and alumina. All the streams coming from the purification section (S9, S9a, S13, S20, and S22) are completely allocated to MMA for a total of 0.39 kg wastewater/kg rMMA.

The wastewater from the alumina calcination (water vapor from char combustion + water vapor from dehydration) are allocated fully to alumina for 0.12 kg wastewater/kg alumina

Finally, the wastewater from cracking (172.7 kg) can be split between MMA and alumina as:

- 0.35 kg wastewater/kg r-MMA, and
- 0.26 kg wastewater/kg alumina.

For a total of:

- 0.39 + 0.35 = 0.74 kg wastewater/kg r-MMA, and
- 0.12 + 0.26 = 0.38 kg wastewater/kg alumina.

14.4.5 Cooling water estimates

We must cool the crude MMA generated at the depolymerization, and the streams at the distillations. The calcined alumina is also heated to high temperature and would have to be cooled. However, in the case of the alumina, we assume that the calcination is done in a counter current furnace. It means that cold air contacts the hot calcined alumina when it enters the furnace. That way, the alumina is cooled and the air is preheated up to the calcination temperature, avoiding water consumption.

For the crude and distilled MMA, the cooling water consumption is proportional to the heat exchanged in the units. For the condensation of the crude MMA from the pyrolysis, one must cool the MMA vapors from 450 °C to the condensation temperature, condensing it, and finally cooling the MMA liquid to storage temperature.

In addition, we must cool water from the first stage (at 250 °C) to 25 °C, and the remaining water heated at depolymerization temperature to storage temperature.

To simplify the calculations, we treat the two streams as pure MMA and pure water, respectively. The total water consumption is then 363 kg water.

For the distillation, we take the energy injected in the reboilers as the energy we have to remove from the unit. This corresponds to 493 MJ and 214 kg of water consumption. The water consumption from cooling is then 3.06 kg water/kg rMMA. Again, this is the same order of magnitude as for the other units processing PMMA scraps, and not composites. However, a major difference here is the high amount of water generated by the process itself, which also must be condensed and cooled.

14.5 CO_2 emissions

The CO_2 emissions in the process are divided into (Figure 14.8):
- CWAO water treatment emissions, from the oxidation of the organic fractions in the wastewater
- Emissions from the thermal oxidation of the odorous streams
- Emissions from the calcination of the crude alumina via char combustion
- Emissions from the combustion of either natural gas, or by-products to supply the energy for the cracking and purification

In each section, we split the emissions by mass allocation into r-MMA and alumina emissions, and finally we calculate the global emissions.

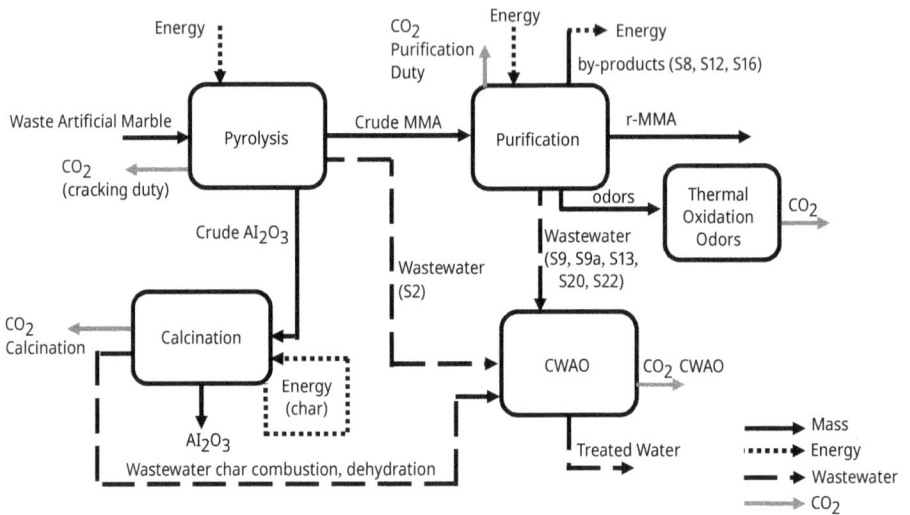

Figure 14.8: Energy, CO_2, and wastewater streams in the process.

14.5.1 Water treatment

The organics in the water treatment will oxidize as follows:
- MMA: $C_5H_8O_2 + 6\ O_2 \rightarrow 5CO_2 + 4H_2O_2$
- MAA: $C_4H_6O_2 + 4.5\ O_2 \rightarrow 4CO_2 + 3H_2O_2$
- Methanol: $CH_3OH + 1.5\ O_2 \rightarrow CO_2 + 2\ H_2O_2$

With the light impurities assumed to behave like MMA, the total amount of CO_2 generated is then 73 kg.

Again, the CO_2 produced by the oxidation of the organic species can be split between alumina and r-MMA by mass allocation. The CO_2 of the stream 2 (38.5 kgCO_2eq) is split into:
- 15 kg MMA
- 23.5 kg alumina

The remaining CO_2 (34.5 kgCO_2eq) coming from the purification section is allocated fully to MMA. Finally, the CO_2 produced in the CWAO section is:
- 23.5 kg CO_2eq allocated to alumina
- 49.5 kg CO_2eq allocated to MMA

14.5.2 Thermal (or catalytic) oxidation of odors

The CO_2 produced by the thermal (or catalytic) oxidation of the odoriferous streams S13a (degassed from 325), and S18 (degassed from 343) comes from the oxidation of methanol and MMA equivalent compounds, following the reaction:
- MMA: $C_5H_8O_2 + 6\ O_2 \rightarrow 5\ CO_2 + 4\ H_2O$
- Methanol: $CH_3OH + 1.5\ O_2 \rightarrow CO_2 + 2\ H_2O$

considering all the other impurities to behave like MMA. To minimize the CO_2 emissions, we consider that a catalytic exhaust treatment would be used; and would then generate 23 kg CO_2, allocated fully to rMMA. The amount is small and a thermal oxidizer would have a limited impact on the process.

14.5.3 CO_2 emissions from calcination

The calcination of crude alumina to Al_2O_3 is fueled by the combustion of the char PMMA residue left on the crude alumina following the reaction:
$$PMMA(C_5H_8O_2) + 6\ O_2 \rightarrow 5\ CO_2 + 4\ H_2O.$$
Thus, 36.7 kg of PMMA generates 81 kg CO_2eq, allocated completely to alumina.

14.5.4 CO_2 emissions from natural gas/by-product combustion

Different cases can be considered for the by-products combustion, with or without energy recovery. The energy recovery minimizes not only the natural gas consumption but also the CO_2 emissions. In case the by-products contribute for part (90%) of the total duty, natural gas (LHV 47.2 MJ/kg) would provide for the rest of the duty at the cracking section, where we would need the highest temperature.

From the by-products, we can recover 2,476 MJ of energy (Section 2.2), while we need to provide 2,742 MJ. The natural gas must cover for the surplus of 267 MJ to be provided at the cracking section.

If we split this surplus by mass allocation, we get:

- 104 MJ MMA, and
- 163 MJ alumina

The 2,476 MJ of by-products fuels the purification (493 MJ completely allocated to MMA), and 1,983 MJ required by cracking.

We can split these 1,983 MJ by mass allocation into:

- 774 MJ MMA, and
- 1208 MJ alumina

Assuming the by-product combustion to be like that of MMA (LHV 27.2 MJ/kg):

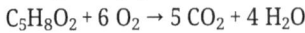

$$C_5H_8O_2 + 6\ O_2 \rightarrow 5\ CO_2 + 4\ H_2O$$

The CO_2 emitted on site, in the case of energy recovery from by-product combustion, is the sum of CO_2 from by-product combustion, CO_2 from natural gas, CO_2 from CWAO, CO_2 from calcination, and CO_2 from thermal oxidation of odors (Table 14.2):

Table 14.2: CO_2 emissions.

Contribution	kg Natural gas	kWh	MJ	kg CO_2 eq	kg CO_2eq/ kg rMMA	kg CO_2eq/ kg alumina
Natural gas MMA contribution (cracking)	2.2	29	101	6	0.03	
By-products MMA contribution (cracking and purification)		353	1267	102	0.55	
1371 MJ is the MMA contribution to the total duty of the process						
By-products Alumina contribution (cracking)		336	1208	98		0.25
Natural gas Alumina contribution (cracking)	3.4	45	163	9.3		0.02
1371 MJ is the alumina contribution to the total duty of the process						
Char combustion (LHV 24.5 MJ/kg) (calcination)				81		0.2
CWAO MMA contribution				49.5	0.26	

Table 14.2 (continued)

Contribution	kg Natural gas	kWh	MJ	kg CO$_2$ eq	kg CO$_2$eq/ kg rMMA	kg CO$_2$eq/ kg alumina
CWAO Alumina contribution				23.5		0.06
Catalytic oxidation				23	0.12	
Total	5.6		2,742	398	0.96	0.53

In the case of energy recovery from by-products, there is an emission of 398 kg CO$_2$eq (221 kg CO$_2$eq from combustion of gas/by-products and 177 from calcination, CWAO and thermal oxidation), with consumption of 5.6 kg of natural gas.

Allocating this CO$_2$ by mass, there will now be 0.96 kgCO$_2$eq/kg r-MMA and 0.53 kgCO$_2$eq/kg alumina.

In the case of no energy recovery from the by-products, and all the energy must be supplied by the extra natural gas consumption, the process would generate 1.48 kg CO$_2$eq/kg r-MMA, and 0.64 kg CO$_2$eq/kg alumina. This shows that there are some incentives to try to valorize these side products.

CO$_2$ emissions could also have been split according to different rules. In the case of alumina substitution, we calculate the amount of energy needed to calcine gibbsite to alumina and the corresponding CO$_2$ emissions that we subtract from the total of the process for the same amount of alumina. When considering the energy recovery case, the impact rises to 1.32 kg CO$_2$eq/kg r-MMA, higher than when using the mass allocation.

We can also allocate all the emissions to the rMMA. Indeed, that would be the case when the co-product would have no marketable value and would have to be landfilled, as it could be the case if the coproduct is glass-fiber. When considering the energy recovery case, the CO$_2$ emissions would rise to 2.11 kgCO$_2$eq/kg r-MMA, but still lower than for virgin MMA.

The last option would be to allocate the CO$_2$ emissions by the market value of the products. Similar alumina could have a market value of about US\$0.75/kg, while the low purity rMMA generated in the process is assumed to have a market value of US\$1.5/kg. When considering the energy recovery case, the CO$_2$ emissions would split into 1.03 kgCO$_2$eq/kg r-MMA and 0.51 kgCO$_2$eq/kg alumina.

The allocations by mass or market value, or by substitution, give similar values. As long as the coproduct of the composite (alumina in the present case) has a market value, the recycling operation could seem attractive. In fact, the capital cost, feedstock, and revenue from products have the biggest impact on the economic viability of PMMA recycling plants [17], and we believe this to hold true for artificial marble recycling. However, there is a limitation to that model. One must keep in mind that in the present case, there is a cluster of solid surface producers in the same region, nevertheless supplying to three different plants. One of the reasons would be to minimize the transportation cost, but also the CO_2 emissions related to the transportation. The longer the distance, the higher are the CO_2 emissions.

A full life cycle analysis would have to include the CO_2 emissions linked not only to transportation, but also to the pretreatment (crushing) of the solid surface scraps. To recover 1 kg of rMMA, about 6 kg of scraps have to be transported. Solid surfaces are very hard materials, so a much higher energy consumption is to be expected for the crushing and grinding of that material. That step is certainly mandatory to promote heat and mass transfer. If the scraps must be transported on a longer distance, the benefit, compared to virgin MMA, would disappear. Of course, it is even worse when the inorganic filler would have no marketable value.

14.6 Conclusions

R&E is a Korean company that collects and recycles 35,000 t/year of waste artificial marble, a mixture of PMMA and $Al(OH)_3$ (64% $Al(OH)_3$, 35% PMMA, 1% additives (by mass)). From their three plants in Korea, they produce 6,600 t/year of r-MMA (96.5% purity (by mass)), and 13,860 t/year low surface area Al_2O_3, with a global r-MMA and alumina mass yield of 0.54 and 0.5, respectively.

Based on their patents [3, 5], we developed a process flowsheet and derived mass and energy balances for the R&E's artificial marble chemical recycling process. They depolymerize the PMMA part in the artificial marble, and concurrently dehydrate the $Al(OH)_3$ in a stirred-tank reactor. The cracked gases, containing MMA, MMA impurities (light and heavy), MAA and other organic acids, methanol, and water, undergo a precleaning, and two distillation steps, to recover a low-quality purified r-MMA (96.5% purity (by mass)). Most of the water released from the $Al(OH)_3$ in the cracking is separated, but the crude MMA still carries a significant quantity of water. The main reason for the complex purification section is the presence of this water (forming a low-boiling azeotrope with MMA), and the high quantity of MMA impurities. Some of these, in particular the acrylates, are generated due to the artificial marble preparing method.

The process produces a high quantity of wastewater (around 0.74 kg wastewater/ kg r-MMA, and 0.38 kg wastewater/kg Alumina) that has to be treated on site because of the high pollutant concentration (11.5% (by mass), or COD > 115,000 mg/L). Such a high concentration of organics in the water would require a rather expensive wastewater treatment, like for instance, catalytic wet air oxidation unit (CWAO), operating at high pressure.

Despite extensive purification steps, MMA quality remains low, and the coproduced alumina has a low surface area alumina, 30–80 m^2/g at best (from BET). Pyrolysis at less than 500 °C would not only yield a higher surface area alumina, but also minimize the losses of PMMA/MMA as carbon residue on the alumina surface. However, looking at the Al_2O_3 datasheet from this process, the pyrolysis step probably operates at 500 °C. A high quantity of charred PMMA on the crude alumina (8–10% (by mass)) remains and makes the calcination temperature more difficult to control.

The process requires around 7.26 MJ/kg purified r-MMA, and 3.45 MJ/kg of alumina, for a total of 2,742 MJ in our case study of 1,000 kg of artificial marble treated. Because of the low r-MMA mass yield, the by-products could theoretically cover for around 90% of this duty, reducing the natural gas consumption.

Considering only the emissions inside the boundaries of the plant, in the case of energy recovery from the by-products, the total on-site CO_2 emissions would go down to 0.96 $kgCO_2eq$/kg r-MMA and 0.53 $kgCO_2eq$/kg alumina. On top of that, emissions have to be added for the transportation and the pretreatment (crushing) of the Solid Surface scraps. To remain attractive versus virgin MMA, the process should collect scraps over a short distance and valorize the coproduct (alumina). Another inorganic filler, like glass fiber, would not necessarily give as good environmental impact.

References

[1]   "R&E Leaflet in English" on their website, https://www.veolia.co.kr/sites/g/files/dvc2646/files/docu ment/2020/06/R%26E%20Leaflet%20English.pdf (accessed June 18, 2024).
[2]   Lee YS, Noh MS, Lee YM. Korean patent application KR101242763, 2013, assigned to R&E Co. Ltd. Accessible at Patent (101242763) < KIPRIS(Korea Intellectual Property Rights Information Service).
[3]   Lee YS, Noh MS, Lee YM. US Patent application US2013/0055926 A1, 2013, assigned to R&E.
[4]   Lee SG, Jeng EK. Korean patent application KR1020080008975, 2008, assigned to R&E. Accessible at Patent (1020080008975) < KIPRIS(Korea Intellectual Property Rights Information Service).
[5]   Jeon MC, Ahn HG, Kim JG, Jung HM, Jeong SH, Park BJ. US patent application US2009/0099290, 2007. Assigned to Lotte Advanced Materials Co Ltd.
[6]   Lee YS, Noh MS, Lee YM. Korean patent application KR1020120017035, 2012, assigned to R&E Co., Ltd. Accessible at Patent (Crushing apparatus of waste scagliola) < KIPRIS(Korea Intellectual Property Rights Information Service).
[7]   Lee YS, Noh MS, Lee YM. Korean patent application KR1020110036934, 2011 assigned to R&E Co., Ltd. Accessible at Patent (101155126) < KIPRIS(Korea Intellectual Property Rights Information Service).
[8]   Lee YS, Noh MS, Lee YM. Korean Patent application KR1020120017037, 2012, assigned to R&E Co., Ltd. Accessible at Patent (Purification apparatus of waste scagliola) < KIPRIS(Korea Intellectual Property Rights Information Service).
[9]   Lee YS, Noh MS, Lee YM. Korean Patent application KR1020120017036, 2012, assigned to R&E Co., Ltd. Accessible at Patent (Decomposition apparatus of waste scagliola) < KIPRIS(Korea Intellectual Property Rights Information Service).
[10] Ishikawa T, Lu BCY. Vapor-liquid equilibria of the methanol-methyl methacrylate system at 313.15, 323.15 and 333.15 K. Fluid Phase Equilibria 1979, 3(1), 23–34.
[11] Kim BR, Chang WK, Yang GS, Young SL. A study on recovery of aluminum oxide from artificial marble waste by pyrolysis. Korean Chemical Engineering Research 2012, 50(3), 567–573.

[12]   Kooi J. The system methylmethacrylate-methanol-water. Recueil des Travaux Chimiques des Pays-
        Bas 1949, 68(1), 34–42.

[13]   Knovel: Yaws' Handbook of Properties for Aqueous Systems, https://app.knovel.com/web/view/it
        able/show.v/rcid:kpYHPAS006/cid:kt00AQ8521/viewerType:eptble//root_slug:solubility-of-organic-
        compounds-in-water-as-a-function-of-temperature/url_slug:yaws-handb-solubility-
        (accessed June 18, 2024).

[14]   Knovel Critical Tables (2nd Edition), Two Component Azeotropes – https://app.knovel.com/kn/resour
        ces/kt004W2YG1/kpKCTE000X/itble/itable?b-q=water±azeotropes±methyl±methacrylate&columns=
        10101010%2C1%2C2%2C3%2C4%2C5%2C6%2C7%2C8%2C9%2C10%2C11%2C12%2C13%2C14&include_
        synonyms=no&q=water±azeotropes±methyl±methacrylate&sort_on=default
        (accessed June 18, 2024).

[15]   Hemingway BS, Robie RA, Fisher JR. Wilson WH heat capacities of gibbsite, $Al(OH)_3$, between 13 and
        480 K and magnesite, $MgCO_3$, between 13 and 380 K and their standard entropies at 298.15 K and
        the heat capacities of calorimetry conference benzoic acid between 12 and 316 K. Journal of
        Research of the U. S. Geological Survey 1977, 5, 6.

[16]   Munro RG. Evaluated material properties for a sintered alpha-alumina. Journal of the American
        Ceramic Society 1997, 80(8), 1919–1928.

[17]   De Tommaso J, Dubois JL. Risk analysis on PMMA recycling economics. Polymers 2021, 13(16), 2724.

Bruno de Caumia, Benjamin Dubois, Gloire Justesse Adolphe Mbou,
David Laplante St-Martin, Gaétan Mercier, Hooshang Pakdel,
and Christian Roy

15 Pyrovac's waste plastic pyrolysis process

The recycling of waste polymers such as PMMA, polyethylene, and polypropylene represents a global challenge. Several recycling processes other than landfilling and incineration are available to make better use of such resources. One such recycling technology has been developed by Pyrovac Inc., Quebec, Canada. The pyrolysis process is already commercially used for the conversion of biomass into useful products. Features of the process include varying feedstock contact time inside the reactor, varying thermal decomposition temperature and pressure, and selective vapor condensing conditions. Preliminary tests were carried out with injection grade PMMA and cast PMMA scraps. These tests were performed at a feed rate of 13–40 kg/h and a decomposition temperature varying from 380 to 420 °C. In both cases, an rMMA monomer with purity of at least 97% was recovered, representing a quality equivalent to the commercially available recycled monomer, especially in the case of the injection-grade PMMA. This technology has also been used to test low- and high-density polyethylene and polypropylene wastes. Pyrolysis of commingled PE and PP derived from urban trash yields hydrocarbon products with commercial value. Trials performed in a demonstration plant at the scale of 250 kg/h showed high conversion of the commingled plastics into liquid and gaseous fuels of high heating value.

15.1 Introduction

Currently, various regeneration techniques are being investigated around the world to achieve upcycling of PMMA (poly-methylmethacrylate). Examples include: the molten metal bath, dry distillation, rotating drum, fluid bed, stirred tank, and twin-screw extrusion [1]. Among these, the most common are the dry distillation and the molten metal processes. However, these methods present serious drawbacks, since the former process leads to low product purity. The molten metal process ends up with a contaminated side-product (char), given the use of lead during the process [2].

Bruno de Caumia, Benjamin Dubois, David Laplante St-Martin, Gaétan Mercier,
Hooshang Pakdel, Pyrovac Inc., 176-2 Damase-Breton, Saint-Lambert-de-Lauzon, QC, Canada G0S 2W0
Gloire Justesse Adolphe Mbou, UNIDO/Centre d'Excellence d'Oyo pour les Énergies Renouvelables et
l'Efficacité Énergétique, République du Congo
Christian Roy, Pyrovac Inc., 176-2 Damase-Breton, Saint-Lambert-de-Lauzon, QC, Canada G0S 2W0,
e-mail: croy@pyrovac.com

https://doi.org/10.1515/9783111076997-015

An upcycling process is under development at Pyrovac (www.pyrovac.com), where a major program is ongoing for the recycling of polyethylene and polypropylene waste plastics into liquid fuels. The process is also commercially operated at the scale of 500 kg/h for the conversion of almond shells in California (www.corigin.co). A larger unit with a feed capacity of 1,500 kg/h is also in development at the Elkem biocarbon plant in Saguenay, QC, Canada, to replace fossil coal by wood-derived biocarbon in the ferrosilicium metallurgy sector. This section of the paper describes preliminary results obtained from a continuous-feed pilot plant equipment fed with injection-grade pellets and cast PMMA.

The Pyrovac process belongs to the category of "continuous stirred tank" reactors [2]. It is a variation of the "dry distillation" process, without the disadvantages of batch processes, such as:

– Daily heating and cooling of the reactor
– Accumulation of solid residues in the reactor
– Poor heat transfer
– Uncontrolled temperature gradient (from bottom to top)
– Safety issues due to cycling temperatures
– Long residence time
– High product hold-up in the reactor

In the Pyrovac process, polymer scraps continuously fall onto a series of heating plates disposed one on top of the other. The top heating plate of the multiple hearth reactor receives the shredded waste plastic scraps where it melts. Such configuration enhances heat transfer between the molten polymer and the fresh solid scraps falling into the well-stirred molten-polymer bath. Depolymerization occurs as the molten polymer moves across the heating plate inside the multiple hearth reactor. Solid residues left are carried outside at the bottom of the reactor. The process enables selection of the right set of feed residence time and decomposition temperature as they can be adjusted independently as a function of the product slates targeted. Multiple heating plates can also be installed to increase productivity.

Another technology in the family of "continuous stirred tank" processes, is the technology developed by R&E in Korea [3]. However, this technology is currently applied for solid surface" recycling, in which a very high volume of solid residues is produced. Solid surface materials are composites of PMMA and aluminum trihydroxide (ATH) ($Al(OH)_3$), which require specific process conditions. The purity of rMMA produced with this process is between 96 to 97 %.

15.2 PMMA upcycling process description

Pyrovac is equipped with different pilot plants for pyrolysis of waste plastics and biomass. The appropriate pilot is selected depending on a case-by-case study. A picture of the pilot used for PMMA scraps pyrolysis is shown in Figure 15.1. The pilot equipment is divided in five subsystems: feeding, pyrolysis reactor, including the molten salts circulation, cooling and recovery of the solid residues, condensation and recovery of the condensed liquids, and combustion of the noncondensable gas. The process diagram is shown below in Figure 15.2 for the large-scale demonstration plant also used in this study. In the feeding section, a small flow of blanket nitrogen ensures that no air enters the pyrolysis unit. The pilot uses electrically heated molten HITEC salts as heat carrier and can process 50 kg/h at a maximum temperature of 525 °C.

Figure 15.1: 50 kg/h pilot unit continuous pilot plant.

Depolymerization trials of the different PMMA scraps were performed using the same unit, although some modifications were required to process the injection grade PMMA. The pyrolysis runs were performed at a feed rate of 13–40 kg/h.

The pyrolysis reactor comprises three heating plates with molten salt circulating in indirect contact with the feed material. The heating plates are installed in an insulated shell. The plates are equipped with agitators to ensure blending and transport of

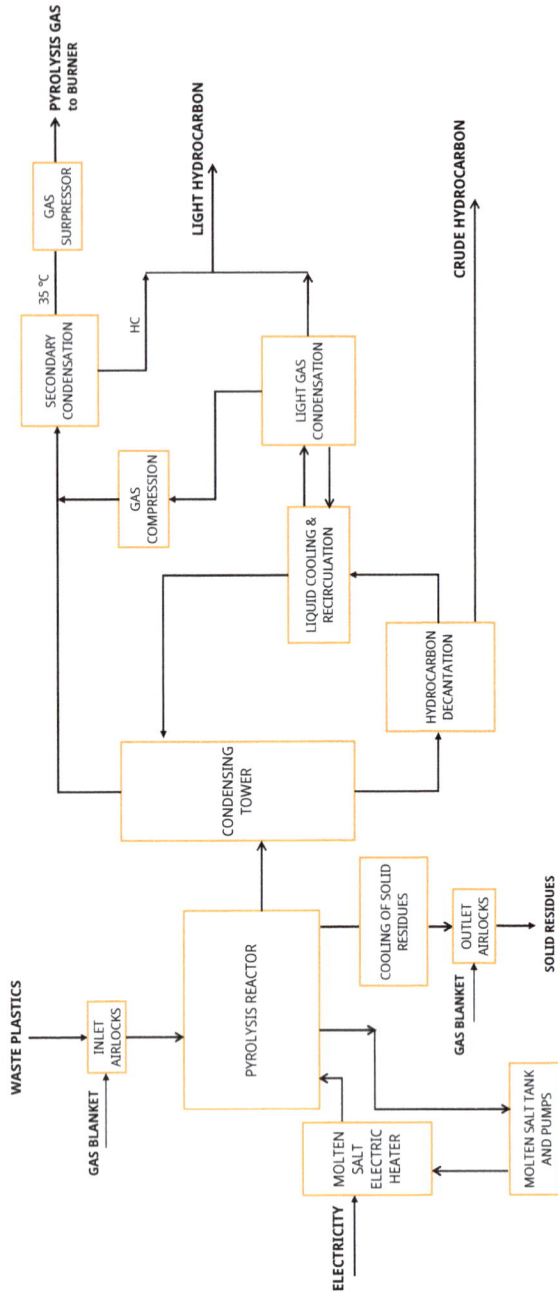

Figure 15.2: Pyrovac's process flow diagram.

the molten material from one plate to the other in a cascade mode. Scraps melt on the first plate, then the viscous liquid is transferred to the following heating plates where depolymerization is achieved. At the bottom of the unit, the solid residue, composed of carbon and mineral impurities, is collected after being cooled.

The Pyrovac process also includes a downstream section for the recovery of the product. The unique and patented process is based on thermodynamic properties [4] and uses two sequential condensation towers to adiabatically cool the organic vapors and water. First, the hot gas from the depolymerization unit is quenched in a first condensation unit with a circulating loop of liquid (Figure 15.3). According to the scenario situation and depending on the raw material processed, different chemical compounds can be used in the liquid loop to obtain an efficient recovery and restitution of the energy [5]. It is also at this point that the non-condensable gases are separated, and the energy contained in hot gas is transferred to the liquid loop. In the industrial unit, the non-condensable gases and other process residues can also be used to service the pyrolysis reactor heat requirement.

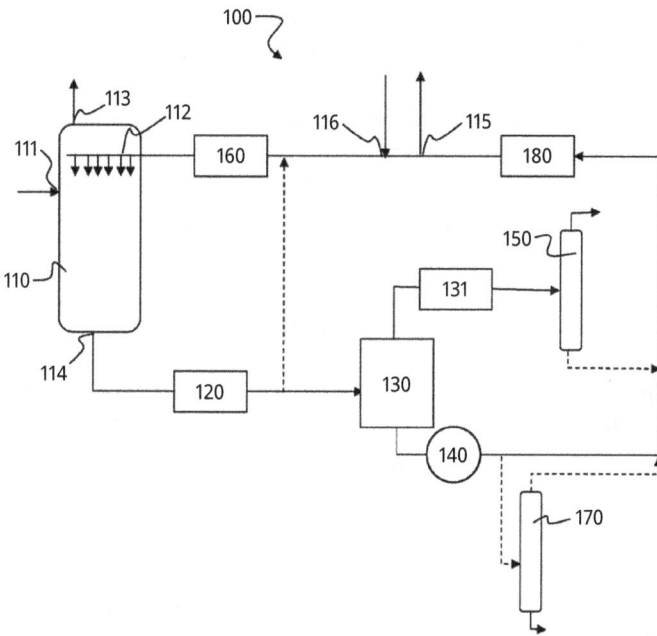

Figure 15.3: Downstream section 110, primary condensation unit; 111, incoming hot gases from the pyrolizer; 112, absorption device; 113, exit for noncondensable gases; 114, circulation loop exit; 116, injection point (additives, inhibitors, and fresh material); 120, separation unit; 130, flash evaporator; 140, recirculation; 131, condenser; 150 and 170, purification devices; 180, analysis device; 160, heat exchanger. Extracted from Patent WO2020079380 [4].

Next, the liquid stream is transferred to a flash evaporation unit under reduced pressure. This thermodynamic cooling unit evaporates the MMA-rich fraction and, at the same time, cools down the liquid, which returns in the first condensation unit. The evaporated fraction is then freed of the heavier impurities and is condensed again at low temperature. Crude MMA obtained through this process must be kept under low temperature to avoid spontaneous polymerization. Otherwise, for transportation and long-term storage, it must be stabilized as described in the Methacrylic Acid Ester Safe Handling Guide [6].

15.2.1 Pyrolysis of injection-grade PMMA and PMMA cast scraps

As explained in the previous chapters, polymerization of MMA is performed in different formulations and processes. Injection and extrusion grades are aimed to be heated again to be injected in molds or extruded as sheets, respectively. Cast sheets are produced either in batch or continuous processes, from a syrup or pre-polymerized MMA, and are characterized by high molecular weight. They are polymerized at "low" temperature and might be later heated to be thermoformed. Cast PMMA starts to depolymerize at lower temperature than injection and extrusion grades and cannot be mechanically recycled, as they start to decompose at the temperature needed to process them. On the other hand, extrusion and injection grades contain comonomers (acrylates), have much lower molecular weight, and can be mechanically recycled. Because of their application, they must be melted to be injected or extruded. One of the roles of the comonomer is then to delay the degradation at higher temperature to allow their process. They are also characterized by a melt flow index.

This difference in physical properties directly impacts the feeding of the pyrolysis reactor. Indeed, cast PMMA scraps remain rigid as the scraps are transferred through the conveyor screws, while injection and extrusion scraps soften and create a plug. Consequently, during pyrolysis performed with the Pyrovac unit, the feeding tube of the injection-grade PMMA pellets had to be cooled to enable continuous feeding. To the contrary, the PMMA cast scraps were heated during the reactor feeding. PMMA depolymerization has been made on both injection-grade PMMA, provided by Altuglas, USA, and cast PMMA, provided by Altuglas, Mexico (Figure 15.4). The injection-grade PMMA were V826 pellets with a size of about 3 mm^3, red-pink color, and an apparent specific gravity of 720 kg/m^3. Cast PMMA scraps, from offcuts in the production, were freed from the protective polyethylene film and from the PVC gasket. The molecular weight of this cast-grade PMMA was between 3,000,000 and 5,000,000 g/mol. The as-received material contained 0.3 wt% moisture. Almost 100% was volatile matter with less than 0.2 wt% ash, as determined by a LECO proximate analysis.

A   B

Figure 15.4: (A) Injection-grade PMMA and
(B) cast PMMA scraps.

15.2.2 rMMA yields and purity

15.2.2.1 Injection-grade PMMA

The continuous feed pilot reactor is a multiple hearth furnace heated with circulating molten salts used as heat carrier. The PMMA material was fed at a mass rate of 13–40 kg/h. The range of temperatures tested was between 380 °C and 420 °C [4].

In the absence of rMMA already produced, pure water was initially used to cool down the vapors. It was found that water and rMMA spontaneously split in two phases in the receiving tank. The water phase also solubilized some of the organic impurities. Condenser #110 on Figure 15.3 was operated between 50 and 65 °C, depending on the test runs. Hydroquinone added as a stabilizer to the circulating liquids resulted in a partition factor favoring water, since hydroquinone is more soluble in water. On the other hand, phenothiazine was a better stabilizer in the condensation circulation loop during the depolymerization process, since it does not require addition of air. It also decomposes at a slower rate than hydroquinone. Nonetheless, hydroquinone or a similar stabilizer is needed in the presence of air or nitrogen when storing the finished product [6].

Analyses of the crude product obtained under such conditions showed that its quality was around 98–99%, as shown in Table 15.1. Pyrolysis temperatures between 380 and 400 °C were better in terms of rMMA purity and product yield. At 400 °C, the feed material residence time inside the reactor was shorter and the rMMA yield was slightly lower at 94.5% versus 97.9% at 380 °C. Impurities present can be removed by distillation to achieve higher purity. The residual solid yield was <0.5 wt% based on a

Table 15.1: Pyrolysis products in pilot-scale pyrolysis of cast- and injection-grade PMMA using Pyrovac process.

Product	CAS no.	Injection	Cast PMMA	Reference
Depolymerization temperature		380 °C	400 °C	
Methyl acrylate	96-33-3	0.11	0.17	[7]
Methyl propionate	554-12-1	0.31	0.67	[8]
Methyl isobutyrate	547-63-7	0.61	1.79	[9]
Dimethyl pentadiene	1000-86-8		0.03	
Ethyl acrylate (ethyl prop-2-enoate)	140-88-5	0.21		[10]
Methyl methacrylate	80-62-6	98.3–98.7	93.0	[11]
Methyl 2,2-dimethylpropanoate (methyl pivalate)	598-98-1		0.23	[12]
2-Methylpropanoyl-2-methylpropanoate (isobutyric anhydride)	97-72-3		≤0.01	[13]
Methyl dihydrofuran	1487-15-6		0.01	
Methyl 2-hydroxy-2-methylpropanoate	2110-78-3		0.04	[14]
2-Butenoic acid methyl ester (methyl crotonate)	623-43-8		0.07	[15]
Methyl 2-methylbut-3-enoate	51747-33-2	0.12	0.20	[16]
Cyclopropane derivatives			0.02	
Butanoic acid 2-methyl methyl ester	868-57-5	0.22	0.83	[17]
2-Ethyl-1-hexene (3-methylene heptane)	1632-16-2		0.07	[18]
3-Pentenoic acid methyl ester	818-58-6	0.02	0.06	[19]
Methyl 2-methylidenebutanoate	2177-67-5	0.03	0.17	[20]
Methyl pentanoate (methyl valerate)	624-24-8	0.02		[21]
Others (including water)		Traces		

Reference column is for additional data on the compound. More information on the chemicals' physicochemical properties are given in references.

material balance after treating 282 kg of inlet PMMA material. The non-condensable gas yield is 1.5 wt% and 4.0 wt% at 380 °C and 400 °C, respectively. Overall, these results showed that the crude rMMA obtained through pyrolysis (Figure 15.5) could compete with distilled rMMA found on the market [2].

15.2.2.2 PMMA cast scraps

The same pilot equipment was used for the cast material industrial scraps. The feed rate was between 15 kg/h and 30 kg/h, with a thermal decomposition temperature of 380–425 °C. An increased concentration of crude rMMA and a higher yield were observed going from 380–400 °C. This could be explained by the fact that there are fewer dimers and oligomers produced under those conditions, which allows more PMMA to completely depolymerize. However, at 425 °C, a decrease in rMMA yield and purity was observed. Therefore, a compromise must be found for the couple "reaction rate and depolymerization temperature." Initially, a reduced feed rate is desirable to

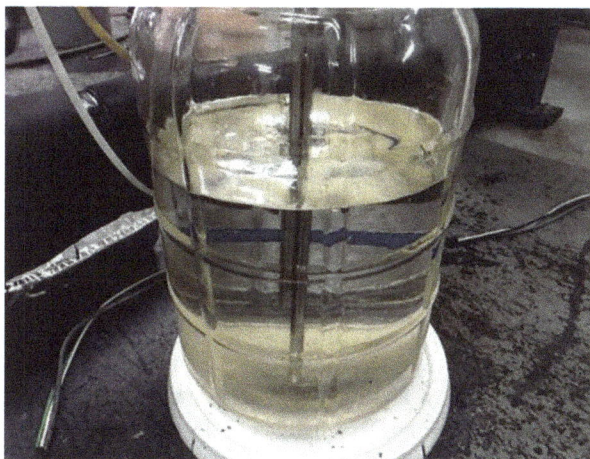

Figure 15.5: Crude MMA from continuous depolymerization (from injection grade).

allow complete decomposition of the molten PMMA inside the reactor. Then, under steady-state conditions, the feed rate should be adjusted to prevent excessive formation of undesirable impurities and limit contact time of solid carbonaceous residues over the heating plates. Presence of residual carbon is a known source of unwanted impurities such as methyl isobutyrate [22]. At 400 °C, average rMMA yield and purity were 81% and 97.9%, respectively. A list and concentration of impurities found in the final product is shown in Table 15.1.

Some 700 kg of Cast PMMA was tested during this series of tests. The overall mass balance closure was between 92.5% and 95.0%, depending on the test run. The lack of balance closure is attributed to non-condensable gases and solid residue. It has not been possible to determine the heat requirement for pyrolysis. Other researchers reported a value of approximately 1.8 MJ/kg for the depolymerization of Cast PMMA or cast-extrusion rich PMMA at 425 °C in a 30–100 kg/h pilot extruder, and 1.7 MJ/kg at 375 °C [23].

15.2.3 Recycled MMA composition

Pyrolysis of PMMA and other acrylate polymers has been studied in laboratory scale micro-pyrolysis coupled to gas chromatography-mass spectrum analysis. Publication in *Techniques and Instrumentation in Analytical Chemistry* [24] already gives a good insight of the impurities generated during the pyrolysis of various PMMA samples. This study showed that there are multiple degradation products with similar physicochemical properties than MMA, which are going to make the separation more difficult.

The GC-MS method used by Pyrovac, involves an HP-5890 gas chromatograph from Agilent, and an HP5-MS column (40 m and 0.25 mm diameter). Helium was used as carrier gas at a flow rate of 1 ml/min. Injection was done in a split mode (1/100) at 250 °C. Column temperature was 35 °C for 2 min, then 100 °C at 5 °C/min, and then 250 °C at 10 °C/min. Methyl cyclopentenone was used as internal standard, and Aldrich MMA (99% pure) was used for calibration. Results obtained using this method are presented in Table 15.1.

15.2.4 Crude MMA distillation

Crude MMA distillation was sent to Speichim Processing, France, where crude MMA from the Cast PMMA sheets scraps were purified. The distilled MMA reached an MMA content of 99%, as shown in Table 15.2. The distilled MMA was tested in cast sheets polymerization trials, confirming it could substitute virgin MMA.

Table 15.2: Crude and distilled MMA from cast PMMA obtained with Pyrovac process (d: detected, nd: not detected).

Product	CAS no.	Crude MMA	Distilled MMA	Reference
Water content			<200 ppm	
APHA color			<10	
Methanol	67-56-1	d		[25]
2-Butene	107-01-7	d		[26]
Methyl acrylate	96-33-3	0.17	0.01	[7]
Methyl propionate	554-12-1	0.67	0.06	[8]
Methyl isobutyrate	547-63-7	1.79	0.50	[9]
Ethyl acrylate	140-88-5	nd	nd	[10]
Methyl methacrylate	80-62-6	93.0	99.0	[11]
3-Butenoic acid 2-methyl methyl ester	51747-33-2	0.20	0.09	[16]
Butanoic acid 2-methyl methyl ester	868-57-5	0.83	0.30	[17]
3-Methylene heptane	1632-16-2	0.07	0.05	[18]
Methyl 2-methylidenebutanoate	2177-67-5	0.17	0.02	[20]

15.3 Pyrolysis of commingled polyolefin plastics

The reaction mechanisms that are leading to the formation of saturated products in PMMA depolymerization (methyl propionate, methyl isobutyrate, methyl pivalate, etc.) and that are lowering the recycled MMA purity or increase the complexity of the purification, are due to hydrogen transfer mechanisms. This is a disadvantage when recycling PMMA, but an advantage when the objective is the recycling of waste plastics into fuels or naphtha. Indeed, hydrogenated products have a much higher energy content than unsaturated molecules, and a good naphtha is also a blend of saturated molecules. Therefore, in such situation one should promote the hydrogen transfer reaction mechanisms. Polystyrene pyrolysis shares the same challenge than PMMA pyrolysis regarding the recovery of an unsaturated monomer (styrene) instead of a blend of molecules. Separation of methyl isobutyrate from MMA and ethyl benzene from styrene are equally difficult, as the molecules have similar physicochemical properties.

To favor the hydrogen transfer reactions, longer residence time and contact between the carbon residue and the molten polymer undergoing thermal cracking should be favored. Coke formation, as in delayed coking units (common in refineries for the treatment of heavy residues), is desirable in this case.

PMMA is now listed as a commodity polymer, but its annual global production is far lower than that of the polyolefins and other packaging plastics. It is then challenging to develop a dedicated technology for a limited market. The Pyrovac process is not only used for biomass pyrolysis, but also for mixed plastics pyrolysis. Therefore, the cost of the technology development and engineering can be split on a larger number of units and applications.

15.3.1 Pyrolysis of PE/PP

A scale-up of the pyrolysis system used for the PMMA wastes was tested for a blend of low- and high-density polyethylene and polypropylene wastes. Unlike PMMA, the pyrolysis of mixed plastics does not produce the original monomers. Instead, it generates hydrocarbon-like liquid products. The largest portion of the condensable organic vapors is obtained at the bottom of the first condensing tower (Figure 15.2) where the so-called crude hydrocarbon product is collected. The lighter oil fraction is collected at a cooler section of the condensing system. Figure 15.6 shows the large-scale industrial demonstration plant used to perform the test program.

Approximately 25,000 kg of shredded materials (Figure 15.7A) were supplied by Sustane Technologies, Inc. in Chester, Nova Scotia, Canada (www.sustanetech.com) and tested in Pyrovac continuous feed reactor. Waste material originates from residential/institutional sources of separated recyclables which are sent to the material recovery facility for sorting. Typically, 20–40% of this material is discarded and sent to landfills. Sustane uses this rejected material and processes it to remove the remaining mixed rigid PP, PE, and PS. Average density of the material processed in our installation was between 455 kg/m^3 and 570 kg/m^3. The average ash content of the feed material was low at 1.1% (anhydrous basis) with only trace amounts of fixed carbon. The volatile matter was high at 98.8%, as determined by LECO proximate analysis. Among the waste material received, there was approximately 6,000 kg of black shredded pans made of PP, which exhibited a higher ash content of 3.3%.

Figure 15.6: 500 kg/h industrial continuous feed pyrolysis demonstration plant.

A   B   C

Figure 15.7: (A) Shredded polyolefins wastes; (B) heavy pyrolytic oil; and (C) diesel fraction.

Temperature of the molten plastic bed ranged between 430 and 455 °C. The overall contact time of the feed material inside the reactor at an operating temperature of 450 °C was estimated to be 100 min. A mass balance was attempted for the PP material thermally decomposed at 435 °C. The lack of mass balance closure was only 5%, a result deemed reasonable for large scale industrial demonstration testing. The total liquid fuel yield was 78.7%, the non-condensable gas phase was 14.8%, and the solid residues yield was 6.5%. The above mass yields were obtained after distributing the lack of mass pro rata of the as-received and weighed products. The product yields for the commingled PE/PP samples were similar.

Gross calorific value of the crude hydrocarbon stream oil was determined to be 37.9 MJ/kg on an as-received basis (0.3% moisture content). The quality of the oil varies depending on the contact time inside the reactor and the maximum operating temperature. Longer residence time results in a lower proportion of long chains of paraffinic hydrocarbons. This can be explained by the fact that the long waxy molecules remain a longer time in the pyrolysis reactor, allowing them to crack further. The influence of thermal decomposition temperature was investigated for the PP waste material. At 445 °C, the relative proportion of light hydrocarbon oils (e.g., gasoline, diesel) was higher than at 430 °C. Conversely, the proportion of heavier oils was lower. The energy consumption should also be higher, although it might be difficult to determine, since only one bond needs to be broken from a C_{50} molecule to make two C_{25} (or two other fragments). The as-received light oil product can be more easily marketed than the crude oil. One possible outlet for the crude oil other than distillation is blending it with heating oil #6. Blends of bunker oil comprising 5–6% of the pyrolytic crude oil were successfully obtained while keeping the flash point of the whole blend above 90 °C. Density of the pyrolytic heavy oil is 802 kg/m³.

The PP-derived solid residue was composed of 50.5% ash. The proximate analysis gave 30.7% volatile matter and 18.8% fixed carbon. The properties of the PE/PP commingled material gave similar values of 55.7%, 24.5% and 19.8%, respectively. Therefore, given the relatively small solid residue carbon content, the waste plastic conversion into useful liquid and gaseous hydrocarbon fuels is higher than 94%.

The non-condensable gas phase is a valuable end product because of its high internal energy content. Table 15.3 shows a typical gas phase composition. The gross heating value of the non-condensable gas obtained at 435–440 °C varies between 44.1 MJ/kg for the PP feed material and 48.5 MJ/kg for the PE/PP. The lower HHV for the PP feed can be partially attributed to its higher ash content.

The pyrolysis process energy requirement was experimentally determined to be approximately 400 Wh/kg (1,440 kJ/kg) at a temperature of 435 °C.

Table 15.3: Noncondensable gas phase composition at 455 °C.

Gas	Volume % (N_2, O_2, and H_2O-free basis)
H_2	Nil
CO	1.2
CH_4	23.3
CO_2	1.4
C_2	16.3
C_3	24.3
C_4	15.9
C_5 isomers	12.0
C_6 isomers	5.6

Table 15.3 (continued)

Gas	Volume % (N$_2$, O$_2$, and H$_2$O-free basis)
Total	100.0
Molecular weight	42.5 g/mol
Gross heating value	48.5 MJ/kg

15.3.2 Crude oil fractioning

Upon thermal decomposition, the high molecular weight polymers are cracked into rather small fragments. A simulated distillation showed that a high percentage of crude pyrolytic oil obtained from plastic waste (Figures 15.7B) is distillable into lighter fractions (Figure 15.8). Figure 15.8 below shows that a high percentage of the crude oil is distillable in lighter fractions. Some 800 L of pyrolytic diesel (Figures 15.7C) obtained during batch distillation of crude pyrolytic oil was blended in successive proportions of 10%, 20%, and 50% v/v with commercial diesel and tested in a public bus in Montreal City, Quebec. The tests were supervised by Parkland Fuels, in collaboration with Polynergy Inc., and took place during an extended period of 33 days during cold winter night conditions in February-March 2021. An important requirement of the STM (Société de Transport de Montréal) was to have a flash point >40 °C and a cloud point <− 22 °C for the blends. Results obtained regarding diesel consumption in relation to bus speed are shown in Figure 15.9. The cumulated distance registered amounted to 5,719 km, and no difference was observed between the different blends of pyrolytic diesel.

Subsequently, 3,000 L of raw oil was fractionated into a 100 L/h pilot-scale distilling column (Figure 15.10). Reprocessing the heavy oil fraction into a 50 kg/h pilot plant pyrolysis reactor was also attempted and enabled further cracking of the heavy oil into lighter oils, mainly diesel. More than half of the heavy oil fed with fresh plastics was cracked after one pass.

15.4 Process economics

15.4.1 Waste PMMA

An industrial pyrolysis plant for the upcycling of waste PMMA would have a throughput capacity of 500 kg/h, or 4000 t/y. This is about the size of current existing capacities in Europe, where distilled MMA at a purity of about 98.5% is produced with the lead-bath process. Elsewhere, lower MMA purities are produced but nevertheless used to make recycled products. A complete and fair benchmark of the economic analysis

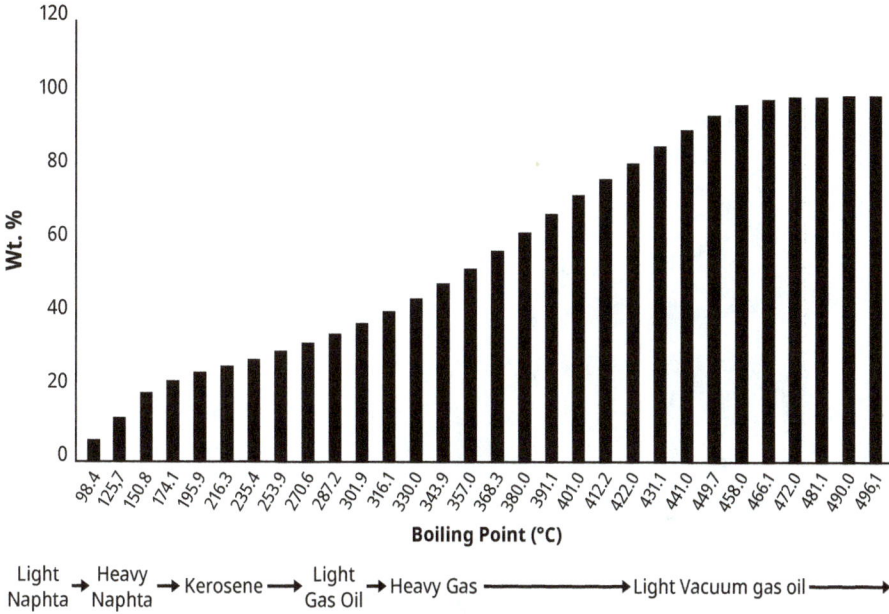

Figure 15.8: Simulated distillation on pyrolytic heavy oil.

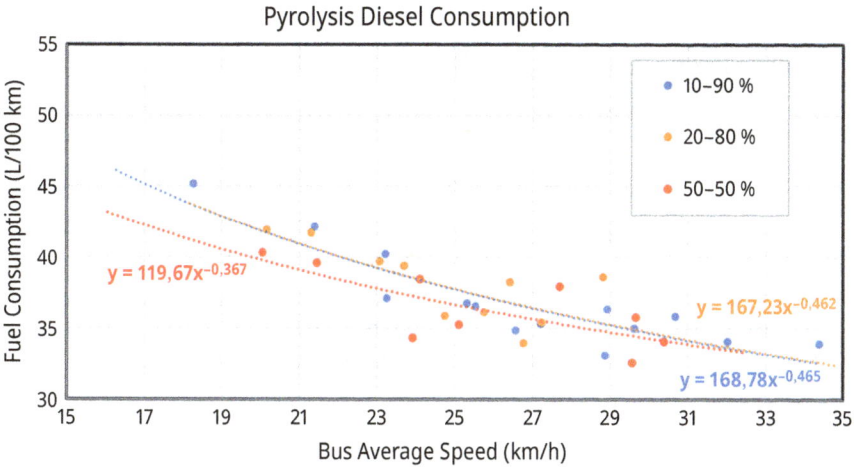

Figure 15.9: Fuel consumption under various conditions. Test run in Montreal Public Transportation bus. Source: Mr. G. Karawani, Parkland Fuels, Montreal, QC.

should be done for the same purity level. With Pyrovac's process, it is possible to reach purity above 99%, where only a few other processes would compete, and where a premium could be expected. More important, the process can easily handle scraps that gen-

Figure 15.10: 100 L/h pilot distillation column.

erate a high amount of solid residue. Another key advantage of the process is scalability. Design of a unit with higher capacity would require a higher heat exchange area, so more heat exchange plates and/or plates of larger diameter. Unlike many of the dry-distillation processes, the scalability would not require the duplication of the reactor. The technology is also versatile, and the same reactor can be used to process other plastics, which could be important for the recyclers who want to minimize their risks.

15.4.2 Commingled plastic wastes

A commercial plant should have a throughput capacity of between 750 kg/h and 2,000 kg/h. The raw oil main product is amenable to a refining process into petroleum fractions. Evolving environmental rules such as the CBAM in Europe [27] will facilitate such opportunity in the future. In the meantime, the process can find its way through simple applications such as using the whole oil in blend with a bunker #6 heating fuel.

15.5 Conclusions

Preliminary tests performed in a Pyrovac multiple hearth pilot-scale pyrolysis reactor was carried out with injection-grade PMMA and cast PMMA scraps. Each heating plate is indirectly heated with molten salt heat carrier, providing even heat distribution throughout the system. Tests performed with injection-grade PMMA in the temperature range of 380–400 °C resulted in the production of high purity rMMA of 99% at a mass yield of nearly 98 wt%. The same temperature range was investigated for Cast PMMA industrial scraps. The best result was obtained at 400 °C with rMMA yield of 81% and purity of nearly 98%. In both cases, the rMMA monomer purity was almost equivalent to the commercially available recycled monomer, especially in the case of the injection-grade PMMA.

This technology was also tested in a large-scale industrial demonstration plant at a throughput capacity 250 kg/h for low- and high-density polyethylene and polypropylene wastes. The commingled PE and PP wastes originated from typical urban and institutional trash streams. Some 25,000 kg of waste material was tested in a continuous feed multiple hearth reactor of 2,400 mm in diameter, also indirectly heated with molten salts at a typical temperature of 450 °C. The waste plastic conversion into hydrocarbon liquids and gases combined was high at 94–97%. The overall liquid yield reached approximately 79%. The gross heating value of this oil was 38 MJ/kg. Such pyrolysis oil can be advantageously distilled into transportation liquid fuels like gasoline, kerosene, and diesel. Yield of distillates reach 75%, more if the heavy oil is recycled back into the reactor. However, multiple large pyrolysis plants are required to economically enable such sophisticated application. Until the day comes when conventional refineries will blend such clean liquid fuels with crude oil, thus contributing to a greener planet, the distillation route is uneconomical. One shorter-term market application for the heavy oil is blending it with heating oil #6 for heating purpose. Another more immediate outlet is cracking the heavy oil into lighter hydrocarbon fractions for which a wider market exists.

15.6 Acknowledgments

The following governmental organizations have partially supported the projects:
- PMMA: Investissement Québec/Innovation Program; MEIE/Programme Nova Science; NRCC/IRAP.
- Waste plastics: Sustainable Development Technology Canada; Programme Technoclimat, Gouv. Québec.

Altuglas provided the PMMA samples for the trials.

Thanks are due to Dr. J.L. Dubois/Trinseo, for fruitful exchanges regarding the depolymerization process chemistry. The technical assistance of Dr. Marie-Odile Guimond and undergraduate Isabelle Dion-Buteau is also acknowledged.

References

[1]   Dubois J-L Guidelines for PMMA depolymerization at pilot and industrial scale. In: MMAtwo's Webinar. 2022. DOI: 10.13140/RG.2.2.31754.75202/1

[2]   Dubois J-L. Chapter 7 – PMMA chemical recycling reactor technologies. In Polymer Circularity Roadmap: Recycling of Poly(Methyl MethAcrylate as a Case Study. De Gruyter STEM, 2023 and this Edition of the book.

[3]   R&E Technology [Internet]. Available from: https://www.veolia.co.kr/en/our-services/re [accessed on Jun 20, 2024].

[4]   Roy C, de Caumia B, Blanchette D, Dubois J-L. Process for Treating a Gaseous Effluent from Pyrolytic Decomposition of a Polymer. PCT International Patent WO 2020/079380 A1, 2020 and EP3866950.

[5]   Roy C, de Caumia B, Blanchette D, Adolphe-Mbou GJ, Pakdel H. PMMA pyrolysis upcycling. In: Proceedings of the ISCRE 27: Chemical Reaction Engineering for Sustainable Development, June 2023, accessed on August 20 2024 at https://www.dropbox.com/scl/fi/vwd30fh2tqi11l0kufctr/Whole-Program.pdf?rlkey=xlq7w56237ags6lspdv5319lp&e=1&dl=0.

[6]   Methacylate Producers Association Inc, Methacrylates Sector Group of the European Chemical Industry Council. Methacrylates Esters – Safe Handling Manual. 2019. Available from https://www.petrochemistry.eu/wp-content/uploads/2019/08/Methacrylate-Esters-Safe-Handling-Manual-Rev-FINAL-8-22-19.pdf.

[7]   https://echa.europa.eu/fr/brief-profile/-/briefprofile/100.002.274 [last accessed on August 20, 2024, as well as for following ones].

[8]   https://echa.europa.eu/fr/brief-profile/-/briefprofile/100.008.238.

[9]   https://echa.europa.eu/fr/brief-profile/-/briefprofile/100.008.118.

[10] https://echa.europa.eu/fr/brief-profile/-/briefprofile/100.004.945.

[11] https://echa.europa.eu/fr/brief-profile/-/briefprofile/100.001.180.

[12] https://echa.europa.eu/fr/substance-information/-/substanceinfo/100.009.055.

[13] https://echa.europa.eu/fr/brief-profile/-/briefprofile/100.002.367.

[14] https://echa.europa.eu/fr/brief-profile/-/briefprofile/100.016.638.

[15] https://echa.europa.eu/fr/substance-information/-/substanceinfo/100.038.636.

[16] https://pubchem.ncbi.nlm.nih.gov/compound/Methyl-2-methyl-3-butenoate.

[17] https://echa.europa.eu/fr/substance-information/-/substanceinfo/100.011.617.

[18] https://echa.europa.eu/fr/substance-information/-/substanceinfo/100.015.125.

[19] https://echa.europa.eu/fr/substance-information/-/substanceinfo/100.011.321.

[20] https://echa.europa.eu/fr/substance-information/-/substanceinfo/100.283.590.

[21] https://echa.europa.eu/fr/brief-profile/-/briefprofile/100.009.853.

[22] German patent DE000002132716, Verfahren zur Herstellung monomerer Ester von Acrylsaeure und substituierten Acrylsaeuren, Application date July 1st 1971. Also available as French patent FR2101684 and British patent GB1350612 Depolymerization of acrylic polymers, available at https://patentscope.wipo.int/search/en/detail.jsf?docId=FR186558175&_fid=GB135519855.

[23] Tojo M, Dubois J-L. Grade-dependent depolymerization process and its use for recycling plastics. European Patent EP4155341 A1, 2023.

[24] Moldoveanu SC. Techniques and Instrumentation in Analytical Chemistry, vol. 25, 1st ed., Moldoveanu SC, ed. Elsevier, 2005, 3–697.

[25] https://echa.europa.eu/fr/brief-profile/-/briefprofile/100.000.599.

[26] https://echa.europa.eu/fr/brief-profile/-/briefprofile/100.003.140.

[27] The EU Carbon Border Adjustment Mechanism (CBAM) applies now [Internet]. 2024 Available from: https://www.dentons.com/en/insights/articles/2024/february/21/the-eu-carbon-border-adjustment-mechanism-cbam-applies-now [accessed on Jun 20, 2024].

Ahmet Turhan Ural, Ezgi Güzel, and Yusuf Uludag
16 A novel method for PMMA recycling: mixed induction reactor

In the last century, plastics emerged as a wonderful alternative to many classical materials in both industry and in our daily lives. It soon became clear that their resistance to biodegradation made them one of the most serious environmental problems worldwide. Unfortunately, with the current trends in their production and consumption, these problems can be expected to get worse as no place on Earth is already left unaffected by plastic wastes. Recycling of plastic wastes is one of the methods to reduce their negative impact on the environment. Here, a new method is presented to recycle polymethyl methacrylate, one of the plastic materials widely used. The method is based on pyrolysis of the polymer in an inert atmosphere. The novelty of this method is associated with the heating of the pyrolysis reactor, which is partly filled with small metal spheres. The spheres are directly heated by induction, and convey necessary energy for the pyrolysis of the plastic particles. Pilot-scale experiments revealed the energy efficiency of the process and monomer of high purity. Comparison with the conventional recycling methods of polymethyl methacrylate demonstrated these key advantages of the presented method.

16.1 Introduction

16.1.1 Importance of MMA recycling

One of the biggest issues facing our planet today, the plastic waste generation, is increasing globally, and the world needs novel solutions contributing to a circular economy.

Polymethylmethacrylate (PMMA), known as acrylic products, is an integral part of our daily lives due to its optical and mechanical properties, like versatility, durability, and aesthetic appeal. Acrylic is applied in various final products, including plastic furniture, construction materials, car headlights, cosmetic bottles, bathtubs, paintings, and more. Its extensive use raises concerns about waste management and recycling. The fate of these products after their primary use often remains hidden; they frequently end up in landfills or incinerators, exacerbating the global plastic pollution crisis.

Ahmet Turhan Ural, M-D2, Engineering Consultancy Contracting Domestic and Foreign Trade Limited Company, N°6 ZK9, ODTÜ-Ostim Teknokent Uzayçağı blv., 06374 Ostim Ankara, Turkey
Ezgi Güzel, M-D2, Ankara, Turkey
Yusuf Uludag, Middle East Technical University, D-112, METU-Department of Chemical Engineering, 06800, Ankara, Turkey

https://doi.org/10.1515/9783111076997-016

Today, only 10% of the acrylic waste is collected in the EU, according to its limited recycling rate. It is estimated that 270,000 tons of the 300,000 tons of annual acrylic waste generation [1] is accumulated in landfill only in the EU.

The urgency of PMMA recycling is underscored by several critical factors:

1. Increased demand for recycling technologies and recycled plastics: Tightening waste trade regulations [2] underline a need for enhanced waste management technologies. With a focus on restricting waste exports and promoting recycling, these regulations contribute to the growing demand for recycled plastics and recycling technologies globally. In 2024, the export of plastic waste from the EU is becoming more challenging, but it can still be sent to OECD countries with permission.
2. Growing PMMA waste volume: The consumption of acrylic products is escalating, yet inadequate recycling infrastructure results in a considerable portion ending up in landfills or incinerators.
3. Negative environmental impacts: PMMA is nonbiodegradable, posing risks to soil and groundwater when disposed of in landfills, releasing toxic fumes when incinerated.

Europe is currently undergoing an ambitious green transition as part of the European Green Deal. Central to this transformation is the aspiration to achieve climate neutrality by 2050. Additionally, the upcoming waste trade regulations reflect a challenge in restricting waste exports and current volumes of plastic waste. Thus, the countries should develop their infrastructure and expertise in response to these regulatory changes.

Acrylic manufacturers are under increasing pressure to adopt sustainable practices due to tightening regulations, while recyclers struggle with inefficient, costly recycling methods that fail to meet industry demands for high-purity materials. These shortcomings of the conventional methods highlight the urgent need for a new acrylic recycling technology.

16.1.2 Recycling methods

Various technologies are presently employed for the regeneration of MMA (methyl methacrylate), including the molten metal reactor (Clementy process), dry distillation, rotating drum process, fluid bed process, stirred tank reactor, auger screw process, and microwaves reactor.

In the molten metal process, PMMA scraps are poured on a metal bath kept above the pyrolysis temperature of PMMA. Lead is the most widely used molten bed material due to its low cost [3, 4]. Once the scraps are in contact with hot metal, a pyrolysis reaction takes place, and MMA vapor is generated. Due to the side reactions, some gaseous molecules, such as CO, CH_4, and H_2, and solid carbon are also produced.

Solid carbon accumulates on the surface and forms a crust that must be periodically removed. Oil burners are used to heat the bath; therefore, temperature control is not accurate. In addition, evaporation of lead takes place, leading to operational and safety problems. Since the melting temperature of lead is very high, it solidifies in evacuation lines and causes blockage. The process must be stopped at short intervals for cleaning. The method is still widely used despite the generation of lead-contaminated hazardous solid waste. It is believed that strict environmental and safety regulations will limit the general use of this method in future.

The dry distillation process, a longstanding method for PMMA recycling, is employed in many countries by heating a stack of PMMA until monomer generation ceases. Major drawback of this method is associated with the inaccurate temperature control. Then, along with appreciable temperature variations in the reactor, side reactions take place. The yield and monomer quality of the dry distillation are, therefore, relatively low. The rotating drum technology, a variation of dry distillation, utilizes a rotating cylinder for scrap placement. This method is mostly used for tire recycling. Similar to dry distillation, temperature control is not very effective. While the stirred tank reactor is not widely adopted, its semicontinuous mode prevents solid residue accumulation, averting catalyzed side reactions. The reasons for its limited use are, first, the difficulty in keeping the polymer in a molten state throughout the reactor, and then effective heat transfer to the polymer for high productivity.

The auger screw process [5, 6], where molten polymer is pushed by an Archimedes screw, is very effective in keeping the polymer in a molten state. But the heat transfer area is limited by the outside surface of the reactor. The size of the Archimedes screw, and hence the size of the reactor cannot be enlarged beyond a certain size due to mechanical limitations. It is still gaining attention because of its robust structure and clean operation compared to the molten metal bed process.

During the last decades, as an alternative technology, the fluid bed process [7–9] has been developed. It was initially promising for its efficient heat transfer. In this method, hot inert gas (generally nitrogen) is passed through a solid bed (generally sand) at sufficiently high velocity to fluidize the solid bed. PMMA scraps introduced to the fluidized hot bed are quickly pyrolyzed by coming in contact with the hot sand particles. MMA vapor leaves the reactor with the fluidizing inert gas. Both are cooled below the dew point to recover MMA as the liquid phase. Inert gas is reheated and returned back to the reactor. Even with the high monomer yield and superior quality of the recovered monomer, only pilot-scale facilities have been established because of its high energy consumption, mainly due to the heating and cooling cycles of inert gas. Additionally, this method is prone to agglomeration of PMMA particles and forms large lumps, which negatively affects the fluidization of the bed. As a result of these two disadvantages, presently, there is no plant at industrial scale.

Another technique based on fluidization, mechanical fluidization [10–12], seems to solve the problem of inert gas heating/cooling energy consumption. Agglomeration of particles and formation of large lumps that greatly diminishes the heat transfer rate is still the most important problem of this method.

Finally, the twin-screw extruder, a continuous process, heats PMMA scraps to depolymerization temperature, condensing vapors at its end. As mentioned above, each method has its merits and challenges [13].

16.2 P2L recycling technology

Pyrolysis is the most important step in production of r-MMA. The efficiency of this step determines the product yield and the purity, as well as energy consumption.

P2L (plant for second life) is a new technique that revolutionizes the safe, cost-effective, and environmentally friendly production of r-MMA. Overcoming the challenge of efficient heating, the method employs inductive heating of small magnetic balls, serving as the pyrolysis medium for PMMA scraps. This unique approach generates heat within the reactor, ensuring exceptional energy efficiency – an approach widely acknowledged in the metallurgical industry with well-established design and manufacturing processes for induction coils.

In order to demonstrate the potential benefits of the technique, a pilot-scale prototype plant was constructed. Picture of the pilot-scale inductively heated pyrolysis reactor is shown in Figure 16.1. Within this reactor, as depicted in Figure 16.2, a ribbon mixer continuously blends magnetic particles, enabling contact with the PMMA chips. The induction field elevates the temperature of these magnetic particles to the pyrolysis threshold, initiating the depolymerization of PMMA chips in direct contact. The mixer plays a critical role in maintaining a uniform temperature and material distribution throughout the reactor.

Temperature distribution in the reactor is crucial for efficiency and product quality. Conventional heat transfer, based on conduction, convection, and radiation, need resistances that yield a temperature gradient in the reactor. Depending on the heat of the reaction, this gradient may be large.

Pyrolysis, inherently an endothermic reaction, continuously draws the required energy from inductively heated magnetic particles. Magnetic particles boast a substantial heat transfer area, enabling high pyrolysis rates at relatively low temperatures. Operating the pyrolysis chamber at lower temperatures provides a key advantage – reducing the production rate of undesired pyrolysis products, compared to high-temperature heating scenarios. This, in turn, enables induction-based heat transfer, boasting an impressive energy efficiency of up to 85%, yielding relatively low carbon footprint. The process is also environmentally clean, eradicating occupational hazards related to lead exposure and significantly reducing energy costs, compared to conventional recycling methods. Another advantage of P2L technology is its operational ease. In the event of a plant malfunction, such as an electricity cutoff leading to

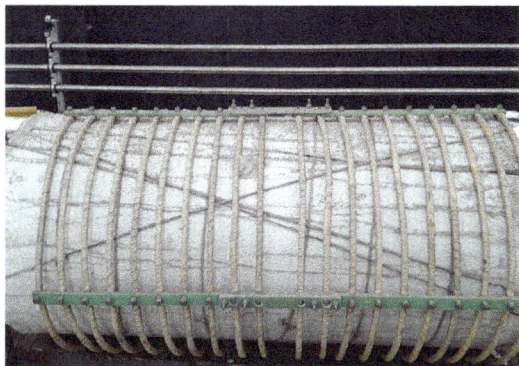

Figure 16.1: Inductively heated pyrolysis reactor.

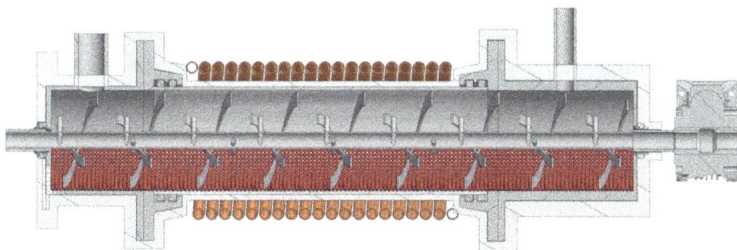

Figure 16.2: 3D model of pyrolysis reactor.

the solidification of polymer inside the reactor, the consequences are less severe, compared to Auger (extrusion)-based reactors and catastrophic scenarios for fluidized beds. Pilot studies have tested and confirmed that after complete solidification, the reactor becomes fully operable within 30 min of heating, requiring no worker intervention. This significant reduction in technical efforts and time-saving enhances the overall efficiency of the system.

The issue of output pipe blockage is a common and troublesome occurrence in pyrolysis processes. Gases generated during pyrolysis carry solid particles (carbon particles, pigments, etc.) and greasy liquid products. When greasy products encounter solid surfaces, they create a sticky layer, where solid particles accumulate, potentially leading to blockage and even the bursting of the reactor, causing a dangerous situation. P2L technology incorporates a self-cleaning system. Solid particles and greasy liquid products are collected in the condensed crude monomer, preventing the risk of blockages and maintaining a smooth operation.

Despite its polluting and hazardous nature, molten lead pyrolysis is the most widely used method, especially in developing countries. P2L distinguishes itself by eliminating the generation of solid side products (wastes), a common occurrence in conventional molten lead bed pyrolysis. In traditional methods, carbon particles, resulting from the chemical breakdown of acrylic polymers, mix with lead and forms a crust on the reactor's surface. This crust necessitates periodic removal and disposal. However, P2L technology allows these particles to exit the reactor with pyrolysis gas, collecting in the crude monomer condenser. Subsequently, carbon is separated through filtration.

The yield of the process, based on the fraction of r-MMA produced/PMMA scrap, is more than 85 wt%. Hence, a maximum 15 wt% of the scraps leaves the plant as a waste stream, which is mainly composed of methanol, ethyl acrylate, MMA dimers, and oligomers.

Energy consumptions of the existing prototype plant are tabulated in Table 16.1. Total consumption is 6.16 MJ/kg r-MMA. Thanks to the high-efficiency chillers used for condensation, cooling energy is relatively low. CO_2 emissions, corresponding to the total energy consumption, is 0.82 kg CO_2 equivalent/kg rMMA (0.44 kg CO_2 is emitted for each kWh of electricity [14]). This result demonstrates compelling advantages. It is more energy-efficient and produces much less CO_2, compared to conventional methods.

A comprehensive comparison is given in Table 16.2. In addition to the low energy consumption of P2L process, its low water consumption should be highlighted, as all cooling operations are carried out with chillers. Efficient recycling plants like P2L are environmentally friendly facilities, since they are using plastic waste as secondary resources.

Table 16.1: Energy consumption of P2L process.

Process step	Energy consumption (MJ/kg MMA)
Pyrolysis heating	2.12
Pyrolysis cooling (raw monomer condensation)	0.09
Distillation heating	2.00
Distillation cooling	0.75
Pumps, control, etc.	1.20
Total	6.16

Table 16.2: Utility consumptions of different methods and carbon footprints.

Properties	Clementi*	Fluidized bed*	Extrusion*	M-D2
Electricity consumption (MJ/kg r-MMA)	14.70	20.40	10.50	6.16
Water consumption (L/kg r-MMA)	2.50	3.00	2.50	0.50
Carbon dioxide emission (kg/kg r-MMA)	1.21	1.67	0.86	0.82

*The utility consumption values for alternative methods are referred to in the *Polymer Circularity Roadmap* book [13].

P2L technology has a potential to expand into recovery of other plastic types. One potential application is the production of recycled styrene from polystyrene (PS) wastes. Although the yield of the PS pyrolysis is inherently lower than that of MMA recycling, only 65% [15], the share of PS waste annually generated is much higher than that of PMMA, making its recycling attractive.

In the near future, research efforts will be concentrated on the recycling of composite (paper/plastic/aluminum) materials by using P2L pyrolysis technology.

16.3 Product characterization

PMMA selectively breaks down to its monomer methyl methacrylate (MMA) on pyrolysis. But other species are also produced. The output of the pyrolysis reactor is called raw r-MMA. GC analysis of a raw r-MMA sample from the pilot reactor is shown in Figure 16.3. Each peak in the figure belongs to a different organic molecule. The largest peak appearing at 105th s (retention time) belongs to MMA. The second-largest peak appeared at 50 s. indicates the presence of methanol. Remaining peaks can be attributed to the impurities. Previous studies [13, 16, 17] revealed that these impurities are mainly ethyl acrylate, methyl isobutyrate, methyl acrylate, and methyl propionate. Our chemical analysis revealed 11 species, in addition to methyl methacrylate. Based on peak areas, the MMA content of reactor product is around 92 wt%.

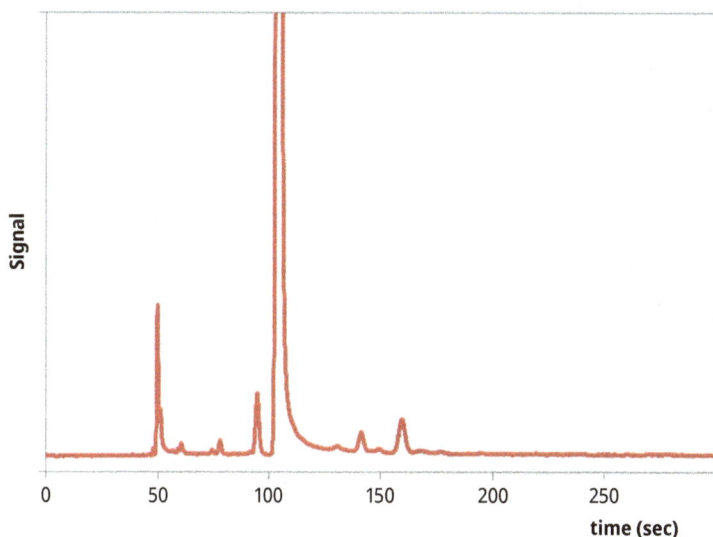

Figure 16.3: GC chromatogram of raw rMMA.

Impurities are separated from the main product, MMA, by two consecutive distillation processes. For this purpose, a 6-meters long packed column is used. The GC analysis of the final product is shown in Figure 16.4. For better comparison, the chromatogram of P2L's product is shown with those of virgin and recycled products available in the market. The analysis results reveal that the MMA purity of the r-MMA produced by the P2L is 99.2%. The success of the process is understood much better when the results are compared with competing products with only 97% purity. A sample PMMA produced by r-MMA from the pilot plant is transparent as shown in Figure 16.5.

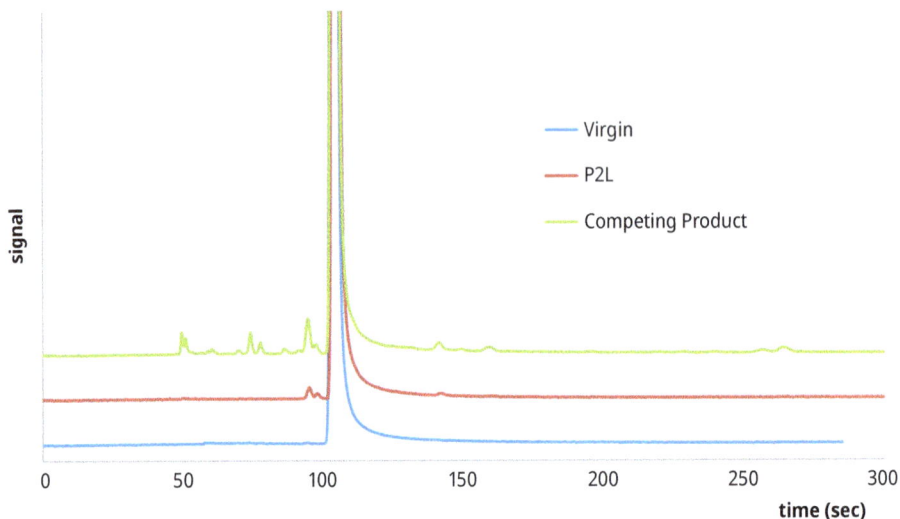

Figure 16.4: GC chromatograms of virgin, competing product, and the product of P2L prototype plant.

Figure 16.5: PMMA produced from r-MMA.

16.4 Conclusion

Based on the results of the pilot-scale application of the presented pyrolysis technique for the recycling of PMMA, the following conclusions can be drawn.

– Inductive heating of metal spheres in the pyrolysis reactor makes this PMMA recycling method highly energy-efficient compared to the other methods used.
– Providing pyrolysis energy through heated metal particles enabled carefully controlled pyrolysis temperature, avoiding exposure of plastic scraps to high temperatures. Therefore, production of unwanted pyrolysis products is kept at low levels as demonstrated by the produced MMA monomers of high purity.
– The method provided a working environment with reduced hazards, as opposed to the conventional methods such as the Clementy process.
– Due to its high energy efficiency, low utility consumption, and safe nature, the presented technique is promising to recycle polymer wastes that can be handled by pyrolysis.

References

[1]   MMAtwo; Today's challenge; https://www.mmatwo.eu/mmatwo-in-a-nutshell/todays-challenge/; Last accessed: 20.11.2024.
[2]   European Union; Waste Framework Directive; https://environment.ec.europa.eu/topics/waste-and-recycling/waste-framework-directive_en; Last accessed: 20.11.2024.
[3]   Newborough M, Highgate D, Vaughan P. Thermal depolymerisation of scrap polymers. Applied Thermal Engineering 2002, 22, 1875–1883.
[4]   Popescu V, et al. The characterization of recycled PMMA. Journal of Alloys and Compounds 2009, 483, 432–436.
[5]   Tokushige H, et al. Method for continuously thermally decomposing synthetic macro-molecule materials, US 3,959,357, 1976.
[6]   Weiss HJ, et al. Method for depolymerizing polymethylmethacrylate, US 6,469,203 B1, 2002.
[7]   Kaminsky W. Recycling of polymers by pyrolysis. Journal de Physique Nov 1993, 3.
[8]   Kang BS, Kim SG, Kim JS. Thermal degradation of poly(methyl methacrylate) polymers: Kinetics and recovery of monomers using a fluidized bed reactor. Journal of Analytical and Applied Pyrolysis 2008, 81, 7–13.
[9]   Vaughan PW, Highgate DJ, Depolymerisation, US 5,663,420, 1997.
[10] Sasaki A, et. al. Recovery method of pyrolysis product of resin, US 8,304,573 B2, 2012.
[11] Newborough M, Highgate D, Matcham J. Thermal depolymerisation of poly-methyl-methacrylate using mechanically fluidised beds. Applied Thermal Engineering 2003, 23(6), 721–731.
[12] Grause G, Predel M, Kaminsky W. Monomer recovery from aluminium hydroxide high filled poly (methyl methacrylate) in a fluidized bed reactor. Journal of Analytical and Applied Pyrolysis 2006, 75, 236–239.
[13] D'hooge D, Marien YW, Dubois J-L, eds. Polymer Circularity Roadmap: Recycling of PMMA as a Case Study, Walter de Gruyter GmbH & Co KG, 2022. ISBN 978-3-11-071649-8 and this current book, PMMA circularity Roadmap.
[14] U.S. Energy Information Administration, https://www.eia.gov/tools/faqs/faq.php?id=74&t=11

[15] Kaminsky W, Predel M, Sadiki A. Feedstock recycling of polymers by pyrolysis in a fluidised bed. Polymer Degradation and Stability 2004, 85, 1045–1050.

[16] Achilias DS. Chemical recycling of poly(methyl methacrylate) by pyrolysis. Potential use of the liquid fraction as a raw material for the reproduction of the polymer. European Polymer Journal 2007, 43(6), 2564–2575.

[17] Esmizadeh E, Khalili S, Vahidifar A, Naderi G, Dubois C. Waste polymethyl methacrylate (PMMA): Recycling and high-yield monomer recovery. In Handbook of Ecomaterials. Springer International Publishing, 2019.

Sara Babo, Anna Micheluz, Eva Mariasole Angelin, Susanne Brunner,
Joana Lia Ferreira, and Marisa Pamplona

17 Identification of additives in historical PMMA: contributions from heritage science

17.1 Introduction

Plastics have become an integral part of our cultural heritage. The advent of these versatile materials has spurred innovation across a multitude of fields including aviation, medicine, food packaging, and information storage, fundamentally reshaping our way of life. The transformation resulting from the introduction of plastics during the twentieth century is so profound that it may be considered a marker for a new historical era, the "Plastic Age" [1, 2]. Consequently, plastics are now a relevant part of museum collections and built heritage.

Poly(methyl methacrylate) (PMMA) is one of these plastics. It can be found in technical heritage collections in diverse items such as canopies of historical airplanes, and windows of cars and omnibuses since the 1930s [3]. Posterior applications include architectural constructions [4], windows of trains, and submarines.

Its properties, such as transparency, lightweight, and molding ability, also made it attractive for artists. In art, the introduction of PMMA is generally attributed to Naum Gabo (1890–1977) and other members of the Russian avant-garde during the late 1930s [5, 6], but it was only during the 1960s that it became popular as an art material [7]. In the building sector, PMMA has been widely used from the 1950s onwards, mainly due to its exceptional outdoor durability and resistance to yellowing over time. Construction elements such as cupolas, roofs, and windows were adapted and further developed from the aviation industry, being often nowadays protected cultural heritage objects. As a material of significant importance in cultural heritage, PMMA has garnered increasing attention from heritage conservation science.

Sara Babo, LAQV-REQUIMTE, Department of Conservation and Restoration, NOVA School of Science and Technology, Universidade NOVA de Lisboa, 2829-516 Caparica, Portugal
Anna Micheluz, Marisa Pamplona, Conservation Science Department, Deutsches Museum, Museumsinsel 1, 80538 Munich, Germany
Eva Mariasole Angelin, Chair of Conservation-Restoration, Art Technology and Conservation Science, Technical University of Munich, Oettingenstr. 15, 80538 Munich, Germany
Susanne Brunner, Professorship of Recent Building Heritage Conservation, TUM School of Engineering and Design, Technical University of Munich, Arcisstraße 21, 80333 Munich, Germany
Joana Lia Ferreira, CIUHCT—Interuniversity Center for the History of Sciences and Technology, Department of Conservation and Restoration, NOVA School of Science and Technology, Universidade NOVA de Lisboa, 2829-516 Caparica, Portugal

https://doi.org/10.1515/9783111076997-017

As it is known, plastics are not pure compounds. Polymers are modified with colorants, fillers, and other additives that can make material processing easier, enhance their physical properties, and improve the environmental stability and performance of the final product [8]. Compared to other plastic polymers, acrylics generally do not need many additives to perform well during production and through time. Common additives include peroxides, which are used as polymerization initiators, plasticizers to improve the elasticity and molding capability, or UV stabilizers and flame retardants, depending on the intended use of the products [9].

These complex mixtures of polymer and additives are designed to be effective during the service lifetime of the plastic objects. However, they might not necessarily remain so throughout their museum life, which is expected to be much longer [10]. This discrepancy is a primary reason why conservation scientists study the detailed composition of historical plastic objects in depth. By understanding the composition of an object, it may be possible to predict degradation mechanisms and, therefore, develop adequate preservation strategies. Additionally, the information related to the objects´ materiality also adds to their historical value and contributes to the history of science and technology. The identification of specific additives may help to date an object, determine its provenance, or appreciate its singularity.

Material characterization of historical objects implies the development of specific approaches that allow obtaining essential information while preserving their integrity and avoiding damage. Hence, techniques suitable for in situ analysis or methods that require only microsampling are preferred. Usually, the polymer and additives in concentrations above 3% are identified using Fourier-transform infrared (FTIR) and Raman spectroscopies [11]. Energy-dispersive X-ray spectroscopy (EDXRF), which provides elemental analysis, may contribute to the identification of inorganic pigments and fillers. Organic additives present in concentrations under 3% typically require gas chromatography/mass spectrometry (GC-MS) techniques for their identification.

Within this framework, the authors have been engaged in several projects [12–14] in which different types of historical PMMA from artworks, buildings, and company swatches were analyzed for the sake of conservation of PMMA-based heritage. The data collected by conservation scientists on historical plastic formulations can also inform modern recycling practices, as the presence of certain compounds affects the process, quality, and toxicity of the reprocessed plastic material. This chapter presents the characterization of inorganic pigments, fillers, and organic additives, focusing on transparent and colored PMMA samples of artistic and historical interest from Portugal, France, and Germany, produced between the 1960s and 2000s. The results highlight how heritage science can contribute to PMMA recycling, as heavy metal pigments and a broad range of organic additives of interest have been identified.

17.2 PMMA samples studied

PMMA samples with varying colors and optical clarities from different producers and dates were selected from our previous studies and revisited for the work presented in this chapter. The diversity of samples (Table 17.1) serves two main purposes: firstly, to demonstrate that heavy metal compounds as inorganic pigments and fillers were widely used to impart color to PMMA formulations across the decades; secondly, to present the variety of organic additives possible to encounter in PMMA produced in the past decades.

Table 17.1: List of analyzed historical PMMA samples for the identification of additives with abbreviation code, color, optical clarity, country of origin, production company, commercial name or company identification code, and approximate production date.

Sample code	Color	Optical clarity	Country of production	Producer	Product name and/or code	Date
RAL	Red	Opaque	France[1]	Altulor[1]	Altuglas[1]	1960s
RPS1			Portugal	Plásticos do Sado	247	1970s
RPS2					9081	2000s
RPS3					9082	2000s
RPA1				Paraglas	6500	2000s
RPA2					6700	2000s
OPS1	orange	opaque	Portugal	Plásticos do Sado	261	1970s
OPS2					9043	2000s
OPA				Paraglas	7900	2000s
YPS1	Yellow	Opaque	Portugal	Plásticos do Sado	353	1970s
YPS2					357	1970s
YPS3					9040	2000s
NWPS	Nacreous white	Semiopaque	Portugal	Plásticos do Sado	n.a	1960s
NYPS	Nacreous yellow	opaque				1960s
NOPS	Nacreous orange					1960s
BWPS	Brilliant white					1960s

Table 17.1 (continued)

Sample code	Color	Optical clarity	Country of production	Producer	Product name and/or code	Date
BYPS	Brilliant yellow					2000s
BRPS	Brilliant red					2000s
TPL1	Colorless	transparent	Germany/France[2]	Röhm und Haas/Alsthom[2]	Plexiglas[2]	1960s
TPL2			Germany	n.a	n.a	1960s
TPL3				Röhm GmbH	Plexiglas® GS215 gereckt	1970s
TPS1			Portugal	Plásticos do Sado	201	1970s
TPS2					n.a	2000s
TPA				Paraglas	n.a	2000s

[1]Sample provided by the artist Lourdes Castro as Altuglas. It is not possible to confirm this information. At that time, Altuglas was produced by Altulor, in France.
[2]Sample provided by the artist Lourdes Castro as Plexiglas. It is not possible to confirm this information. Since during the 1960s acrylic sheet traded under the name Plexiglas was produced both in Germany, by Röhm und Haas, and in France, by Alsthom, and the artist lived and worked in both countries, it is not possible to know either the producer or the country of production.

17.2.1 Colored samples

To display the range of inorganic pigments, including heavy-metal compounds, two groups of colored samples were considered: i) the first consisting of red, orange, and yellow opaque samples; ii) the second containing special effect pigments such as nacreous. With one exception, all samples in these groups were produced by two Portuguese companies, Plásticos do Sado and Paraglas. Both these companies produced cast PMMA sheets, although by different methodologies. Plásticos do Sado used recycled monomer and polymerization was performed exclusively in water tanks. Instead, Paraglas used pure MMA and the sheets were produced through the Rostero process (in chambers with controlled temperature and pressure), including a post-polymerization step at 120 °C [15, 16].

The colored samples of the first group comprise swatches that were produced by Plásticos do Sado from the 1970s and from the 2000s (RPS, OPS, and YPS samples), and samples by Paraglas from the 2000s (RPA and OPA). Additionally, a red sample from a cast sheet of PMMA, produced by Altuglas (a French company) during the 1960s (RAL) was also considered. This sample was offered for study by Lourdes Castro (1930–1922), a Portuguese artist who extensively used acrylic sheets in her artworks during the 1960s [17].

The nacreous and brilliant samples of the second group were taken from acrylic sheets found in the studio of the Portuguese artist Ângelo de Sousa (1938–2011). These included three sheets from the 1960s with a nacreous effect in three different colors (white, yellow, and orange, identified as NWPS, NYPS, and NOPS, respectively); a brilliant sheet of white color also from the same decade (BWPS); and two brilliant sheets from the 2000s of yellow (BYPS) and red (BRPS) colors. Previous research [15, 18] has shown that the artist likely bought these sheets from Plásticos do Sado, which had produced them for the button and cutlery industries.

17.2.2 Transparent colorless samples

Transparent and colorless samples were selected from Germany and Portugal, from different dates: two samples from the 1960s, two from the 1970s, and two from the 2000s. The first sample considered (TPL1) was also offered by the artist Lourdes Castro and identified by her as Plexiglas bought during the 1960s. The PMMA samples produced in Germany were originally applied on buildings. One (TPL2) was taken from the original large bent windows (1 × 2 m^2), made for the facade of the former Felix Wankel Institute building (Lindau), dated from 1961. This PMMA was produced as cast sheet and then thermoformed over a wooden mold [19]. The other sample (TPL3) derives from the Roof of the Olympic Sports Facilities in Munich, originally dated from 1972. This PMMA was also produced as cast sheets (1.5 × 1.5 × 0.12 m^3) and then biaxially stretch-formed under heat in four directions at the same time, thus gaining a final shape of 3 × 3 × 0.04 m^3. In a later step, the sheets were cold bent into aluminum frames. These transparent and colorless PMMA sheets cover ca. 20% (16,000 m^2) of the roof [20]. In contrast, the Portuguese samples selected were all produced as 3-mm transparent cast sheets and not subjected to further processing. These consist of a Plásticos do Sado swatch sample from the 1970s (TPS1) and samples from transparent sheets from 2000s by Plásticos do Sado (TPS2) and Paraglas (TPA). Their casting processes are the same as described above for the colored samples produced by these two companies.

17.3 Characterization technique for additives in PMMA

17.3.1 X-ray fluorescence (XRF)

Elemental analysis is used to identify additives [8], such as fillers and inorganic pigments, supporting preliminary discrimination of plastics possibly containing toxic heavy metal components in the plastic matrices. Among the elemental analytical methods, X-ray fluorescence spectrometry allows the characterization of the elemental composition in situ, respecting the integrity of the materials without sampling and sample preparation.

Micro-energy-dispersive X-ray fluorescence spectrometry (μ-EDXRF) was used in the current investigation (ArtTAX, Intax GmbH) and the experimental parameters were 25 kV voltage, 300 μA of intensity, and 360 s acquisition time under a Helium gas flux to reduce the bremsstrahlung effect ("braking radiation") due to the polymeric (organic) matrix.

17.3.2 Raman microscopy (μ-Raman)

Raman microscopy proved to be a valuable technique for the characterization of pigments in historical plastics [13] overcoming their low concentration (ca 0.1%-3%) and the micro-size of their particles/agglomerates (ca 0.02–30 μm). Taking advantage of the confocal microscope system, it is possible to selectively analyze the particles/agglomerates dispersed in the matrix nondestructively with no sample preparation by simply focusing the laser beam on their surfaces.

The Raman investigation presented in this chapter was carried out using a μ-Raman Labram 300 Jobin Yvon spectrometer equipped with three laser sources: a 17-mW He–neon laser operating at 632.8 nm, a 100-mW diodes laser operating at 785 nm, and a solid state 50- mW laser operating at 532 nm. The three excitation lines were employed to explore the resonance Raman spectra and maximize sensitivity avoiding fluorescence. The selected laser beam was focused either with a 50× or a 100× Olympus objective lens. The laser power at the surface of the samples was controlled with neutral density filters.

17.3.3 Thermal desorption-gas chromatography/mass spectrometry (TD-GC/MS)

Thermal desorption-gas chromatography/mass spectrometry (TD-GC/MS) analyzes the volatile part of polymers, which can be composed by additives, residues of solvents, and unreacted monomers. Volatiles evolve during the heating program, starting from ca. 40 °C generally until 250–300 °C, being separated from the polymeric matrix, which simplifies their identification. The analysis can be followed by a subsequent step of Py-GCMS for the complete characterization of the polymers at higher temperatures. The thermal desorption is a selective method that enables the detection of additives in plastics [16, 21–25], which normally are present at a very low concentration (0.1–1% w/w) [26] and are undetectable by other conventional method, e.g., FTIR [16]. It uses a solvent-free direct sample injection without requiring any complicated pretreatment procedures [27] and this is an advantage compared to conventional GCMS method, which requires prolonged, e.g., solvent extraction and clean-up procedures, usually expensive and time-consuming [28]. To enhance the selectivity of the analysis, a cryo-trap cooled by liquid nitrogen was used to focus temporarily the volatiles (i.e., gather their small amount) at the beginning of the chromatographic separation column.

In this study, around 100 µg of each PMMA sample was analyzed by thermal desorption with a multi-shot pyrolyzer EGA/PY-3030D (Frontier Lab.) connected to a 7890B GC gas chromatograph/5977B MS mass spectrometer (Agilent) [16].

17.4 Additives identification: results and discussion

XRF, µ-Raman, and TD-GCMS analyses detected several additives in the PMMA samples, including heavy metal pigments, fillers and organic compounds used as initiators, plasticizers, UV absorbers, fire retardants, and lubricants. The results are summarized in Table 17.2 by the identified additive.

Table 17.2: Inorganic pigments, fillers, and organic additives identified in the PMMA samples.

Additive	Compound, formula, CAS number	Period	PMMA sample
Inorganic pigment	Cadmium sulfo-selenide, Cd(S,Se), CAS 58339-34-7	1960s	RAL
		1970s	RPS1; OPS1
		2000s	RPS2; RPA1; OPS2
	Cadmium-zinc sulfide, (Cd,Zn)S, CAS 8048-07-5	1970s	OPS1; YPS2
		2000s	OPS2; YPS3
	Mercury-cadmium sulfide, (Cd,Hg)S, CAS 1345-09-1	2000s	RPS3
	Cadmium sulfide, CdS, CAS 1306-23-6	2000s	OPS2
	Lead chromate, $PbCrO_4$, CAS 1344-37-2	1970s	YPS1
	Zinc sulfide, ZnS, CAS 1314-98-3	1970s	RPS1; OPS1; YPS2
		2000s	RPS2; YPS3
	Titanium dioxide, TiO_2, CAS 13463-67-7	1970s	YPS1
		2000s	RPA2; OPA
Filler	Barium sulfate, $BaSO_4$, CAS 7727-43-7	1970s	RPS1; OPS1
		2000s	RPS2; RPS3; OPS2; YPS3

Table 17.2 (continued)

Additive	Compound, formula, CAS number	Period	PMMA sample
Initiator	Azobisisobutyronitrile (AIBN), $[(CH_3)_2C]_2N_2$, CAS 78-67-1	1960s	TPL1; RAL; NWPS, NYPS, NOPS, BWPS
		1970s	TPS1, RPS1, YPS1, YPS2, OPS1, TPS1
		1990s	BYPS, BRPS
		2000s	TPA, RPA1, RPA2, OPA TPS2, RPS2, RPS3, YPS3, OPS2
	Dibenzoylperoxide (DBPO), $C_{14}H_{10}O_4$, CAS 94-36-0	1960s	TPL2
	Dilaurylperoxide (LP), $C_{24}H_{46}O_4$, CAS 105-74-8	1970s	TPL3
Plasticizer	Diethyl phthalate (DEP), $C_{12}H_{14}O_4$, CAS 85-66-2	1960s-2000s	All samples
	Dibutyl phthalate (DBP), $C_{16}H_{22}O_4$, CAS 84-74-2	1960s	RAL TPL2 BW, NW, NY, NO
		1970s	RPS1, YPS1, YPS2
		1990s	BYPS, BRPS
		2000s	RPA RPS2, OPS2
	Bis(2-ethylhexyl) phthalate (DEHP), $C_{24}H_{38}O_4$, CAS 117-81-7	1960s	TPL1 BWPS, NWPS, NYPS, NOPS
		1970s	TPS1, RPS1, YPS1, YPS2, OPS1
		2000s	TPA, RPA2, OPA TPS2, RPS2, RPS3, YPS3, OPS2
	Diisooctyl phthalate (DOP), $C_{24}H_{38}O_4$, CAS 27554-26-3	1960s	BWPS
	Diisononyl phthalate (DINP), $C_{26}H_{42}O_4$ CAS 28553-12-0	2000s	TPA, RPA2, OPA
	Dibutyl adipate (DBA), $C_{14}H_{26}O_4$, CAS 105-99-7	1970s	OPS1
	Bis(2-ethylhexyl) adipate (DEHA), $C_{22}H_{42}O_4$, CAS 103-23-1	1960s	NYPS

Table 17.2 (continued)

Additive	Compound, formula, CAS number	Period	PMMA sample
UV absorber	2-(2-Hydroxy-5-methylphenyl)benzotriazole (Drometrizole), $C_{13}H_{11}N_3O$, CAS 2440-22-4	1970s	TPL3[1]
		1990s	BYPS, BRPS
		2000s	RPA1, RPA2, OPA TPS2
	Ethyl-2-cyano-3,3-diphenylacrylate (Etocrylene), $C_{18}H_{15}NO_2$, CAS: 5232-99-5	2000s	TPS2
Fire retardant	Phosphonic acid, $C_{14}H_{28}Cl_5O_9P_3$, CAS 4351-70-6	1970s	TPL3
Lubricant	Palmitic acid, methyl ester, $C_{17}H_{34}O_2$, CAS 112-39-0	1960s–2000s	All samples, with the exception of: TPA, RPA1, RPA2, OPA, TPL3
	Stearic acid, methyl ester, $C_{19}H_{38}O_2$, CAS 112-61-8	1960s–2000s	All samples, with the exception of: TPA RPA1, RPA2, OPA, TPL3

[1]Confirmed by the company Röhm GmbH to be the product Tinuvin® P [29].

17.4.1 Inorganic pigments and fillers

PMMA polymer formulations can be mass-colored with pigments and dyes to reach respectively opaque and transparent appearance, as well as with special colorant types (e.g., metallic, pearlescent, fluorescent, phosphorescent, thermochromic, and photochromic colorants) to impart specific effect appearance. Instead, inorganic fillers are used to reinforce the polymeric resin or reduce the cost by replacing part the pigment and behaving, in most cases, as scattering opacifiers.

Important requirements of the colorants for a successful coloring process are the chemical compatibility with the acrylic resin, thermal stability (as it must be capable of surviving the temperatures at which the polymer is processed), migration resistance during the full cycle of plastic life, satisfactory dispersibility during processing, good lightfastness, and weatherability [30]. Colorants stable up to ca. 230 °C are advisable for the mass-coloring of PMMA [31]. Several inorganic pigments containing heavy metal elements meet this requirement. Table 17.2 provides an overview of the cadmium (Cd)-, mercury (Hg)-, chromium (Cr)-, lead (Pb)-, zinc (Zn)-, and titanium (Ti)-containing pigments identified in the historical PMMA samples.

17.4.1.1 Cadmium pigments

The basic composition of cadmium pigments is cadmium sulfide (CdS) (PY 37), which can be coprecipitated with increasing amounts of selenium Cd(S,Se) (PR 108) to produce shades ranging from orange to deep red, or with increasing amounts of zinc (Cd, Zn)S (PY 35) to achieve greener hues of yellow. Another variation involves substituting some of the cadmium with mercury in the crystal lattice, yielding mercury-cadmium sulfide (Cd,Hg)S (PR 113:1), which produces colors from deep orange to maroon. Cadmium-based pigments are heat-resistant above 400 °C, making them suitable for coloring PMMA. Additionally, they are known for their lightfastness, chemical stability, ease of dispersion, and good migration resistance [32].

These pigments have been used in the plastic industry since the 1950s [33]. Despite their excellent properties, specific regulations and restrictions due to potential toxicological and environmental concerns have limited their use. [30, 33–35]. Recently, the EU has proposed new restrictions on the use of cadmium compounds in plastic products [36, 37].

Even so, all the mentioned pigments (cadmium sulfide, cadmium sulfo-selenide, cadmium-zinc sulfide, and mercury-cadmium sulfide) were identified as the main coloring agents, either alone or in combination with each other or with lead- or titanium-based pigments, in some of the historical PMMA samples (Table 17.2 and Figure 17.1a). The identification of mercury-cadmium sulfide is particularly notable, as it was considered a very rarely used pigment by some authors [38]. Primarily used in plastic applications [39, 40], its production was limited after the 1980s [41]. The presence of this pigment in a sample from the 1990s indicates its continued use as a plastic colorant.

17.4.1.2 Lead chromate

Besides cadmium pigments, lead chromate, $PbCrO_4$ (PY 34), was also identified (Table 17.2). Such pigment typically produces warm yellow tonalities and has been a plastics mainstay due to its very good heat stability, lightfastness, weatherability, and bleed resistance [30, 34, 42]. Widely used in the past, this pigment is also regulated by strict guidelines because of toxicological considerations [36, 37]. Chromium was also detected by XRF in samples RPA2 and However, the lack of lead suggests the presence of another Cr-based additive such as spinel pigments, inorganic or organometallic compounds for the production and processing of PMMA.

17.4.1.3 Fillers and other inorganic pigments

Inorganic additives such as barium sulfate, $BaSO_4$, and zinc sulfide, ZnS, were identified in the samples (Table 17.2). They could have been used as extender ($BaSO_4$, PW 21) and lightener (ZnS, PW 7), respectively, or as an intimate mixture to produce lithopone ($ZnS/BaSO_4$, PW 5), a white pigment for plastic coloring. Titanium dioxide TiO_2 (PW 6), is the pigment of choice for producing whites, pastel tints, and opacifying plastic products still today. Due to its very high refractive index ($n > 1.7$), titanium dioxide is indeed considered a white pigment rather than a filler [42].

17.4.1.4 Pearlescent pigments

µ-Raman analysis confirmed unequivocally the presence of two different types of pearlescent pigments (Figure 17.1b), namely plumbonacrite and bismuth oxychloride, respectively, in the lustrous acrylic sheets of 1960s and 2000s found in the studio of Ângelo de Sousa [43, 43].

Plumbonacrite ($Pb_5(CO_3)_3O(OH)_2$) is a monocrystalline pigment that, together with hydrocerussite $Pb_3(CO_3)_2(OH)_2$ (lead white), is described as being pearlescent and falls under the definition of basic lead carbonates pigments. Those pigments were manufactured in the 1930s [44, 45] and, by the 1950s and 1960s, represented the most common synthetic pearlescent pigment category employed in the production of artificial pearls, bijouterie, and acrylic and polyester buttons. The distinction of the two basic lead carbonate pigments is critical as they differ in stability and notoriety [43]. In contrast to hydrocerussite, plumbonacrite is rarely used as a pigment [46] and no longer commercially available [47]. The identification of plumbonacrite in the acrylic samples evidences for the first time the use of this compound as pearlescent pigment by the plastic industry, which was never specifically reported in the coloring technology literature.

Bismuth oxychloride BiOCl (PW 14) is a rare example of a nontoxic heavy metal compound, in use since the early 1960s in cosmetics and later in that decade also in the plastics industry [48]. It is a pearlescent pigment currently used in coatings for cars and in cosmetics, printing, buttons, bijouterie, and plastics [30, 36, 49]. It is acknowledged that bismuth oxychloride started to replace lead-based effect pigments from the late 1960s due to growing concerns regarding toxicity [42, 48, 50, 51].

Summarizing, heavy metal containing pigments were identified in the Portuguese PMMA samples from all decades considered (1960s–2000s). Recently, the authors analyzed red, orange, and yellow PMMA samples from Plexiglas produced in Germany in 2002–2003, and no pigments based on heavy metals were identified. This indicates that depending on the origin of the material (country, manufacturer, period of production), different colorants may have been used, highlighting the importance of their identification in heritage science and recycling fields. In addition, the detection of plumbonacrite and, in more recent samples, bismuth oxychloride, reflects that there has been an evolution over time in the selection of pearlescent pigments due to toxicity and environmental concerns.

17.4.2 Organic additives

17.4.2.1 Initiators

PMMA can be produced using a variety of polymerization mechanisms. The most common technique is the free radical polymerization of MMA, a chain polymerization process, which involves the double bond of the monomer. It generally starts with the production of radicals due to the reaction with other labile compounds, through radiation, heat, or chemical agents [52]. The following three initiators were identified: azobisisobutyronitrile (AIBN), dibenzoylperoxid (DBPO), and dilaurylperoxide (LP). AIBN was found in the Portuguese and French samples from the 1960s to 2000s, while DBPO and LP were identified in transparent PMMA produced in Germany, respectively, in the 1960s and 1970s. AIBN is a thermally labile compound which, upon heating, forms free radicals initiating the polymerization of MMA [52]. DBPO is both a photochemically and thermally labile compound that can be used either for the radiation or for the heat-initiated polymerization of MMA. It can be also used as a free radical initiator for the suspension polymerization of PMMA [52]. As the previous peroxide, LP can be used as initiator for a free radical polymerization of PMMA [53].

17.4.2.2 Plasticizers

Plasticizers are chemical additives that provide durability, elasticity, and flexibility to polymeric products. Phthalate esters (PEs) are the dominant class of plasticizers used in plastic formulations. In particular bis-2-ethylhexyl phthalate (DEHP) has been the most widely used substance [54]. Alternative plasticizers that can be used instead of PEs are adipates, benzoates, citrates, cyclohexane dicarboxylic acids, epoxidized vegetable oils, glycerol-acetylated esters, phosphate esters, sebacates, terephthalates, and trimellitates [55]. Plasticizers can be distinguished by their concentration. Typically, primary plasticizers are those used alone or as main plasticizer when mixed with others, enhancing the elastic behavior and softness of the plastic material. In contrast, secondary plasticizers are used in smaller concentrations and are often combined with the primary plasticizers to improve specific performance features, such as reduced flammability [56, 57].

Different PEs were identified in the historical PMMA samples (Table 17.2). Dibutyl phthalate (DBP) and DEHP were identified as the primary plasticizers, while all other identified phthalates can be considered as secondary. Diisononyl phthalate (DINP) was found exclusively in the PMMA samples produced by the Portuguese company Paraglas (Figure 17.1c).

It is worth noticing that several PEs have been shown to cause negative health effects to animals as well as to humans, and therefore, their use began to be regulated in the European Union. This is the case of DBP and DEHP. DEHP is also classified as toxic to reproduction [58] and has thus been banned in toys, children's articles, cosmetics, and medical devices [37]. DBP is also listed as environmental hazard [59].

Some PMMA samples, such as TPL3 from the 1970s, are characterized by relatively low amounts of plasticizers. On one hand, this highlights the relatively pure plastic formulation; on the other hand, makes it harder to distinguish between primary and secondary plasticizers. Interestingly, the transparent sample from Germany (TPL2) used for windows also possesses an extremely simple PMMA formulation, lacking stabilizers and characterized by very low amounts of additives.

17.4.2.3 UV absorbers

Drometrizole was found in several PMMA samples such as TPL3 from the 1970s and especially in samples produced in the 1990s and 2000s (Table 17.2). Drometrizole is an odorless benzotriazole compound that appears as an off-white to yellow crystalline powder. It is used as an UV absorber and stabilizer in plastics, polyesters, celluloses, acrylates, dyes, rubber, synthetic and natural fibers, waxes, detergent solutions, and orthodontic adhesive. It is also categorized as a sunscreen ingredient used in a variety of cosmetics and personal care products with no significant toxicity properties [60].

Furthermore, in one sample (TPS1) etocrylene was found in combination with drometrizole (Table 17.2). It has been listed as a UV absorber used in cosmetics and plastics [61, 62].

17.4.2.4 Fire retardants and lubricants

A fire retardant agent was only found in one sample, TPL3, and identified as phosphonic acid, known as Phosgard C22R. This sample comes from plates applied on the roof of the Olympia Park in Munich, which required a special composition for their use as part of public external infrastructure. The usage of a flame retardant in combination with a UV-absorbent for outdoor acrylics was particularly relevant because it led to the development of Plexiglas GS215 Gereckt, a material that became part of Röhm's commercial catalogue from 1972 onward [63, 64].

In contrast, traces of palmitic and stearic acid methyl esters were found in almost all samples (Table 17.2). These are commonly used as lubricants and are classified as irritant substances [65].

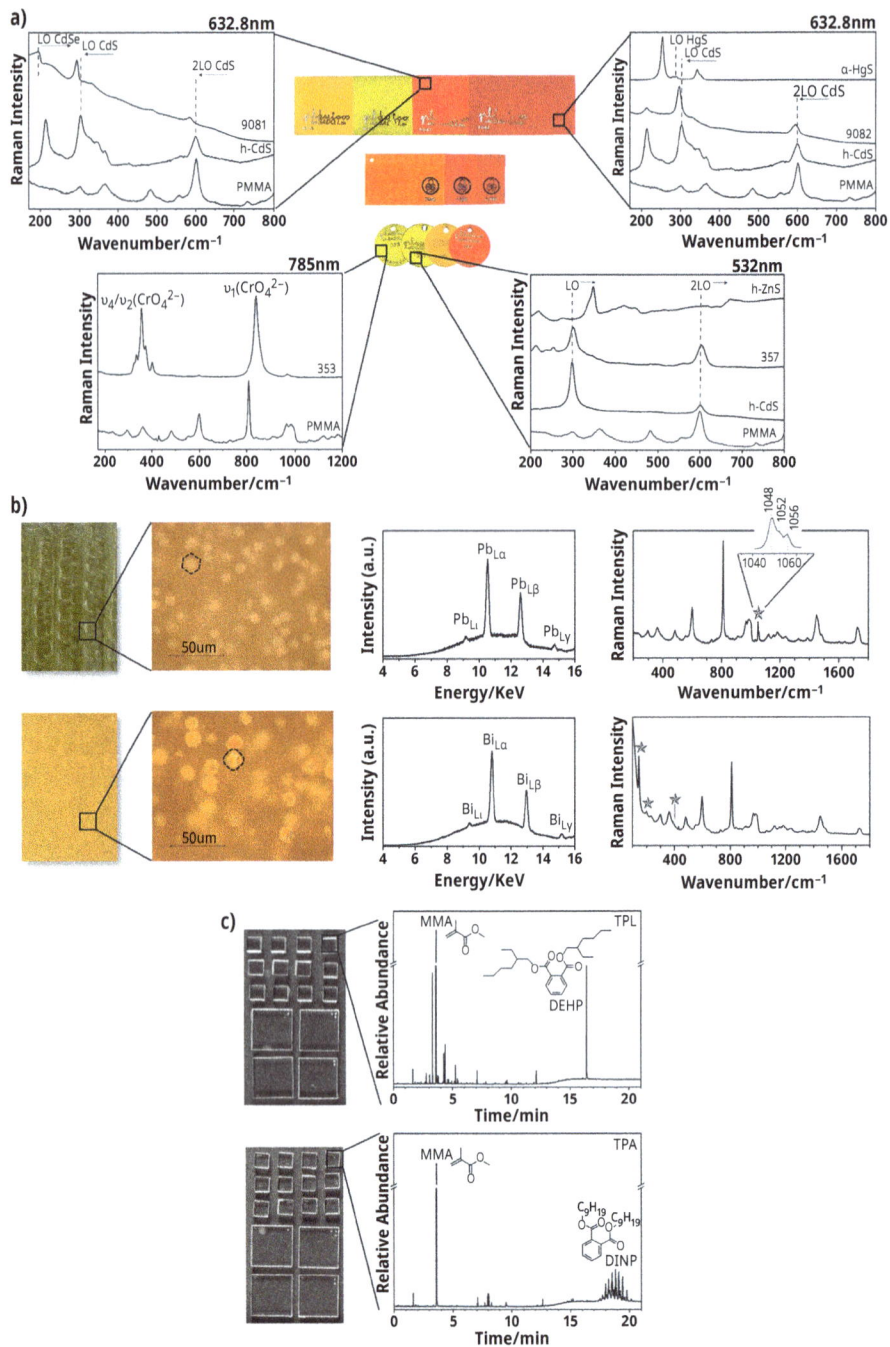

Figure 17.1: (a) μ-Raman spectra of red, yellow, and orange samples recorded under resonance and out-of-resonance conditions for the identification of cadmium and lead chromate pigments, respectively.

17.5 Conclusions

The investigation into historical PMMA samples from the 1960s to the 2000s, produced in Portugal, France, and Germany, has yielded significant insights into the complex composition of these materials, revealing the presence of a wide range of additives. These additives include inorganic pigments and fillers, and various organic compounds used as initiators, plasticizers, UV absorbers, fire retardants, and lubricants. Understanding these additives is crucial for heritage conservation as well as for recycling practices.

One of the key findings of this study is the identification of heavy metal pigments, such as several cadmium and lead-based compounds in the PMMA samples. These pigments, known for their excellent thermal stability, lightfastness, and weatherability, were commonly used in the plastics industry, despite toxicological and environmental concerns. The presence of these pigments across different decades and producers underscores their widespread use in the recent past. Regarding the organic additives, it is worth noticing the continuous use of phthalate esters (PEs) as plasticizers, including dibutyl phthalate (DBP) and bis-2-ethylhexyl phthalate (DEHP). The current regulatory restrictions on these compounds due to their adverse health effects emphasize the importance of identifying and managing these substances in recyclable materials.

Contrary to what was expected considering the analyzed timeframe, as well as the known toxicity of some of the compounds, the authors could not find a systematic evolution of the additives used through time. An important exception is the case of the lead white pigment plumbonacrite that has been replaced by nontoxic bismuth oxychloride, as shown by the lustrous samples from the 2000s. Following current legislation on human and environment toxicity restricting the usage of chemical substances, this scenario is changing, and a follow-up study would certainly show a different trend.

The analytical methodology used to identify additives in cultural heritage objects made of PMMA is similar to that employed by industrial deformulation; however, the former works with very small samples and noninvasive methods to respect the value of heritage. Studies dedicated to the identification of additives in historical PMMA are relevant not only for heritage professionals, who need to understand their role in deg-

---

Figure 17.1 (continued)

References of PMMA, h-CdS, α-HgS, h-ZnS pigments are also reported for comparison. (b) Identification of plumbonacrite and bismuth oxychloride in the yellow pearlescent acrylic sheets from the Ângelo de Sousa studio from the 1960s (up) and 2000s (bottom). μ-EDXRF identified counts of lead (Pb) and bismuth (Bi), respectively. The vibrational fingerprint of the pigments was obtained by observing some Raman bands not assignable to the PMMA Raman spectrum: three subbands at 1,048, 1,052, and 1,056 cm^{-1} due to the factor-group splitting of the v1 symmetric stretching mode of the CO_3^{2-} ion in plumbonacrite, bands at 142, 198, and 394 cm^{-1} for bismuth oxychloride. (c) TD chromatograms of two historical PMMA samples TPL1 (top) and TPA (bottom). Both samples are characterized by the MMA fragment (RT 3.61 min). DEHP (RT 16.4 min) and DINP (RT 17–19 min) were identified as plasticizers, respectively, in TPL1 and TPA.

radation for preservation purposes, but also for other professionals, like those working in the recycling industry.

References

[1] Thompson RC, Swan SH, Moore CJ, vom Saal FS. Our plastic age. Philosophical Transactions of the Royal Society B: Biological Sciences 2009, 364(1526), 1973–1976. 10.1098/rstb.2009.0054.

[2] Madden O. Balancing ingenuity and responsibility in the age of plastic. In Madden O, Charola AE, Cobb KC, DePriest PT, Koestler RJ, eds. The Age of Plastic: Ingenuity and Responsibility: Proceedings of the 2012 MCI Symposium. Smithsonian Institution; Museum Conservation Institute, Washington, D.C., 2017, 1–17.

[3] Röhm, Haas AG. What is happening here? Hanau: Evonik industries AG, corporate archives. Record Group Röhm 1936, 10–10.

[4] Brunner S. Transparent acrylic constructions before and after 1950 – From the 1935 Opel Olympia to the 1972 Olympic roof. In History of Construction Cultures. CRC Press, London, 2021, 275–282. 10.1201/9781003173359-36

[5] Pullen D. Sculpture. In Quye A, Williamson C, eds. Plastics Collecting and Conserving. NMS Publishing Limited, Edinburgh, 1999, 48–50.

[6] Waentig F. Plastics in Art, Michael Imhof Verlag, Petersberg, 2008.

[7] Williamson C. Foreword. In Lavédrine B, Fournier A, Martin G, eds. Preservation of Plastic Artefacts in Museum Collections. Paris: Éditions du comité des travaux historiques et scientifiques. 2012, 15–21.

[8] Bart JCJ. Additives In Polymers: Industrial Analysis And Applications, John Wiley & Sons, Chichester, 2005.

[9] Vieweg R, Esser F. Kunststoff-Handbuch, Carl Hanser, Munich, 1975.

[10] Williams RS. Composition implications of plastic artifacts: A survey of additives and their effects on the longevity of plastics. In Grattan DW, ed. Proceedings of a Conference Symposium 91 – Saving the Twentieth Century Ottawa, Canada 15 to 20 September 1991. Canadian Conservation Institute, Ottawa, 1993, 135–153.

[11] van Oosten T (coord.). Identification and characterisation of plastic artefacts. In Lavrédrine B, Fournier A, Martin G, eds. Preservation of Plastic Artefacts in Museum Collections. Paris: Éditions du comité des travaux historiques et scientifiques. 2012, 29–105.

[12] Babo S. From industry to artworks by Lourdes Castro and Ângelo de Sousa: conservation studies on cast acrylic sheet [PhD Thesis]. Universidade NOVA de Lisboa, 2021.

[13] Angelin E. The fate of colours in the 20th – 21st centuries: preserving the organic colourants in plastic artifacts [PhD thesis]. Universidade NOVA de Lisboa, 2021.

[14] Brunner S. Abschlussbericht des Forschungsvorhabens Zur Erhaltung historischer Acrylgläser, Erhaltungsstrategien für transparentes Polymethylmethacrylat (PMMA). Architektur und musealem Kulturgut im Außenraum 2023.

[15] Babo S, Ferreira JL, Melo MJ, Ramos AM. Back to the origin: Understanding the history of production and its influence on the properties of acrylic sheet. In Bechthold T, ed. Future Talks 015–Processes the Making of Design and Modern Art. Materials, Technologies and Conservation Strategies. Die Neue Sammlung – The Design Museum, Munich, 2017, 160–170.

[16] Babo S, Ferreira JL, Ramos AM, Micheluz A, Pamplona M, Casimiro MH, Ferreira LM, Melo MJ. Characterization and long-term stability of historical PMMA: Impact of additives and acrylic sheet industrial production processes. Polymers 2020, 12(10). 10.3390/POLYM12102198.

[17]   Restany P. Lourdes Castro: A presença da ausência. In: Castro L, Zimbro M, eds. Lourdes Castro Além da Sombra, Exhibition Catalogue. Lisbon: Centro de Arte Moderna, Fundação Calouste Gulbenkian; 1992. p. 36–38.

[18]   Angelin EM, Babo S, Ferreira JL, Melo MJ. Raman microscopy for the identification of pearlescent pigments in acrylic works of art. Journal of Raman Spectroscopy 2019, 50(2). 10.1002/jrs.5431.

[19]   Brunner S, Putz A. Acrylglas im Bauen – Materialgerechte Konstruktion oder konstruktionsgerechtes material? In Materialgerecht konstruiert!?! Fünfte Jahrestagung der Gesellschaft für Bautechnikgeschichte 2021. Michael Imhof Verlag, Petersberg, 2023, 241–257.

[20]   Wittig E, Eizenhöfer D. 100 Jahre Zukunft : die Röhm GmbH von 1907 bis 2007, Röhm, Darmstadt, 2007.

[21]   Jansson KD, Zawodny CP, Wampler TP. Determination of polymer additives using analytical pyrolysis. Journal of Analytical and Applied Pyrolysis 2007, 79(1–2), 353–361. 10.1016/j.jaap.2006.12.009.

[22]   Yuzawa T, Watanabe C, Freeman RR, Tsuge S. Rapid and simple determination of phthalates in plastic toys by a thermal desorption-GC/MS method. Analytical Sciences 2009, 25(9), 1057–1058. 10.2116/analsci.25.1057.

[23]   Micheluz A, Angelin EM, Lopes JA, Melo MJ, Pamplona M. Discoloration of historical plastic objects: New insight into the degradation of β-Naphthol pigment lakes. Polymers 2021, 13(14), 2278. 10.3390/polym13142278.

[24]   Elsässer C, Micheluz A, Pamplona M, Kavda S, Montag P. Selection of thermal, spectroscopic, spectrometric, and chromatographic methods for characterizing historical celluloid. Journal of Applied Polymer Science 2021, 138(21). 10.1002/app.50477.

[25]   Micheluz A, Angelin EM, Sawitzki J, Pamplona M. Plastics in robots: A degradation study of a humanoid skin mask made of soft urethane elastomer. Heritage Science 2022, 10(1), 4. 10.1186/s40494-021-00636-8.

[26]   Hunt TP. Polymer additives: Supercritical fluid chromatography. In Wilson ID, ed. Encyclopedia of Separation Science. Academic Press, San Diego, 2000, 3901–3906.

[27]   Yanagisawa H, Maruyama F, Fujimaki S. Verification of simultaneous screening for major restricted additives in polymer materials using pyrolyzer/thermal desorption gas–chromatography mass spectrometry (Py/TD-GC-MS). Journal of Analytical and Applied Pyrolysis 2019, 137, 37–42. 10.1016/j.jaap.2018.11.004.

[28]   Yanagisawa H, Kudo Y, Nakagawa K, Miyagawa H, Maruyama F, Fujimaki S. Simultaneous screening of major flame retardants and plasticizers in polymer materials using pyrolyzer/thermal desorption gas chromatography mass spectrometry (Py/TD–GC–MS). Molecules 2018, 23(4), 728. 10.3390/molecules23040728.

[29]   Brunner S. Outdoors! Transparent acrylics in cultural heritage exposed to weather – Case studies. Degradation phenomena and their possible causes. In Bechthold T, ed. Future Talks 021, Smart Solutions in the Modern. Die Neue Sammlung – The Design Museum, Munich, 2023, 44–51.

[30]   Charvat RA, ed. Colouring of Plastics: Fundamentals, 2nd edition, John Wiley & Sons Inc., New Jersey, 2004.

[31]   Webber TG, ed. Coloring of Plastics, John Wiley & Sons Inc., New York, 1979.

[32]   Angelin EM, Ghirardello M, Babo S, Picollo M, Chelazzi L, Melo MJ, Nevin A, Valentini G, Comelli D. The multi-analytical in situ analysis of cadmium-based pigments in plastics. Microchemical Journal 2020, 157, 105004. 10.1016/j.microc.2020.105004.

[33]   Dunning P. Cadmium pigments. In Faulkner EB, Schwartz RJ, eds. High Performance Pigments, 2nd edition. Wiley-VCH, Weinheim, 2009, 13–26.

[34]   Buxbaum G, ed. Industrial Inorganic Pigments, 2nd edition, Wiley-VCH Verlag GmbH, Weinheim, 1998.

[35] Turner A. Cadmium pigments in consumer products and their health risks. Science of the Total Environment 2019, 657, 1409–1418. 10.1016/j.scitotenv.2018.12.096.

[36] Etzrodt G. Industrial Colouration of Plastics, Hanser, Munich, 2022.

[37] ECHA. Substances restricted under REACH (Annex XVII to REACH). 2024.

[38] Bacci M, Baronti S, Casini A, Lotti F, Picollo M, Casazza O. Non-destructive spectroscopic investigations on paintings using optical fibers. MRS Proceedings 1992, 267, 265. 10.1557/PROC-267-265.

[39] Fiedler I, Bayard MA. Cadmium yellows, oranges and reds. In Feller RL, ed. Artist's Pigments, A Handbook of Their History and Characteristics, vol. 1. National Gallery of Art, Washington, 1986, 65–108.

[40] Huckle WG, Swigert GF, Wiberley SE. Cadmium pigments. Structure and composition. I&EC Product Research and Development 1966, 5(4), 362–366. 10.1021/i360020a016.

[41] De Keijzer M. The history of modern synthetic inorganic and organic artists' pigments. In Mosk JA, Tennet NH, eds. Contributions to Conservation: Research in Conservation at the Netherlands Institute for Cultural Heritage (ICN). James & James (Science Publishers), London, 2002, 42–54.

[42] Müller A. Coloring of Plastics. Fundamentals – Colorants – Preparations, Carl Hanser Verlag GmbH & Co., Munich, 2003.

[43] Angelin EM, Babo S, Ferreira JL, Melo MJ. Raman microscopy for the identification of pearlescent pigments in acrylic works of art. Journal of Raman Spectroscopy 2019, 50(2). 10.1002/jrs.5431.

[44] Brossman PD. Pigment and process of preparing same, 1934.

[45] Thompson GW, Roselle AS. White Lead, 1937.

[46] Corbeil M-C, Sirois PJ. A note on a modern lead white, also known as "Synthetic Plumbonacrite. Studies in Conservation 2007, 52(4), 281–288. 10.1179/sic.2007.52.4.281.

[47] Franz KD, Emmert R, Nitta K. Interference pigments. Kontakte (Darmstadt) 1992, 2, 3–14.

[48] Greeinsten LM. Properties and economics. In Lewis PA, ed. Pigment Handbook, vol. 1, 2nd edition. J. Wiley & Sons, New York, 1988, 829–858.

[49] Suzuki EM. Infrared spectra of U.S. automobile original finishes (1998–2000). IX. Identification of Bismuth Oxychloride and Silver/White Mica Pearlescent pigments using extended range FT – IR spectroscopy, XRF spectrometry, and SEM / EDS analysis. Journal of Forensic Sciences 2014, 59(5), 1205–1225. 10.1111/1556-4029.12414.

[50] Pfaff G. Special effect pigments. In Faulkner EB, Schwartz RJ, eds. High Performance Pigments, 2nd edition. Wiley-VCH Verlag GmbH & Co, Weinheim, 2009, 77–104.

[51] Pfaff G, Franz K-D, Emmert R, Nitta K. Luster pigments. In Buxbaum G, ed. Industrial Inorganic Pigments, 2nd edition. Wiley-VCH Verlag GmbH & Co., Weinheim, 1998, 211–228.

[52] Rasmussen WL. Novel Carbazole Based Methacrylates, Acrylates, and Dimethacrylates to Produce High Refractive Index Polymers, Virginia Polytechnic Institute and State University, 2001.

[53] Wilson GO, Henderson JW, Caruso MM, Blaiszik BJ, McIntire PJ, Sottos NR, White SR, Moore JS. Evaluation of peroxide initiators for radical polymerization-based self-healing applications. Journal of Polymer Science Part A: Polymer Chemistry 2010, 48(12), 2698–2708. 10.1002/pola.24053.

[54] Murphy J, ed. Additives for Plastics Handbook, Elsevier, 2001. 10.1016/B978-1-85617-370-4.X5000-3.

[55] Bui T, Giovanoulis G, Cousins AP, Magnér J, Cousins IT, de Wit CA. Human exposure, hazard and risk of alternative plasticizers to phthalate esters. Science of the Total Environment 2016, 541, 451–467. 10.1016/j.scitotenv.2015.09.036.

[56] Godwin A. Plasticizers. Applied Plastics Engineering Handbook. Elsevier 2017, 533–553. 10.1016/B978-0-323-39040-8.00025-0.

[57] Krauskopf LG, Godwin A. Plasticizers. In Wilkes CE, Daniels CA, Summers JW, eds. PVC Handbook. Hanser Publications, Cincinnati, 2005.

[58] ECHA. Substance Infocard: Bis(2-ethylhexyl) phthalate. https://echa.europa.eu/substance-information/-/substanceinfo/100.003.829, 2023.

[59] ECHA. Substance Infocard: Dibutyl phthalate. https://echa.europa.eu/substance-information/-/sub stanceinfo/100.001.416, 2023.

[60] Lee JK, Kim K-B, Lee JD, Shin CY, Kwack SJ, Lee B-M, Lee JY. Risk assessment of drometrizole, a cosmetic ingredient used as an ultraviolet light absorber. Toxicological Research 2019, 35(2), 119–129. 10.5487/TR.2019.35.2.119.

[61] Cosmile Europe. Ingredient: Etocrylene. https://cosmileeurope.eu/inci/detail/5508/etocrylene/, 2024.

[62] ECHA. Substance Infocard: Etocrilene. https://echa.europa.eu/es/substance-information/-/substan ceinfo/100.023.663, 2023.

[63] Röhm GmbH. Olympiade unter „Plexiglas". Röhm Spektrum 1971, 7.

[64] Röhm GmbH. Lieferprogramm Halbzeug, 1968.

[65] ECHA. Substance Infocard: Palmitic acid. https://echa.europa.eu/substance-information/-/substan ceinfo/100.000.284, 2023.

Simon van der Heijden and Jean-Luc Dubois
18 Conclusions and future outlook

18.1 Conclusions

The new edition of this book is more focused on PMMA recycling. Mechanical and chemical recycling have their own specificities, but both need efficient collection and sorting, which are also addressed in the initial chapters. The "chemical recycling process" is a "thermolysis" step; but it is also described by many other words that may also convey a bad image. Of late, terms such as "advanced recycling" and "pyrolysis" have been associated with the recycling of mixed plastics (mostly, mixed polyolefins) with low mass yields to produce a pyrolysis oil that needs further treatment, to be used as a recycled "naphtha" in steam crackers. The polymer-to-polymer yields of these processes are around 25%. However, with PMMA, heat treatment in the absence of oxygen triggers a "thermolysis" or unzipping mechanism that gives back the monomer in high yields. When yields can be as high as 90 wt%, it means that the material can be recycled six times, while still keeping 50% of the carbon in the economy.

PMMA depolymerization has been selected as the main case study throughout the book, with results and examples inspired by the MMAtwo project, which experimented with the twin-screw extruder technology and which can deliver the highest regenerated monomer purity (virgin-like) for closed-loop applications.

New chapters have been added in this second edition. Technologies under development have been included: the continuous stirred tank process from Pyrovac, the inductively heated process from M-D2, and an analysis of the technology to recycle artificial marble/solid surface. In addition, a new chapter illustrates several examples generated with the MMAtwo Footprinter, benchmarking the environmental impacts of various virgin and recycling processes. When looking at recycled end-of-life PMMA scraps, we gathered data on the composition of PMMA-based materials, which have been put on the market more than 20 years ago. Indeed, in the sectors of building and construction, and also in cars, planes, and other transportation systems, the life expectancy of the products is of several decades. Even in flat-screen TVs, the life is much longer than initially anticipated, as the product would be reused or repurposed. So, for the products collected, we have typically lost track of the producer and the recipes. Additives could have been used at the time of production, which are no longer allowed today, and they compromise mechanical or physical recycling processes that are unable to remove those additives. A very valuable source of information has been found in museums. To be able to preserve art master pieces and cultural heri-

Simon van der Heijden, Heathland B.V., Arkansasdreef 8, 3565 AR Utrecht, The Netherlands
Jean-Luc Dubois, France Trinseo France SAS – Altuglas International SAS Tour CB21,16, place de l'Iris, 92400 Courbevoie, France

https://doi.org/10.1515/9783111076997-018

tage, conservation and restauration experts have the need to know what they are made of, and in their search of this information, they have developed analysis techniques that provide very useful data for recyclers. During the MMAtwo project, such analytics were used to gather data on old PMMA car taillights pieces, and therefore on the production methods of that time.

The activities of the MMAtwo project under the European Union's Horizon 2020 research and innovation funding scheme (grant agreement no. 820687) have been illustrated by the contribution of many of its partners in writing and preparation of this book. PMMA depolymerization is already implemented in several industrial sites throughout the world, including in Europe. Current technologies have major disadvantages and struggle to achieve a virgin-like quality, justifying the relevance of the MMAtwo project and the long-term impact of this book.

The most important conclusions of each chapter (Chapters 2–17) are outlined in Table 18.1. To facilitate interpretation, the scope of each chapter is provided.

Table 18.1: Scope and main conclusions of each chapter.

Chapter	Scope	Conclusions
Chapter 2	Vinyl polymer waste market for chemical recycling	– Five end-use market segments account for about 90% of the global demand. – Europe and governmental rules open the window to a market share for recycled products from polymer waste. – REACH is still in development, which may result in short-term resistance to adoption of regenerated raw material in the market.
Chapter 3	Collection, sorting, and pretreatment	– Sorting can be done by size, density, magnetism, rigidity, and melting point. – NIR and MIR are important technologies regarding sorting as well as color detection. – The feedstock to either the recycling process or the sorting process might need to be pretreated to facilitate subsequent processing. – PMMA exists as various grades. The most common grades are cast, extrusion, and injection grades. Cast-grade waste cannot be recycled mechanically, and to remove colors, the depolymerization process is necessary for all grades.
Chapter 4	Mechanical recycling	– Presents, in principle, the lowest environmental impact among all plastic recycling alternatives. – Low capital expenditure and low operational complexity. – Limited number of reprocessing cycles, as each cycle leads to a loss of quality. – Transparent clear injection and extrusion-grade PMMA can be recycled mechanically.

Table 18.1 (continued)

Chapter	Scope	Conclusions
Chapter 5	Thermochemical recycling: reactions	– Radical routes have many side reactions. – Structural defects are points of initiation of degradation. – Coupled matrix-based kinetic Monte Carlo simulations are a leading kinetic tool.
Chapter 6	Thermochemical recycling: lab-scale experiments	– The combination of TGA data and lab-scale experiments targeting isothermicity and the measurement of the MMD evolution can provide detailed mechanistic and kinetic information. – Polymer feedstock design is relevant to obtain kinetic insights. – One should strive for the determination of Arrhenius parameters for elementary, instead of overall, lumped reactions.
Chapter 7	Thermochemical recycling: reactor technologies	– Several PMMA depolymerization reactor technologies are in operation to date. – Depending on the operating conditions and types of wastes to be processed, yields above 50% and rMMA purity above 90% are routinely obtained. – Best performers can expect yields up to 90 wt% at purities of 99.8%: – Process intensification technologies using electricity as energy source (reactive extrusion, microwave heating, inductive heating, etc.) have the potential to reduce the carbon footprint of the processes.
Chapter 8	Chemical recycling: process design	– The hierarchy decomposition method allows for process design at several levels. – A straightforward comparison of the material efficiency of the virgin and rMMA processes can be performed using the heat value. – Design choices vary when it comes to continuous processing of scraps, primarily in the means of feeding and heating the media.
Chapter 9	Purification of regenerated monomers	– Purification of rMMA from continuous twin-screw extruder depolymerization of PMMA feedstocks may require the use of a two-step continuous distillation process. – Thermal depolymerization products are divided into three fractions, namely, gas, liquid (after condensation), and solid residues. – GC-MS/FID (composition) and dynamic olfactometry (odor) are important analysis techniques to qualify the process.

Table 18.1 (continued)

Chapter	Scope	Conclusions
Chapter 10	Industrial repolymerization	– Viscosity control is relevant to control the interplay of chemistry and diffusional limitations. – Exothermicity control requires dedicated cooling procedures, supported by measurement- and model-guided design. – Testing of final parts is recommended.
Chapter 11	Applications based on regenerated monomer	– rMMAs from extruder-assisted depolymerization showed fully comparable optical properties as their virgin reference during accelerated weathering/aging tests. – MMA and PMMA in current commercial recipes for composite acrylic kitchen sinks can be fully replaced by rMMA and rPMMA, while maintaining all physical, chemical, and mechanical requirements, resulting in a sufficiently low odor concentration for indoor applications. – rMMA thermoplastic infusionable resins are promising to increase the sustainability of wind turbine blades.
Chapter 12	Life cycle approaches	– Goal and scope definition are key elements of LCA. – Life cycle costing is generally defined as an analysis of all costs related to a certain product or process over the product life cycle. – An approach linked to LCA is footprinting. The LCA approach takes various impact categories (e.g., environmental and health categories, energy consumption, and emission) into account, whereas the footprinting approach focuses on one environmental impact.
Chapter 13	MMAtwo's Footprinter	– MMAtwo developed a tool to benchmark alternative solutions. – Most of the existing recycling technologies have a lower carbon footprint than the virgin MMA processes. – The difference between the technologies lies in the purity that can be achieved. A fair comparison should be made at the same purity level.
Chapter 14	Recycling of artificial marble	– Solid surface or artificial marble is a composite of PMMA and $Al(OH)_3$, which can be recycled into MMA and alumina. – Although more energy intensive than from PMMA scraps, the process remains attractive compared to virgin MMA, but for low demanding applications.
Chapter 15	Pyrovac's process	– Continuous stirred tank technology with high heat transfer area. – Same technology can be used for other plastics and also biomass pyrolysis.

Table 18.1 (continued)

Chapter	Scope	Conclusions
Chapter 16	M-D2's process	Inductive heating process uses small metallic balls to transfer the heat required by the depolymerization process. Energy-efficient process can minimize the heat losses.
Chapter 17	Contribution from PMMA heritage	Preservation of art pieces made with PMMA requires detailed knowledge on how the products were made several decades ago. Same information is needed for an efficient recycling of post-consumer PMMA scraps. Additives such as heavy metal pigments and plasticizers were common in the past.

LCA, life cycle analysis; rMMA, NIR, near-infrared; MIR, mid-infrared; PMMA, poly(methyl methacrylate); TGA, thermogravimetric analysis; GC-MS/FID, gas chromatography-mass spectrometry/flame ionization detector.

Since the first edition of this book and the end of the MMAtwo project, recycling of PMMA entered a new phase. Trinseo, which joined the MMAtwo project during the last year, took over Altuglas and also Heathland, to have access and to secure the required supply of PMMA waste feedstock for a depolymerization plant. During the first semester of 2024, Trinseo started its demonstration plant in Rho, Italy, using the technologies developed during the MMAtwo project.

We have also seen other companies becoming more active with projects involving PMMA depolymerization, a recycling concept that is clearly gaining traction in the market. Trinseo introduced its regenerated MMA under the name of R-Life MMA, but there are other names circulating that we summarize in Table 18.2. When a product deserves new commercial names, this means that it is now sufficiently mature.

18.2 Future outlook

This book (first and second editions) can serve as a guideline for future developments regarding thermochemical recycling, specifically utilizing twin-screw extruder technology. For the PMMA market, industrial activities are ongoing in the complete recycling value chain, starting from collection and pretreatment to chemical recycling, and ultimately to rMMA production and utilization in the final application. The MMAtwo project is a leading example in the realization of circularity in our society regarding the treatment of PMMA waste, as it demonstrates that low-grade waste PMMA can be used to make high quality optical-grade PMMA. This book also serves as a general roadmap to tackle other (vinyl) polymer waste streams.

Table 18.2: Alternative names for the recycled MMA currently in use, and also some ideas.

Name	Meaning	Companies
R-Life MMA	In line with brand names for recycled sheets and resins that also use R-Life	TRINSEO
rMMA	Regenerated MMA, because of the higher purity achieved	MMAtwo
MAM-R	MAM for French "MethAcrylate de Methyle" and R for "Recyclé"	ALTUGLAS
D-MMA	Depolymerized MMA	Asian most often
c-MMA	Circular MMA	MITSUBISHI
Reborn MMA	Recycled or reborn MMA (translated from Russian)	ACROLAB
vMMA	Greek letter Nu, may sound like new . . . but sounds like nu . . .	
βMMA	Greek letter Beta, since that's the second cycle	

However, the reader should not underestimate the difficulties in implementing such a technology. There are many challenges to be solved:
–   Collection of waste: a sufficient amount of waste has to be secured to justify construction of a plant. Initially, the waste may be free or even come with a gate fee, but as soon as the demand increases, prices will follow. Waste is an interesting resource that is available close to the consumption point, but one should not expect that the amount of waste will increase in the future. Instead, one should expect optimized production systems reducing waste, and longer life-reuse-repurpose, which will delay the time at which the product enters the recycling loop. At the same time, however, one may expect that "eco-conception" and "design-for-recycling" rules will enhance the amounts of end-of-life products that can be collected for recycling.
–   Plant size: to minimize the transportation distances and its impact on the carbon footprint, it may be beneficial to target multiple small-sized plants rather than a single large one. When the depolymerization and purification processes are energy efficient, it means that transportation and pretreatment will have a major share of the carbon footprint. The challenge is to make a small plant sufficiently economically attractive.
–   Parallelization versus up-scaling: there is limited economy of scale in parallelization because the same equipment has to be purchased to double the capacity. But in thermolysis processes, the limiting factor is often the heat transfer, and so the surface area through which heat is transferred to the material. Increasing the area for heat transfer is what dominates the plant design, and so the equipment cost.
–   rMMA purity level: introducing a high purity recycled product with a different impurity profile is also a challenge. The market knows either the virgin product or a low purity recycled product, which has been tried in the past and failed in numerous applications. That is also a reason why recycling was not developed

further in many past instances. Many applications do not necessarily need "virgin" quality, because anyway co-monomers are deliberately added in the process, or the product is polymerized in water, excluding certain impurities, but nevertheless it is an enormous effort for customers to dedicate a specific tank or line for a segregated recycled product. Even if the new recycled product has lesser impurities than the previous generation, no one (customers or producer) has a precise idea of the role of each one of the individual impurities. This may lead to over-quality at a higher cost, or a too low a quality to address the market pricing requirements. The difficulty comes when one needs to know for which rMMA quality level to design the plant and process, while the customer still needs to identify what rMMA quality level fits with its own product applications.

- Pricing: for most people, recycled product means lower quality, and so should be sold at lower price, even when it has the same technical properties. There is this idea that the waste was available at a low price, so the product should be cheap. The economics, however, are such that for non-marginal products, the feedstock price will increase as demand increases, until a new equilibrium is reached. Recently, what has changed is the social pressure/demand for recycled products, but also and most importantly, the regulations that are forcing companies to recycle and/or use recycled plastics. In that situation, premium quality recycled material can expect a price premium, especially as long as the supply of top-quality recycled product remains limited.

In order to motivate companies and consumers, and facilitate the transition toward recycled plastics, new legislations that will enforce the collection of post-industrial and post-consumer wastes are needed. These materials are available at the point of consumption, and that's what justify recycling them locally. Legislation that also enforces the use of recycled product, preferably closed-loop recycled product, will not only help to access high performance markets but also promote the use of "eco-conception" and "design-for-recycling" rules, making product easier to collect, dismantle, sort, and recycle.

Index

https://doi.org/10.1515/9783111076997-019

www.ingramcontent.com/pod-product-compliance
Lightning Source LLC
Chambersburg PA
CBHW061345210326

41598CB00035B/5884